Environmental Effects of Offshore Oil Production

THE BUCCANEER GAS AND OIL FIELD STUDY

MARINE SCIENCE

Coordinating Editor: Ronald J. Gibbs, *University of Delaware*

Recent Volumes:

Volume 8 - *Ocean Wave Climate*
Edited by Marshall D. Earle and Alexander Malahoff

Volume 9 - *Marine Geology and Oceanography of the Pacific Manganese Nodule Province*
Edited by James L. Bischoff and David Z. Piper

Volume 10 - *Ecological Processes in Coastal and Marine Systems*
Edited by Robert J. Livingston

Volume 11 - *Estuarine and Wetland Processes: With Emphasis on Modeling*
Edited by Peter Hamilton and Keith B. Macdonald

Volume 12 - *Ocean Dumping of Industrial Wastes*
Edited by Bostwick H. Ketchum, Dana R. Kester, and P. Kilho Park

Volume 13 - *Oceanography From Space*
Edited by J. F. R. Gower

Volume 14 - *Environmental Effects of Offshore Oil Production: The Buccaneer Gas and Oil Field Study*
Edited by Brian S. Middleditch

Environmental Effects of Offshore Oil Production

THE BUCCANEER GAS AND OIL FIELD STUDY

Edited by
Brian S. Middleditch
University of Houston
Houston, Texas

PLENUM PRESS • NEW YORK AND LONDON

Library of Congress Cataloging in Publication Data

Main entry under title:

Environmental effects of offshore oil production.

(Marine science; 14)
"Proceedings of a symposium on the buccaneer gas and oil field study, held October 8-9, 1980, during Expochem '80 in Houston, Texas" — T.p. verso.
Sponsored by the Environmental Protection Agency et al.
Includes bibliographical references and index.
1. Oil well drilling, Submarine — Environmental aspects — Mexico, Gulf of — Congresses. I. Middleditch, Brian S. II. Expochem '80 (1980: Houston, Tex.) III. United States. Environmental Protection Agency. IV. Series.

TD195.P4E55 333.91'7'097641 81-11934
ISBN 0-306-40826-0 AACR2

Proceedings of a symposium on The Buccaneer Gas and Oil Field Study, held October 8-9, 1980, during EXPOCHEM '80 in Houston, Texas

A Division of Plenum Publishing Corporation
233 Spring Street, New York, N.Y. 10013

Printed in the United States of America

PREFACE

The Buccaneer Gas and Oil Field Study has been the most comprehensive research project to date concerned with assessing the ecological effects of offshore production activities. It took nearly five years to complete and involved almost 30 individual research groups. All of the raw data have been archived with NOAA's Environmental Data and Information Services, and detailed technical reports have been deposited with the National Technical Information Service so the interested investigator should be able to gain access to them. However, we felt that it would be desirable to present a distillation of our more significant findings in a form that was more readily available to the scientific and lay community. Thus, we conducted a symposium on the study during EXPOCHEM '80 at the Astrohall, Houston, Texas during October, 1980. This volume comprises the proceedings of that symposium.

All but two of the papers presented are included in this book. Manuscripts were not received from Dr. D. A. Wiesenburg (Texas A&M University: Volatile Hydrocarbons) or Dr. J. Tillery (Southwest Research Institute: Trace Metals), but these topics are adequately covered by other authors. An introductory chapter was added to place the study in its proper perspective and to provide some background material on the Buccaneer Field, a brief chapter on biocides was inserted since this topic generated much discussion at the symposium, and a bibliography is provided to direct the interested reader to sources of additional published information on the Study.

Each of the authors was encouraged to avoid the rather dry format characteristic of a technical report and most have succeeded. We hope that our book will appeal to a wider readership than the technical reports that are normally produced following studies of this type.

Several of the administrative and support personnel involved in this study are not given adequate recognition elsewhere in the book, and it is appropriate to acknowledge their contributions here. Dr. Joseph W. Angelovic, formerly the Director of the NMFS Gulf Coastal Fisheries Center, Galveston, Texas was instrumental in initiating involvement of the Galveston Laboratory in environmental impact research through Interagency Agreements with other federal agencies. The first project, through Interagency Agreement with the Department of Interior, Bureau of Land Management, involved fisheries investigations associated with the South Texas Outer Continental Shelf. The second was the study of the Buccaneer Field, supported by Interagency Agreement with the Environmental Protection Agency. Dr. Angelovic is currently Deputy Director of NOAA's Ocean Programs Office in Washington, D.C., which provided the NOAA program overview of the Buccaneer Gas and Oil Field Study, with Mr. W. Lawrence Pugh as Program Manager. The EPA Program Manager was Dr. Alfred A. Galli of EPA's Office of Energy, Minerals, and Industry in Washington, D.C. Mr. John T. Hinman, Ms. Joan Blaylock Sylvester and Mr. Joseph L. Gray of the NMFS Southeast Regional Office in St. Petersburg, Florida were Contracting Officers for the Study.

When the project first began, Dr. Angelovic was the Contracting Officer's Technical Representative and Dr. Richard Berry the Project Manager at the NMFS Galveston Laboratory. When Dr. Angelovic left for Washington, the former task was passed to Dr. Berry, and Dr. Ernest D. McRae (now deceased) became Project Manager. Early in 1977, Dr. Edward F. Klima was appointed as Director of the Galveston Laboratory and assumed the role of Contracting Officer's Technical Representative. At that time, Dr. Charles W. Caillouet became Chief of the Galveston Laboratory's Environmental Research Division and Project Manager of the Study. It was Dr. Klima and Dr. Caillouet who were most directly responsible for the transition from a multidisciplinary survey to the interdisciplinary study that the project became.

Mr. William B. Jackson (Senior Advisor, Contracts and Deliverables) of the Galveston Laboratory was the Report Editor during most of the Study. Mr. Gregg R. Gitschlag was Acting Senior Advisor, Field Operations and Logistics. Other members of the technical staff at the Galveston Laboratory who participated in the project were Dr. Robert M. Avent (Oceanographer, Field Party Chief), Mr. E. Peter Wilkens (Fishery Biologist), Mr. Gary Faw (Fishery Biologist), Mr. Dennis Koi (Computer Programmer), and Ms. Leesa M. Young (Biological Aid).

Mr. John A. Burgbacher of Shell Oil Company's Offshore Division in New Orleans, Louisiana deserves special mention for providing information about operations in the Buccaneer Field, arranging for access to the structures to be granted, and providing helicopter transportation on many occasions. Other Shell employees in the Field were equally cooperative.

We are also indebted to Dr. Albert Zlatkis of the University of Houston for allowing us to hold our symposium as a component of EXPOCHEM '80.

The prompt publication of these proceedings was achieved with the cooperation of the authors and the able assistance of Mr. James Busis of Plenum.

Brian S. Middleditch
University of Houston

DISCLAIMER

The Buccaneer Gas and Oil Field Study was sponsored by the Environmental Protection Agency and Department of Commerce, National Oceanic and Atmospheric Administration, National Marine Fisheries Service, Southeast Fisheries Center, Galveston Laboratory under Interagency Agreement EPA-IAG-D5-E693-EO. The contents of this publication do not necessarily reflect the views and policies of the Environmental Protection Agency or the Department of Commerce. Mention of trade names or commercial products herein does not constitute endorsement or recommendation for use.

CONTRIBUTORS

JOHN B. ANDERSON, Geology Department, Rice University, Houston, Texas 77001.

REED S. ARMSTRONG, Atlantic Environmental Group, National Marine Fisheries Service, National Oceanic and Atmospheric Administration, Narragansett, Rhode Island 02882.

GLENN D. AUMANN, Department of Biology, University of Houston, Houston, Texas 77004.

E. WILLIAM BEHRENS, University of Texas Marine Science Institute, 700 The Strand, Galveston, Texas 77550.

GREGORY S. BOLAND, LGL Ecological Research Associates, Inc., 1410 Cavitt Street, Bryan, Texas 77801.

JAMES M. BROOKS, Department of Oceanography, Texas A&M University, College Station, Texas 77843.

JACK CALMAN, Environmental Research and Technology, Inc., 1111 Third Avenue, Suite 2770, Seattle, Washington 98101. Present address: Applied Physics Laboratory, Johns Hopkins University, Johns Hopkins Road, Laurel, Maryland 20810.

ROBERT J. CASE, Texas A&M Marine Laboratory, Building 311, Fort Crockett, Galveston, Texas 77550.

LARRY J. DANEK, Hazleton Environmental Sciences, 1500 Frontage Road, Northbrook, Illinois 60062. Present address: Environmental Science and Engineering, Inc., P.O. Box ESE, Gainesville, Florida 32602.

GEORGE D. DENNIS, LGL Ecological Research Associates, Inc., 1410 Cavitt Street, Bryan, Texas 77801.

ERNIE L. ESTES, Department of Marine Sciences, Texas A&M University, Galveston, Texas 77843.

NICK FOTHERINGHAM, Department of Biology, University of Houston, Houston, Texas 77004. Present address: Dames & Moore, 4321 Directors Row, Houston, Texas 77092.

KENNETH W. FUCIK, Science Applications, Inc., 2760 29th Street, Boulder, Colorado 80301.

BENNY J. GALLAWAY, LGL Ecological Research Associates, Inc., 1410 Cavitt Street, Bryan, Texas 77801.

DONALD E. HARPER, JR., Texas A&M Marine Laboratory, Building 311, Fort Crockett, Galveston, Texas 77550.

RICHARD P. HERBST, Environmental Research and Technology, Inc., 1111 Third Avenue, Suite 2770, Seattle, Washington 98101. Present address: Regulatory Affairs Department, Exxon Minerals Co., P.O. Box 2180, Houston, Texas 77001.

RANDALL L. HOWARD, LGL Ecological Research Associates, Inc., 1410 Cavitt Street, Bryan, Texas 77801.

CHI-HSIN HSU, Department of Biology, University of Houston, Houston, Texas 77004.

REBECCA L. JASCHEK, Texas A&M Marine Laboratory, Building 311, Fort Crockett, Galveston, Texas 77550.

LARRY R. MARTIN, LGL Ecological Research Associates, Inc., 1410 Cavitt Street, Bryan, Texas 77801.

BRIAN S. MIDDLEDITCH, Department of Biophysical Sciences, University of Houston, Houston, Texas 77004.

KURT D. OLSEN, Department of Biology, University of Houston, Houston, Texas 77004.

DEBORAH L. POTTS, Texas A&M Marine Laboratory, Building 311, Fort Crockett, Galveston, Texas 77550.

CURT D. ROSE, Environmental Sciences Division, Energy Resources Company, Inc., 185 Alewife Brook Parkway, Cambridge, Massachusetts 02138.

ROBERT R. SALZER, Texas A&M Marine Laboratory, Building 311, Fort Crockett, Galveston, Texas 77550.

CLAUDE R. SCHWAB, Department of Oceanography, Texas A&M University, College Station, Texas 77843.

RUDOLPH R. SCHWARZER, Chemistry Department, Texas Southern University, Houston, Texas 77004.

ROBERT F. SHOKES, Science Applications, Inc., 1200 Prospect Street, La Jolla, California 92038.

IVAN T. SHOW, Science Applications, Inc., 476 Prospect Street, La Jolla, California 92038.

RONALD K. SIZEMORE, Department of Biology, University of Houston, Houston, Texas 77004.

GERALD W. SMEDES, Environmental Research and Technology, Inc., 1111 Third Avenue, Suite 2770, Seattle, Washington 98101.

MICHAEL S. TOMLINSON, Hazleton Environmental Sciences, 1500 Frontage Road, Northbrook, Illinois 60062. Present address: Environmental Science and Engineering, Inc., P.O. Box ESE, Gainesville, Florida 32602.

CONSTANCE M. WALKER, Texas A&M Marine Laboratory, Building 311, Fort Crockett, Galveston, Texas 77550.

TIMOTHY J. WARD, Environmental Sciences Division, Energy Resources Company, Inc., 185 Alewife Brook Parkway, Cambridge, Massachusetts 02138.

RICHARD B. WHEELER, Geology Department, Rice University, Houston, Texas 77001. Present address: Exxon Production Research Company, Houston, Texas 77005.

DENIS A. WIESENBURG, Department of Oceanography, Texas A&M University, College Station, Texas 77843.

CONTENTS

PROLOGUE

Brian S. Middleditch and Benny J. Gallaway

HYDROCARBONS AND SULFUR

Brian S. Middleditch

TOTAL ORGANIC CARBON AND CARBON ISOTOPES OF SEDIMENTS

E. William Behrens

DISTRIBUTION AND ABUNDANCE OF MACROBENTHIC AND MEIOBENTHIC ORGANISMS

Donald E. Harper, Jr., Deborah L. Potts, Robert R. Salzer, Robert J. Case, Rebecca L. Jaschek, and Constance M. Walker

EFFECTS ON ARTIFICIAL REEF AND DEMERSAL FISH AND MACROCRUSTACEAN COMMUNITIES

B. J. Gallaway, L. R. Martin, R. L. Howard, G. S. Boland, and G. S. Dennis

ACUTE TOXICITY AND AQUATIC HAZARD ASSOCIATED WITH DISCHARGED FORMATION WATER

Curt D. Rose and Timothy J. Ward

ENVIRONMENTAL SYNTHESIS USING AN ECOSYSTEMS MODEL

Kenneth W. Fucik and Ivan T. Show

TRANSPORT AND DISPERSION OF POTENTIAL CONTAMINANTS

Reed S. Armstrong

EPILOGUE

Brian S. Middleditch

PROLOGUE

Brian S. Middleditch

Department of Biophysical Sciences
University of Houston
Houston, Texas 77004

Benny J. Gallaway

LGL Ecological Research Associates, Inc.
1410 Cavitt Street
Bryan, Texas 77801

INTRODUCTION

Debate over the environmental effects of offshore oil production has been continuing ever since the concept of drilling at sea was introduced. On the one hand, there are those who wish to protect the environment from unreasonable assault and, on the other hand, there are pressures to produce more oil and to produce it as inexpensively as possible. In the United States, the latter concern has been compounded by a realization of the problems which can result from overdependence upon imported oil.

For a reasonable debate to proceed to a resolution of differences of opinion, reliable and appropriate data must be made available. Many studies have been made of the environmental effects of major oil spills and well blowouts. Much data has been obtained concerning the toxicity of oil and its components on marine organisms. At the inception of this project, however, little was known about the environmental effects of routine offshore oil production.

To place our research in a proper perspective, we are first reviewing the existing literature on inputs of petroleum hydrocarbons to the marine environment. This is followed by a description of the Buccaneer Gas and Oil Field Study.

MAJOR SPILLS

Major oil spills have resulted from tanker wrecks and drilling accidents.

Tanker Wrecks

One of the first major oil spills to receive world-wide attention, from the press and scientific community alike, was the stranding of the "Torrey Canyon" in the western approach to the English Channel on March 18, 1967. During March and April her 29.4 million gallon load of Kuwait crude oil was released, and much of it reached the beaches of England and France. Unfortunately, conclusions on the biological effects of this well documented spill remain somewhat confused since large quantities of toxic dispersants were used during the clean-up process (Nelson-Smith, 1973).

When the oil barge "Florida" ran aground just off Little Island, West Falmouth, Massachusetts in 1969, approximately 175,000 gallons of Number Two fuel oil leaked into Buzzards Bay and the adjacent Wild Harbor Marsh. Scientists from the nearby Woods Hole Oceanographic Institution were able to study the incorporation of polluting hydrocarbons into the marsh ecosystem. Massive kills of marine and marsh life were reported immediately after the spill and oysters and scallops were found to be contaminated with fuel oil (Blumer et al., 1970). Analyses of sediments and marsh organisms collected a year later showed an accumulation of petroleum hydrocarbons. Chromatograms of pentane extracts revealed the loss of lower boiling components of the oil, but retention of cyclic and aromatic hydrocarbons (Burns and Teal, 1971; Blumer et al., 1973). Sediments examined two years after the spill still demonstrated the persistence of fuel oil hydrocarbons (Blumer and Sass, 1972).

When the tanker "Argo Merchant" went aground near Nantucket Island, Massachusetts on December 15, 1976 more than 7.7 million gallons of number six fuel oil were lost. Immediately after the wreck scientists from throughout the U.S. began to make an assessment of the impact of spilled oil on the ecology of the area. Analysis of sediments collected within two months of the spill did not indicate the presence of oil in significant quantities, except within 10 miles of the bow section, where concentrations up to 100 ppm were measured in a few samples. There was evidence of oil contamination in fish, shellfish, ichthyoplankton, and other zooplankton in the area of the spill. Benthic filter-feeders such as mussels, scallops and quahogs, however, did not then show an accumulation of hydrocarbons (Grose and Mattson, 1977).

In the largest tanker disaster recorded, the "Amoco Cadiz" released 65 million gallons of crude oil and bunker fuel just 1.5 km from the northwest coast of France on March 16, 1978. During the 15 days following the wreck, 140 km of the Brittany coast was heavily impacted by oil. High mortalities of some intertidal organisms, birds, and other marine organisms were observed as well as incorporation of oil in sediments up to 70 cm in depth (Hess, 1978). Just one year after the wreck, however, a return survey of the area by biologists from the U.S. and France reported a substantial increase in Odum's diversity index and attendant recolonization of coastal areas (Cross, 1979), thus indicating movement toward ecological recovery of the damaged area.

Toward the end of 1979, a collision occurred between the tanker "Burmah Agate" and the freighter "Mimosa" in the Gulf of Mexico off Galveston, Texas. Some of the tanker's cargo spilled into the sea, but much of it burned over a period of several weeks. The environmental impact of this incident is still being assessed.

Drilling Accidents

The estimated total volume of oil spilled from the blow-out during January, 1969 of Union oil well A-21 in the Santa Barbara Channel was in excess 3.25 million gallons. While the volume is certainly comparable to that spilled by tankers, it appears that the ecological effects were not as substantial. Phytoplankton productivity (essential to any marine ecosystem) after the blow-out was well within the established pattern observed under normal conditions of nutrient availability. Reproduction of the copepod *Labidocera trispinosa* was successful, thus indicating minimal ecological stress (McGinnis, 1971). "Blooms" of a tunicate were also observed as well as a normal variety and abundance of fish eggs and larvae (Straughan, 1970). It has been argued, however, that had the sophisticated analytical chemical techniques now used in evaluating spills been applied to the Santa Barbara spill a more serious picture of the ecological damage may have been obtained (Easton, 1972).

The largest oil spill due to a drilling accident was the Ixtoc-I blowout on June 3, 1979 on the Mexican continental shelf of the Gulf of Mexico. Non-toxic dispersants were used to prevent oil from reaching Mexican beaches (except on a few occasions), but some South Texas beaches were heavily impacted until the prevailing currents changed direction in the fall. Although the well was not capped until the March of 1979, only small amounts of additional oil reached the Texas coast. The ecological effects of this incident are still being studied.

OTHER SOURCES OF OIL

Natural Seeps

The presence of oil seeps in the Santa Barbara Channel, California has been documented as early as 1543 (Easton, 1972). In order to assess the consequences of this geological phenomenon, a series of field surveys was conducted from 1972 through 1974. Coal Oil Point, an area of natural oil seepage in the Santa Barbara Channel, was compared with nearby control sites that were not chronically exposed to natural oil seepage. The studies were designed to encompass most segments of the environment, invertebrate groups, and habitat types. Species examined were sedentary (non-migratory) to ensure that specimens had actually been exposed to conditions in these areas. Some of the most significant results were (Straughan, 1970):

(1) Animals are inhabiting areas of Coal Oil Point which have relatively high levels (up to 9%) of petroleum hydrocarbons in the sediments.

(2) The distribution of petroleum hydrocarbons is patchy throughout the environment in terms of time, location, and tissue type.

(3) Total hydrocarbon values alone cannot be used as indicators of petroleum contamination due to the variability in composition of both petroleum and biogenic hydrocarbons.

(4) No malformations were observed in the thousands of organisms examined from Coal Oil Point.

(5) Specimens of abalone (*Haliotis* sp.), sea urchin (*Stronglylocentrotus purpuratus*), and mussel (*Mytilus californianus*) from Coal Oil Point are more tolerant of exposure to Santa Barbara crude oil and seem to be hardier in terms of survival and adjustment to laboratory conditions than animals from control locations.

It has been estimated that the petroleum contribution of natural seeps is 0.6 million tons annually (Wilson et al., 1973). While these figures have been criticized as being too high, petroleum geologists generally agree that a more accurate figure would be extremely difficult to estimate.

Anthropogenic Sources

Other sources of oil in the sea include those from shipping operations, oil ports and terminals, offshore drilling and production, refineries, runoff, and sewage. The only source of continuous low-level chronic pollution which had received any great amount of attention prior to our study was refinery effluents (Cowell, 1976). However, these results are not comparable to those obtained for effluents from offshore production platforms since the toxicity of the refinery effluents is not attributable solely to oil (Baker, 1975).

Relative Significance

Several attempts have been made to provide an inventory of the various inputs of oil to the marine environment. One such tabulation is reproduced as Table 1. In most of these estimates, offshore production is credited with only a small proportion of the total oil input. Wilson et al. (1973) estimated that 80,000 metric tons per annum derived from this source. The National Academy of Sciences (1975) evaluated these data and indicated a probable range of input to be 80,000 to 150,000 metric tons per annum. They concluded also that the total oil input was approximately 6.113 million metric tons, so that offshore production would account for 1.3-2.5% of the total input.

Other estimates are difficult to compare. Travers and Luney (1976) estimated that only 0.006% of the oil handled during offshore operations was spilled (compared with 0.016% of the oil carried by tankers). The Massachusetts Institute of Technology Study Group on Criticial Environmental Problems reported that, along with ocean shipping and accidents, offshore drilling may have accounted for a "significant" portion of about 2 million tons of oil introduced into the oceans annually (Wilson, 1970). McCaslin (1972) calculated that 1,500 bbl of oil per year were spilled in U.S. waters and that 10,000 metric tons per year were spilled world-wide.

It is clear that a global estimate of oil input to the sea from offshore operations is difficult, if not impossible, to obtain. Moreover, since the contribution of oil from natural seeps is particularly difficult to determine, the relative amount of oil from offshore production losses is even more elusive. Most estimates agree that this input is relatively low in comparison to the total global input of oil to the sea. The question which remains unanswered is whether a significant quantity of oil is released in areas (such as the Gulf of Mexico) where intensive offshore production is taking place.

TABLE 1. BUDGET OF PETROLEUM HYDROCARBONS INTRODUCED INTO THE OCEANS[a]

	Input Rate (million tons per year)	
	Best Estimate	Probable Range
Natural seeps	0.6	0.2-1.0
Offshore production	0.08	0.08-0.15
Transportation		
LOT tankers	0.31	0.15-0.4
Non-LOT tankers	0.77	0.65-1.0
Dry docking	0.25	0.2-0.3
Terminal operations	0.003	0.0015-0.005
Bilge bunkering	0.5	0.4-0.7
Tanker accidents	0.2	0.12-0.25
Nontanker accidents	0.1	0.02-0.15
Coastal refineries	0.2	0.2-0.3
Atmosphere	0.6	0.4-0.8
Coastal municipal wastes	0.3	-
Coastal industrial wastes	0.3	-
Urban runoff	0.3	0.1-0.5
River runoff	1.6	-
TOTAL	6.113	

[a]From National Academy of Sciences (1975).

PREVIOUS STUDIES IN THE GULF OF MEXICO

Studies designed to provide information enabling an assessment of the effects of offshore oil and gas development activities on marine ecosystems in the northwestern Gulf of Mexico have historically consisted of broad multidisciplinary ecological surveys performed over large geographical areas. Two previous studies have been performed. The initial effort was a multidisciplinary study performed in 1973-1974 off Louisiana in the intensively developed "oil patch" (Gulf Universities Research Consortium, 1974). This investigation was predicated upon synoptic comparisons of impacted (platform) sites with control sites (no platforms) which had not experienced petroleum activity. As might have been expected, one of the major findings of the GURC study was the "reef effect" of the platforms. However, seasonal variation in most response variables was found to have been substantially greater than treatment variation, such that few significant differences could be delineated. The program yielded a wealth of descriptive data (about 1.5 million data points), but most of the data were inadequate to provide a meaningful assessment of impacts because of (1) the inadequacy of the controls (comparison of reef to non-reef habitats), and (2) the lack of depth of the investigations (study was multi- as opposed to interdisciplinary). More recently (1978-1979), the GURC study was essentially repeated by the Bureau of Land Management Study, "Ecological Investigations of Petroleum Production Platforms in the Central Gulf of Mexico." Again, the "reef effect" of platforms was documented and measurable environmental and ecological change was found to be restricted to the immediate vicinity of the platforms (Gallaway et al., 1979).

The second historical approach to obtaining data for evaluating effects of offshore oil and gas development has been based upon the baseline concept. Undeveloped areas subject to lease to industry for development have served as study areas for large multidisciplinary surveys intended to provide an "environmental benchmark" against which comparisons could be made following development (e.g., the Bureau of Land Management "Environmental Studies of the South Texas Outer Continental Shelf"). Again, a wealth of descriptive data has been provided by these studies. However, due to annual and other natural variations as well as to the general lack of depth in most of the disciplinary efforts attributable to the breadth of the surveys, the resulting data hold little immediate promise in terms of making realistic impact projections.

THE BUCCANEER GAS AND OIL FIELD STUDY

Rationale

The foregoing survey reveals many gaps in our knowledge of the environmental effects of offshore activities. Given the time and the level of funding available for this study, it was impractical to gather sufficient data to assess the overall impact of these activities, even in a limited geographic region such as the Gulf of Mexico. The approach selected was to perform an intensive study of a single offshore field so that we could fully document whatever effects might be ascribed to its operations. It was fully realized at the outset that the findings of such a study might not be legitimately applied by extrapolation to other fields in determining the overall effects of offshore operations in the Gulf of Mexico. It was anticipated, however, that sufficient knowledge might be obtained to recognize the parameters which

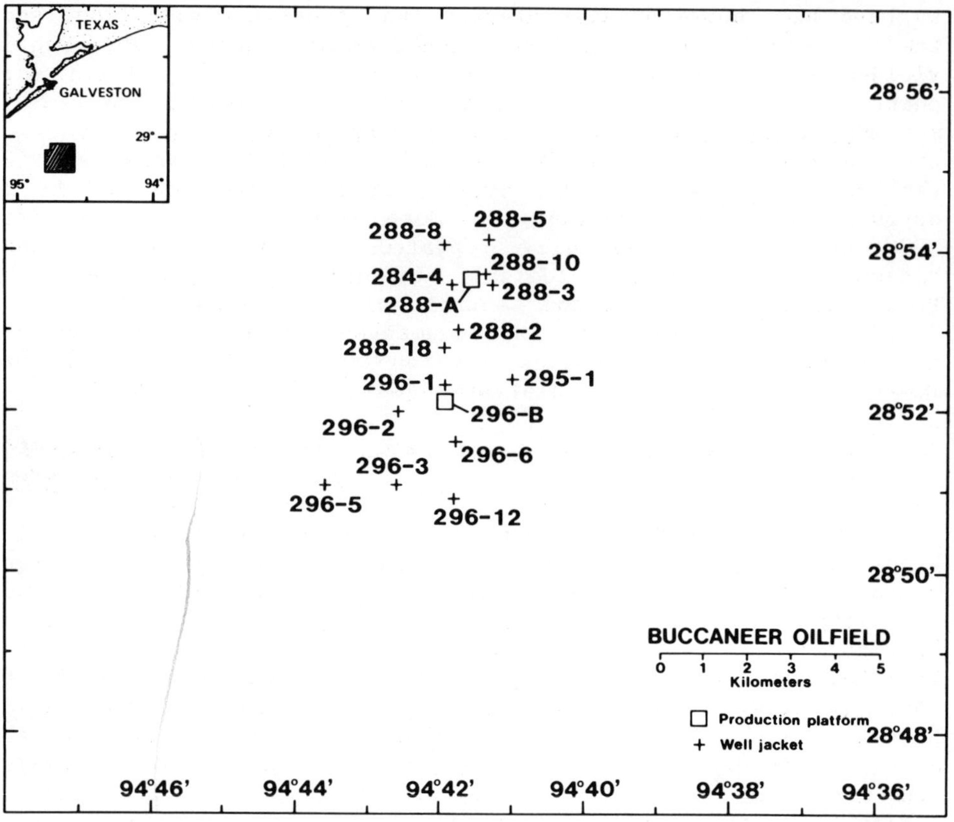

Figure 1. Map showing locations of production platforms and well jackets in the Buccaneer Field.

might be responsible for the individual effects so that they could be monitored and controlled.

The area selected for study was the Buccaneer Field located 50.5 km south of Galveston, Texas (Figure 1). Rationale for selection of this field included:

(1) The field had been in production for about 15 years, which time had allowed full development of the oil field associated climax marine communities.

(2) It is isolated from other fields, which facilitated the selection of an unaltered area (for comparison) within a reasonable distance of the field.

(3) It produces both oil and gas that create a major portion of the pollutants from marine mineral extraction.

(4) Its location simplified logistics and reduced the cost of the research.

(5) The Texas offshore area has not as yet been fully developed for gas and oil production but will probably experience accelerated exploitation in the near future.

Objectives

The Buccaneer Gas and Oil Field Study was inaugurated with the following specific objectives:

(1) to identify and document the types and extent of biological, chemical, and physical alterations of the marine ecosystem associated with the Buccaneer Gas and Oil Field,

(2) To determine specific pollutants, their quantity, and effects, and

(3) To develop the capability to describe and predict fate and effects of Buccaneer Gas and Oil Field Contaminants.

The project was divided into a number of different work units: some were performed by the National Marine Fisheries Service, while others were carried out under contract. Most of the individual components of the study are described in this volume.

The approach of the Buccaneer Gas and Oil Field Study has been to sacrifice breadth of spatial and disciplinary coverage for depth in selected studies specifically designed to measure impacts. Following a brief pilot study in 1975 (Harper et al., 1976), the first year's investigation (1976-1977) consisted of a multidisciplinary ecological survey comparing the relatively isolated Buccaneer Field to control, or undeveloped, areas. Following the initial descriptive effort of 1976-1977, a series of iterative project reviews held on an annual basis served not only to reduce the disciplinary scope as appropriate, but also to transform the multidisciplinary effort to an interdisciplinary one. In this manner, the finite project resources were committed to in-depth investigations of key processes deemed suitable for study in terms of their promise for yielding data enabling a quantitative assessment of impacts. In 1977-1978, investigations were intensified within the Buccaneer Field, comparing conditions around production platforms (source of contaminant discharge) to those around well jackets (structures with no contaminant discharges). In 1978-1979, studies focused more on production platforms, particularly in terms of amounts, fates, and effects of contaminants being discharged and the processes accounting for the observed dispersal and distribution of contaminants in the environment (circulation, hydrography, trophic linkages). During the final year (1979-1980), field studies were reduced to fill gaps in areas of remaining uncertainty, and the major effort was devoted to preparing integrated milestone reports evaluating the effects of the Buccaneer Field on ecological and environmental systems.

The Buccaneer Field

A detailed description of the Buccaneer Field was compiled early in the study (Gallaway et al., 1976; National Marine Fisheries Service, 1977); a brief summary is given here.

History. Initial seismic reflection surveys were conducted in Galveston Block 288 from 1953 through 1959. Based upon the results of this survey and regional geological studies, Shell acquired some 43,000 acres at the federal lease sale in February 1960 and proceeded with the drilling of a discovery well. This well reached a total depth of 16,088 feet in November 1960, but mechanical difficulties caused it to be abandoned and no tests were obtained on the potential pay sands.

During 1961 and 1962, additional seismic surveys were performed in Block 288 in order to better assess the size, shape, and complexity of the structural trap. In October 1963, another well was drilled at a location approximately one mile east of the original well. This test well led to a multi-well delineation program. Within a year, 21 additional wells were drilled, 12 of which were cased and completed from 4-pile satellite jackets.

The delineation program indicated that the field reserves were primarily gas, and that those reserves could be developed most economically from two drilling-production platforms. Platform 288-A was installed in September 1964 and 15 wells were drilled from the structure between January and August of 1965. Platform 296-B was placed down in May 1965 and 15 development wells were drilled from it between October 1965 and July 1966.

As the field developed, an onshore gasoline plant was constructed near Freeport, Texas, and the plant was connnected to the field (platform 288-A) by a 20-inch diameter flowline some 43 miles in length. The pipeline is operated by Blue Dolphin

Pipeline Company, a Shell subsidiary. Upon completion of the platform 288-A wells, gas sales commenced on January 1, 1966 at the contract rate of 100 MMCF/day of residue. This was increased in January 1967 to 175 MMCF/day as 296-B began production.

Additional wells were drilled in the field between 1966 and 1971 for the purpose of oil rim development, for geopressured gas and oil, and for untested fault blocks located by seismic reviews. Four wells in those categories have been drilled in Block 289, which was purchased at the May 1968 lease sale. Through June 1971, 81 wells had been drilled in Block 288; of these, 56 were completed and the remainder were plugged. In 1977 there were 34 active wells including 31 gas completions and 3 oil competions. Both production platforms and 6 of the 13 satellite jackets were active. The total fluid stream to the production platforms was approximately 1000 bbl/day from which about 65 MMCF of gas was extracted. The fluid typically contained about 600 bbl of produced water and about 400 bbl of crude oil and condensate. Shell's current lease comprises portions of Galveston Blocks 288, 289, 295, 296, and 322 for a total of 20,070 acres.

Structures that remain in the Buccaneer Field are of five types; production platforms, quarters platforms, satellite well platforms, flare stacks, and pipelines. A total of 18 platforms have been constructed in the field: 2 production platforms, 2 quarters platforms, and 14 satellite platforms.

Production Platforms. The two production platforms in the Buccaneer Field each have two decks approximately 150 ft by 150 ft and supported by piles or legs. A boat landing is located near the water level below the lower deck. The oil and gas lines of the production platforms and from satellite wells enter the structures at the lower deck level.

Produced water flows to "skim tank" vessels, also at atmospheric pressure, located on the lower decks. Any remaining petroleum products are separated there and are put back into the oil-stream flow. Remaining water is discharged overboard. All drains on the production platform, including those from "drip pans" and gutters, also flow into the skim tanks. When the study commenced, produced water was discharged from both production platforms. Since November 1978, however, routine discharge has been from platform 296-B only.

Also located on the lower decks of production platforms are seawater pumps. Seawater goes to pressurized lines for fire-fighting systems, and to desalinization plants on the lower decks of the nearby quarters platforms. The pumps are fitted with back-pressure valves such that excess seawater is discharged when desalinization plants and fire-fighting water lines are at capacity and/or pressure.

Quarters Platforms. Quarters platforms contain in addition to kitchens, living, and recreational quarters, a gas-fired electrical generator, a desalinization plant, and a sewage treatment plant. A helicopter landing pad is located on the upper deck. An upper and lower platform deck are each about 40 ft by 40 ft. All garbage except for food scraps is compacted and transported to shore for disposal; food scraps are collected and dumped overboard. In 1977, about a dozen persons resided in the field, with the work force supplemented as neccesary.

Water needs are normally met by the desalinization plant, although lines are present to take fresh water onto the platform from boats. Raw seawater is pumped continuously through the fresh water condenser, with most of the seawater being discharged overboard.

Flare Stacks. Each of the production platforms has an associated flare stack sited a short distance away from the platform. This is an emergency safety device and no discharges are normally associated with these simple structures. In an emergency, gas is shunted to the flare stacks and ignited to dispose of the discharge.

Sportfishing and Diving. The Buccaneer Field is a focal point for offshore sport fishing and diving activity in the area. Sportfishing boats range in size from 16-ft-long boats to large yachts and sailboats. Many of the craft that come to the area are only marginally seaworthy and are in various states of operating efficiency. Fuel is sometimes lost overboard from the vessels; also, bottles and paper containers are frequently thrown overboard from pleasure craft. Raw sewage and food wastes are commonly discharged overboard.

Oil Spills. The Buccaneer Field has a good record in terms of oil spills, with a total of about 250 gallons (6 bbl) reported as spilled. Twenty-one gallons (0.5 bbl) were lost from platform 296-B on September 28, 1973; 126 gallons (3 bbl) from an unknown source on November 26, 1973; 21 gallons (0.5 bbl) from platform 296-B on July 21, 1974; 70 gallons (1.7 bbl) from a leak in a line on platform 296-B on August 6, 1974; and 15 gallons (0.4 bbl) from platform 288-A on September 30, 1975. No spills were reported during our study.

We have no evidence that oil from the Ixtoc-I blowout or the "Burmah Agate" reached the Buccaneer Field. Most of our sampling had been completed before these incidents.

REFERENCES

Baker, J. M., 1975, "Biological monitoring: principles, methods and difficulties," in: J. M. Baker (ed.), "Ecology and Oil Pollution." Applied Science, London.

Blumer, M., and Sass J., 1972, "Oil pollution: persistence and degradation of spilled fuel oil," Science, 176:1120.

Blumer, M., Souza, G., and Sass J., 1970, "Hydrocarbon pollution of edible shellfish by an oil spill," Marine Biol., 5:195.

Blumer, M., Ehrhardt, M., and Jones, J. H., 1973, "The environmental fate of stranded crude oil," Deep Sea Res., 20:239.

Burns, K. A., and Teal M., 1971, "Hydrocarbon Incorporation into the Salt Marsh Ecosystem from the West Falmouth Oil Spill." Bureau of Commercial Fisheries, Government Printing Office, Washington, D.C.

Cowell, E. B., 1976, "Oil pollution of the sea," in: Johnston, R. (ed.), "Marine Pollution." Academic Press, New York, p. 353.

Cross, F. A., 1979, "Biological effects observed along the northwest coast of Brittany as a result of the Amoco Cadiz oil spill," presented at the 145th National Meeting of the American Association for the Advancement of Science, Houston, Texas, January, 1979.

Easton, R., 1972, "Black Tide: The Santa Barbara Oil Spill and Its Consequences." Delacorte Press, New York.

Gallaway, B. J., Johnson, M. F., Boland, G. S., Lewbel, G. S., Martin, L. R., Margraf, F. J., and Howard, R. L., 1979, "The artificial reef studies," in: "Ecological Investigations of Petroleum Production Platforms in the Central Gulf of Mexico." Final Report to Southwest Research Institute for the Bureau of Land Management, Washington, D.C., 108 pp. plus appendices.

Grose, P. L., and Mattson, J., 1977, "The "Argo Merchant" Oil Spill." NOAA, Boulder, Colorado.

Gulf Universities Research Consortium, 1974, "The Offshore Ecology Investigation." Gulf Universities Research Consortium, Galveston, Texas, 9 pp.

Harper, D. E., Jr., Scrudato, R. J., and Giam, C. S., 1976, "Pilot Study of the Buccaneer Oil Field (Benthos and Sediments). A Preliminary Environmental Assessment of the Buccaneer Oil/Gas Field." Report to NMFS, Galveston, Texas from Texas A&M University, College Station, Texas, 63 pp. plus appendices.

Hess, W. N., 1978, "The "Amoco Cadiz" Oil Spill." NOAA, Boulder, Colorado.

McCaslin, J, 1972, "Worldwide offshore oil output nears 9 million B/D," Oil Gas J., 70:196.

McGinnis, D. R., 1971, "Observations on the zooplankton of the eastern Santa Barbara Channel from May 1969 to March 1970, "in: Straughan, D. (ed.), "Water Pollution by Oil." Institute of Petroleum, London.

National Academy of Sciences, 1975, "Petroleum in the Marine Environment. Workshop on Inputs, Fates and the Effects of Petroleum in the Marine Environment." Washington, D.C.

National Marine Fisheries Service, 1977, "Environmental Assessment of an Active Oil Field in the Northwestern Gulf of Mexico. A Report of NMFS/LGL Workshop I, 1977-1978." NOAA, Galveston, Texas.

Nelson-Smith, A., 1973, "Oil Pollution and Marine Ecology." Plenum Press, New York.

Straughan, D., 1970, "Breeding and larval settlement of certain intertidal invertebrates in the Santa Barbara Channel following pollution by oil," in: Straughan, D. (ed.), "Water Pollution by Oil." Institute of Petroleum, London.

Travers, W. B., and Luney, P. R., 1976, "Drilling, tankers, and oil spills on the Atlantic outer continental shelf," Science, 194:791.

Wilson, C. L., 1970, "Man's Impact on the Global Environment." MIT Press, Cambridge, Massachusetts.

Wilson, R. D., Managhan, P. H., Asanik, A., Price, L. C., and Rogers, M. A., 1973, "Estimated annual input of petroleum to the marine environment from natural marine seepage," in: "Background Papers, Workshop on Petroleum in Marine Environment." Ocean Affairs Board, National Academy of Science, Washington, D.C.

HYDROCARBONS AND SULFUR

Brian S. Middleditch

Department of Biophysical Sciences
University of Houston
Houston, Texas 77004

INTRODUCTION

A detailed study of the Buccaneer Gas and Oil Field was preceded by a pilot study conducted by scientists from Texas A&M University. One of the conclusions of the pilot study was that there was sufficient evidence for petroleum hydrocarbons in the environment in the vicinity of the Buccaneer Field to warrant inclusion of a hydrocarbon study in the main project (Giam, 1976). Contracts awarded to the University of Houston enabled it to perform the hydrocarbon study during all four years of the project (1976-1980).

Our study plan was modified many times as we learned more about the hydrocarbon ecology of the Buccaneer Field, and some additional tasks were added as the need arose. We were primarily interested in the petroleum hydrocarbon components of the produced water discharge (brine in Figure 1), and their distribution in the environment. Very early on in the study we found moderate concentrations of squalene in the produced water discharges from the production platforms. This was indicative of the presence of bacteria in the effluents. Discussions with the platform operators revealed their use of biocides to suppress the populations of sulfur-bacteria, which produce corrosive metabolites. We also noted that the effluents contained comparatively high concentrations of elemental sulfur. We accordingly included analyses of the biocides and sulfur in our protocol. We also performed a limited number of analyses of selected lipids in some samples and occasionally encountered such substances as antioxidant food additives in some samples. All of these findings are described

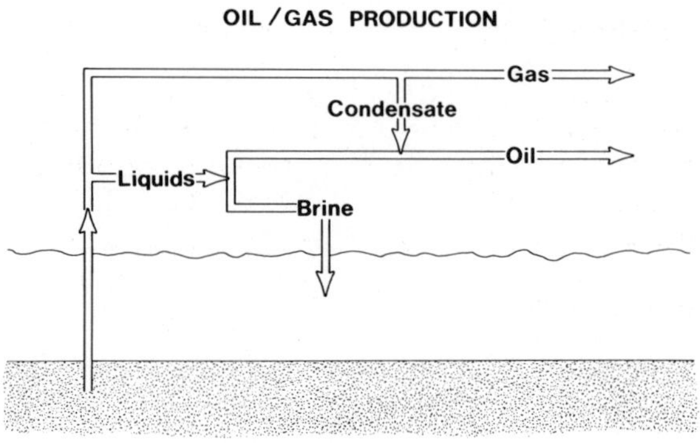

Figure 1. Schematic representation of the separation of oil and gas from produced water ("brine").

here in appropriate detail, although the title of the chapter reflects only the main focus of the later years of the study.

EXPERIMENTAL PROCEDURES

Our primary consideration in conducting this study was producing data which were both appropriate and reliable. The program was designed and managed in such a manner that we had adequate flexibility in the sampling protocol to be able to modify it in response to preliminary findings. No arbitrary sampling grid was imposed upon us. Moreover, we were free to use any analytical procedure that we could justify in terms of accuracy, precision, and cost-effectiveness. We were not expected to use any "standard" procedure: consistency with other studies was deemed less important than producing reliable data.

Sampling Design

The types of samples collected, and their locations, are dealt with in detail in the "Results" section, which provides a narrative of the evolution of the project. There were some constraints on the sampling design, but they were not major impediments.

During the first year of the project, facilities were available for the collection of samples only from boats or the production platforms. Sediment samples were, therefore, collected using coring devices, dredges, or grabs. It was not possible to collect sediments from locations close to pipelines because of the risk of rupturing them. Moreover, the boat could not be stationed for sampling close to the platforms.

During subsequent years of the project, divers were available for sample collection. This enabled us to obtain samples from locations very close to and below the platforms. A constraint here was in the number of samples that a team could collect during a desired time period. This problem was exacerbated when synoptic sampling for other studies was performed. On one occasion when sampling scheduled for a single day was extending into a second day, however, we discovered that there was a large day-to-day change in sediment hydrocarbon concentration.

Sampling Procedures

Seawater and discharge samples were collected directly into glass bottles with Teflon-lined screw caps. Sediment samples collected by divers were scooped directly into wide-mouth glass bottles with ground glass stoppers. Other sediment samples were transferred to similar bottles in the field. Some biota samples were collected by hand by divers while others were collected by trawling or by hook-and-line methods. Plankton samples were collected according to MARMAP II protocols.

All samples were frozen in the field, whenever practical, to minimize bacterial contamination during transit and storage. Freezers were available both on the boats and on platform 296-B.

Analytical Procedures

Particular care was taken to minimize contamination of the samples and to obtain the highest degree of accuracy and precision.

Contamination. Contamination by bacterial metabolites was minimized by freezing the samples as soon as possible after collection and maintaining them in this state until they were analyzed. Some sulfur-bacteria produce hydrocarbons, so this type of contamination would have been particularly serious.

Contamination in the laboratory was minimized by judicious choice of apparatus, solvents, and reagents. Samples were allowed to come into contact with apparatus constructed only of glass, Teflon, and stainless steel. Most plastics contain plasticizers (usually phthalate esters) which interfere with hydrocarbon analyses, so these were not used. All solvents were "distilled-in-glass" grade: less expensive grades contain impurities which interfere with hydrocarbon analyses. Periodic blank analyses were performed, and the validity of analyses performed by gas chromatography (GC) was checked using combined gas chromatography - mass spectrometry (GC-MS).

Some contamination was occasionally encountered, but this was recognized as such and the validity of the data was not compromised. Some batches of cyclohexane were contaminated with bicyclohexyl (Middleditch and West, 1979) which interfered with the determination of n-tridecane. Phthalate esters, one of which had the same retention time as n-pentacosane, were sometimes detected. It was not always possible to determine whether the phthalate esters were contaminants (introduced during sampling or analysis) or whether they were actual components of the samples. We were never able to eliminate chloroform from our blank analyses. However, it did not interfere with our analyses of volatiles.

A related problem is that squalene (a polyunsaturated hydrocarbon common in tissues of marine organisms) interferes with the analysis of n-octacosane if it is not completely removed from the alkane fraction prior to gas chromatography.

Accuracy and Precision. Complete extraction of hydrocarbons is required to obtain an accurate estimate of their concentration. If extraction is incomplete but reproducible, accuracy will suffer but precision will be maintained. It is impossible to determine whether complete extraction is being obtained from a "real" sample. Our procedures achieve the recovery of more than 95% of added hydrocarbon, but this does not prove that a similar proportion of endogenous hydrocarbon is being extracted. If the residue of a sample is further extracted after our routine extraction procedure, no more than 5% of additional hydrocarbon can be obtained, even using exhaustive methods. Again, this does not prove that at least 95% of the total hydrocarbon content is recovered: some endogenous hydrocarbon may not be amenable to extraction.

To optimize the recovery of hydrocarbons from tissue and sediment samples, they were homogenized by a combination of mechanical and ultrasonic action and then saponified prior to solvent extraction. The objective was to break up the sample as much as possible so that a large surface area could be exposed to solvents. Thus, the saponfication step was employed to degrade cell membranes as much as to hydrolyze polar lipids.

The small quantities of hydrocarbons present in many of the seawater samples could have been lost by adsorption onto glass surfaces if precautions had not been taken to minimize this effect. Seawater samples were collected directly into 1-liter glass bottles. Only about 800 ml was collected in each bottle so that there was room for expansion during freezing. This also left sufficient room in the same bottles for addition of the solvent used for extraction. Solvent was tumbled with the sample in these

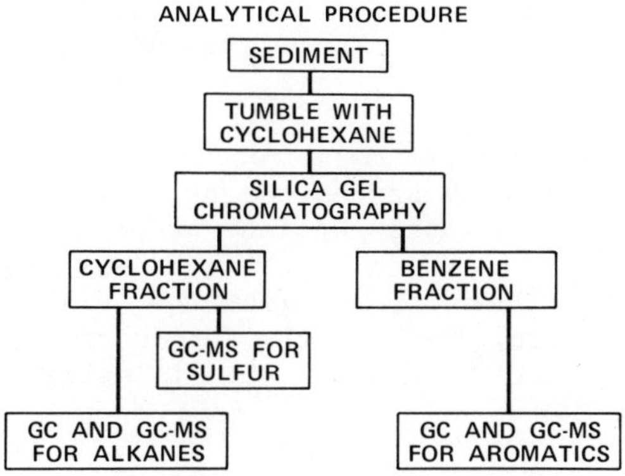

Figure 2. Integrated analytical procedure for alkanes, aromatic hydrocarbons, and sulfur, illustrated for sediments.

bottles for 24 hours to extract hydrocarbons both from the water and from the glass itself.

Even if complete extraction is obtained, the accuracy of the method can suffer if there are losses of hydrocarbons before the final step of the analysis. If these losses vary from sample to sample, precision will also be impaired. A simple method of avoiding this problem involves the use of isotope-labeled internal standards, a method that has been used in biomedical research for more than a decade. Addition to the sample of an isotope-labeled analog of the substance of interest prior to extraction will compensate both for losses of the material after extraction and for injection errors in gas chromatography.

We used isotope-labeled internal standards for all of the analyses. In most instances a mass spectrometer is required as the gas chromatographic detector to differentiate between the labeled and unlabeled species (Sweeley et al., 1966). For n-alkanes, however, each perdeuteriated compound is completely separated from the corresponding unlabeled n-alkane. Thus, no mass spectrometer is required to differentiate these pairs of compounds, even when conventional packed gas chromatographic columns are employed (Middleditch and Basile, 1976).

The various compounds employed for hydrocarbon analysis were n-[$^2H_{66}$]dotriacontane, [$^2H_{10}$]pyrene, [$^2H_{12}$]benzanthracene, [$^2H_{10}$]-1-methylnaphthalene, [$^2H_{10}$]-2-methylnaphthalene, and [$^2H_{12}$]benzo[*a*]pyrene. An appropriate quantity of the deuteriated hydrocarbons was added to the sample before analysis. For example, 2 l of a 0.1% solution added to 1 l of seawater afforded a concentration of 2 ppb.

We used sulfur enriched in sulfur-34 as the internal standard for the analysis of elemental sulfur (Middleditch and Basile, 1979).

In summary, we are no more certain of the accuracy of our analytical procedures than anyone can be, but we were using state-of-the-art methods to maintain optimum accuracy and precision. If the precision of our analytical techniques was inadequate we would not, for example, have been able to discern concentration gradients in the vicinity of the production platforms. Our integrated analytical procedure is illustrated for sediment samples in Figure 2.

RESULTS

Oil and Condensate

Oil and condensate were examined, not because they are discharged from the platforms but because a knowledge of their composition was an aid to interpreting other observations. Figure 1 shows the product flow on the production platforms. The oil was sampled before it was mixed with condensate. A third type of sample (not indicated in Figure 1) is designated "separator tank oil." Some oil floats to the surface of the produced water in the final skim tank; the composition of this is similar to that of the produced water, and provides additional useful data.

Crude Oil. The alkane composition of the crude oil is consistent with reports on the composition of other crude oils. Alkanes from C_{12} to C_{36} comprise approximately 18% of the oil, and they show a decreasing concentration with increase in chain length and an OEP average of 1.02. The n-heptadecane/pristane ratio and the n-octadecane/phytane ratio, often used as indicators for the source of oil, are 0.90 and 2.44, respectively. Crude oils from both production platforms have similar compositions.

Condensate. The condensate contains a lower concentration of alkanes in the C_{12} to C_{36} region than the crude oil (approximately 14%), and larger amounts of the lighter alkanes. The n-heptadecane/pristane ratio is 1.30, and the n-octadecane/phytane ratio is 3.33. The difference between these ratios and those found for crude oil probably reflects the greater volatility of the C_{17} and C_{18} n-alkanes, compared with that of the C_{19} and C_{20} branched alkanes. Condensate from both platforms was similar in composition.

TABLE 1. EXTRACTABLES IN SEPARATOR TANK OIL: JUNE 24, 1979.

Peak	I[a]	Component	ppm
1	1100	n-Undecane	5,840
2	1105	C_4-Benzene	57
3	1133	C_4-Benzene	216
4	1156	Naphthalene	537
5	1200	n-Dodecane + C_5-Benzene	8,790
6	1214	Branched tridecane	2,400
7	1267	2-Methylnaphthalene	519
8	1277	Branched tetradecane	3,250
9	1281	1-Methylnaphthalene	486
10	1300	n-Tridecane	10,890
11	1317	C_7-Benzene	40
12	1347	Acenaphthene or Biphenyl	162
13	1363	C_2-Naphthalene	94
14	1374	C_2-Naphthalene	203
15	1287	C_2-Naphthalene	400
16	1390	C_2-Naphthalene	61
17	1400	n-Tetradecane	13,850
18	1405	C_2-Naphthalene	193
19	1417	C_2-Naphthalene	81
20	1441	?	45
21	1450	Methylbiphenyl	245
22	1463	Branched pentadecane	6,220
23	1469	?	41
24	1492	C_3-Naphthalene	114
25	1500	n-Pentadecane	13,750
26	1513	C_3-Naphthalene	164
27	1530	C_3-Naphthalene	234
28	1543	C_2-Biphenyl	44
29	1550	C_3-Naphthalene	178
30	1557	C_2-Biphenyl	1,970
31	1563	Branched alkane	75
32	1565	?	41
33	1582	?	41
34	1600	n-Hexadecane	11,690
35	1618	C_4-Naphthalene	45
36	1638	C_4-Naphthalene	57
37	1644	?	55
38	1651	C_4-Naphthalene + branched alkane	3,310
39	1671	?	61
40	1694	C_4-Naphthalene	111
41	1700	n-Heptadecane	11,260
42	1709	Pristane	6,630
43	1763	?	1,090

(continued)

TABLE 1 (continued)

44	1800	n-octadecane	8,720
45	1811	Phytane	1,290
46	1900	n-Nonadecane	6,500
47	2000	n-Eicosane	3,850
48	2100	n-Heneicosane	2,850
49	2200	n-Docosane	2,700
50	2300	n-Tricosane	2,420

[a]I = Retention index.

Separator Tank Oil. The major components of the separator tank oil were, again, the n-alkanes. Their relative concentrations also decreased with chain length, and the heaviest alkanes seen were usually around C_{30}. A typical analysis of the nonvolatile components is given in Table 1. The n-heptadecane/pristane ratio is 1.70, and the n-octadecane/phytane ratio is 6.76.

The n-heptadecane/pristane and n-octadecane/phytane ratios are frequently used to characterize spilled oil. The values reported here for the crude oil would be appropriate for this purpose if a spill of crude oil were to occur. However, the different ratios for condensate or separator tank oil should be used if these materials were to be spilled. Moreover, if there were to be a spill from a leaking pipeline, a different ratio might be anticipated. In such a case, it would be prudent to obtain a sample of the oil at its source for comparison with environmental samples.

Produced Water

Many analyses of produced water were performed during our study. During the early years of the survey, we used alkanes as "markers" for tracing the distribution of oil field contaminants throughout the area, so we concentrated on the alkane content of the produced water. We later added methylnaphthalenes and benzo[*a*]pyrene as markers for the light and heavy aromatic hydrocarbons, respectively. We also determined the concentrations of elemental sulfur. Several detailed analyses were carried out to obtain a more complete inventory of the composition of the produced water. Finally, the biocide content of these discharges was investigated.

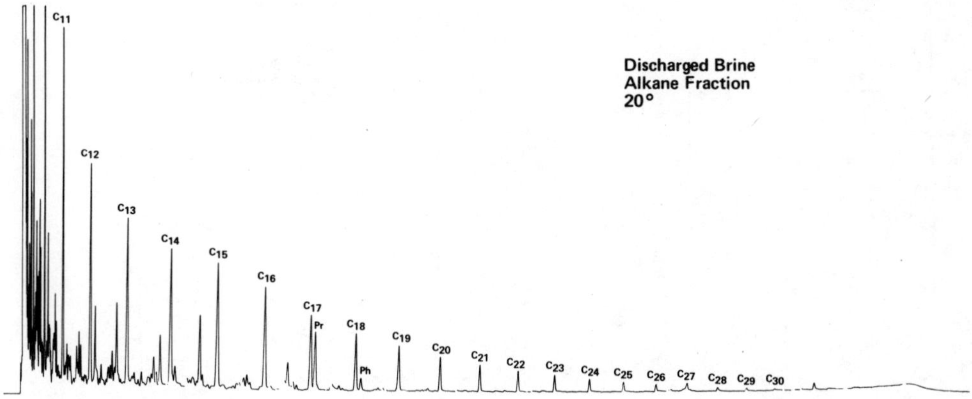

Figure 3. Gas chromatogram for alkane fraction of produced water discharged from platform 296-B on March 8, 1977 at 20°C.

Heavy Alkanes. The protocol used for this analysis provided quantitative data for alkanes greater in molecular weight than dodecane. n-Alkanes with chain lengths to C_{38} were found, with odd-even preference values of around unity. The n-heptadecane/pristane ratios for the produced water range from 0.89 to 1.25 (mean, 1.05), while value for the n-octadecane/phytane ratios range from 1.0 to 5.26 (mean, 2.44). These ratios are closer to those for the crude oil than to those for the condensate and separator tank water.

Most of the produced water samples exhibited concentration maxima below n-C_{23}. It is possible that differences in maxima are due to varying degrees of evaporation of the lighter alkanes. Thus, a sample collected at 20°C from platform 296-B on March 8, 1977 exhibited a concentration maximum below C_{11} and an alkane concentration of 6.2 ppm (Figure 3), whereas a sample collected at 28°C on the same day from platform 288-A exhibited a concentration maximum at C_{17} and an alkane concentration of 570 ppb (Figure 4).

Moreover, the concentration of the alkanes in the produced water varies with time. Ten samples collected over a 30-hour period on September 2-3, 1978 had a mean alkane concentration of 2.98 ppm, but concentrations in individual samples ranged from 0.9 to 4.8 ppm.

An additional variable is the rate of discharge of produced water. Information provided by Shell Oil Co. for the period November 1973 to April 1978 indicates discharge rates from 118 to 2000 bbl/day (1 bbl = 159 l). Measurements reported for August 1979 by Gallaway (1980) corresponded to a mean discharge rate of 1000 bbl/day.

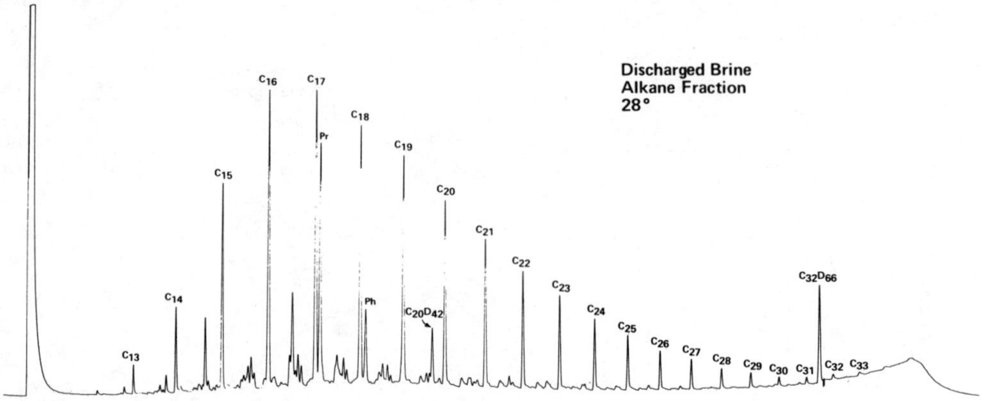

Figure 4. Gas chromatogram for alkane fraction of produced water discharged from platform 288-A on March 8, 1977 at 28°C.

Given these variations in discharge rate, alkane concentration, and composition of the alkane fraction, it is not easy to estimate the mean rate of discharge of individual alkanes. However, a concentration of 3.37 ppm of total alkanes in 1000 bbl/day corresponds to a daily discharge of 560 g, or an annual discharge of approximately 200 kg.

Aromatic Hydrocarbons. Particular emphasis was given during this study to the methylnaphthalenes (1-methylnaphthalene and 2-methylnaphthalene) and to benzo[*a*]pyrene. The former compounds were chosen as being characteristic of the lighter aromatic hydrocarbons (which are relatively soluble and exhibit toxicity toward many marine organisms), while benzo[*a*]pyrene is typical of the less soluble polycyclic aromatic hydrocarbons (some of which are carcinogens). Concentrations for the individual methylnaphthalenes were determined. The ratio of 2-methyl isomer to 1-methyl isomer was approximately 5:4 and relatively constant. Only the total methylnaphthalene concentrations are discussed here.

Methylnaphthalenes were detected in most samples of produced water. Their mean concentration was 43 ppb, while concentrations up to 170 ppb were encountered. This corresponds to a mean discharge rate of about 7.1 g/day or 2.6 kg/year. This is much less than for the alkanes. In determining their significance, however, relative toxicities should be considered.

Concentrations of benzo[*a*]pyrene as high as 5 ppb were occasionally encountered. The mean concentration was only 1.2 ppb which corresponds to a mean discharge rate of 0.2 g/day or 72 g/year.

TABLE 2. VOLATILES IN PRODUCED WATER DISCHARGE: JANUARY 14, 1979.

Peak	Component	ppb
1	2-Methylbutane	960
2	Chloroform (artifact)	100
3	2-Butanone	300
4	n-Pentane	720
5	Cyclohexane	520
6	?	660
7	Methylcyclopentane	460
8	3-Methylpentane	80
9	?	80
10	Benzene	6100
11	?	60
12	2-Pentanone	160
13	Heptane	80
14	n-Hexane	460
15	1,3-Dimethylcyclopentane	240
16	Methylcyclohexane	1080
17	Octadiene	40
18	Heptane	80
19	Heptane	100
20	Heptane	80
21	Octene	380
22	3-Methylhexane	200
23	Heptane	240
24	Octane	80
25	Toluene	5460
26	2,2-Dimethyl-3-hexene	160
27	Dimethylcyclohexane	120
28	Octane	40
29	Dimethylcyclohexane	80
30	Trimethylcyclohexane	80
31	Ethylbenzene	1200

Sulfur. Elemental sulfur was visible in most samples of produced water, and it was determined by a gravimetric procedure. The maximum concentration of elemental sulfur was 1200 ppm, and the mean concentration was 460 ppm. This corresponds to 73 kg/day or 27 metric tons/year. This is more than 100 times the discharge rate for alkanes. As with the alkanes, the toxicity of sulfur is generally considered to be low. In assessing the significance of the very large amounts of sulfur being discharged, however, its potential value as a nutrient should also be considered.

TABLE 3. EXTRACTABLES IN PRODUCED WATER: JUNE 1979.

Peak	I[a]	Component	ppb
1	1018	C_3-Benzene	12.5
2	1039	C_3-Benzene	11.7
3	1062	C_4-Benzene	1.7
4	1065	C_4-Benzene	0.4
5	1068	C_4-Benzene	1.8
6	1073	C_4-Benzene	1.5
7	1084	C_4-Benzene	3.6
8	1088	C_4-Benzene	2.4
9	1092	C_4-Benzene	2.4
10	1102	C_4-Benzene	1.7
11	1113	C_5-Benzene	1.8
12	1118	C_4-Benzene	3.2
13	1133	C_5-Benzene + $C_{5:1}$-Benzene	0.4
14	1130	C_4-Benzene	5.2
15	1148	C_5-Benzene	0.4
16	1156	Naphthalene	11.1
17	1158	$C_{5:1}$-Benzene	0.5
18	1159	C_5-Benzene	0.6
19	1160	C_5-Benzene	0.8
20	1162	C_6-Benzene + $C_{6:1}$-Benzene	0.7
21	1172	C_6-Benzene	1.3
22	1186	C_6Benzene	0.3
23	1193	?	0.5
24	1243	C_6-Benzene	0.4
25	1249	C_6-Benzene	0.5
26	1255	?	0.7
27	1281	1-Methylnaphthalene	7.2
28	1307	?	0.5
29	1320	C_7-Benzene	0.9
30	1338	?	0.3
31	1348	Biphenyl or Acenaphthene	2.8
32	1364	C_2-Naphthalene	1.0
33	1376	C_2-Naphthalene	1.8
34	1390	C_2-Naphthalene	4.0
35	1393	C_2-Naphthalene	1.0
36	1398	C_8-Benzene	0.5
37	1400	n-Tetradecane	22.5
38	1403	C_2-Naphthalene	1.7
39	1418	C_2-Naphthalene	0.9
40	1424	C_8-Benzene	0.4
41	1442	C_8-Benzene	0.7
42	1452	Methylbiphenyl	2.3
43	1458	?	0.5

(continued)

TABLE 3 (continued)

44	1476	C_3-Naphthalene	0.6
45	1488	$C_{8:1}$-Benzene	0.5
46	1494	C_2-Biphenyl + C_3-Naphthalene	0.6
47	1500	C_3-Naphthalene + n-Pentadecane	b
48	1516	C_3-Naphthalene	1.1
49	1526	?	0.4
50	1532	C_3-Naphthalene	1.2
51	1535	C_9-Benzene	1.0
52	1542	C_9-Benzene	0.5
53	1552	C_3-Naphthalene	1.4
54	1558	C_2-Biphenyl	0.4
55	1584	C_{10}-Benzene	0.5
56	1600	n-Hexadecane	35.7
57	1641	C_{10}-Benzene + C_4-Naphthalene	0.5
58	1662	C_{10}-Benzene + C_4-Naphthalene	0.5
59	1697	C_4-Naphthalene + C_5-Naphthalene	0.4
60	1700	n-Heptadecane	29.6
61	1764	C_{11}-Benzene	0.3
62	1800	n-Octadecane	31.0
63	1868	?	0.5
64	1900	n-Nonadecane	35.4
65	2000	n-Eicosane	32.7
66	2100	n-Heneicosane	50.0
67	2200	n-Docosane	77.4
68	2300	n-Tricosane	197.1
69	2400	n-Tetracosane	248.0
70	2500	n-Pentacosane	208.6
71	2600	n-Hexacosane	170.7
72	2700	n-Heptacosane	128.5
73	2800	n-Octacosane	103.2
74	2900	n-Nonacosane	105.1

[a]I = Retention index, [b]a C_3-naphthalene is obscured by n-pentadecane.

The concentration of sulfur in the produced water changes throughout the day and is unrelated to the alkane concentration.

Other Contaminants. One of the tasks that we undertook during this study was to identify and quantitate as many as possible of the components in the produced water discharge.

We were not satisfied with the degree of accuracy that could be attained using existing procedures for the analysis of volatiles. Indeed, as of December 1979 the EPA had not evaluated the accuracy and precision of the "priority pollutant" method for volatiles. Accordingly, we developed a procedure employing [$^{2}H_6$]benzene as internal standard. Results of a typical analysis are given in Table 2. Most of the volatile components were expected petroleum hydrocarbons. Chloroform was an artifact which appeared in blank analyses and could not be eliminated. The 2-butanone and 2-pentanone were probably from solvent employed by painters on the platforms. In July 1979 a 55-gallon drum of "methyl ethyl ketone" (2-butanone) was in use on Platform 296-B for this purpose. 2-Pentanone is a known contaminant of technical grade methyl ethyl ketone. Runoff from the platform is channeled to the skim tanks, so this is the probable source of these compounds.

Responsibility for the analysis of volatile components of the discharges passed to Dr. James Brooks of Texas A&M University in August 1979: his data are presented elsewhere (Brooks et al., 1980).

A typical analysis of the produced water extractables is given in Table 3. All of these compounds are expected petroleum constituents.

The total number of components identified in produced water is:

Volatiles (Table 2)	30
Extractables (Table 3)	74
Other n-alkanes (C_{30} to C_{38})	9
Benzo[*a*]pyrene	1
Sulfur	1
TOTAL	115

Many additional compounds could be found if specific analyses were performed, as for benzo[*a*]pyrene.

Priority Pollutants. Attention is given here to the organic priority pollutants because they are featured in the new Ocean Discharge Criteria (Federal Register, February 12, 1980). At the detection limits of the priority pollutant protocol (generally, 10 ppb), only the following would be reported:

Acenaphthylene
Benzene
Ethylbenzene
Naphthalene
Toluene

Benzo[*a*]pyrene is discussed above, and acrolein in the following Chapter. The recently issued Consolidated Permit Regulations

(Federal Register, May 19, 1980) list 79 "hazardous pollutants" in addition to the priority pollutants. Of those which are organic compounds, we detected only cyclohexane. Xylenes were almost certainly present but were too volatile for analysis as extractables and insufficiently volatile for detection by the protocol that we employed for volatiles. Isoprene may also have been present, but it is too volatile for analysis by the standard purge-and-trap procedure.

Other Platform Discharges

The rate of treated sewage discharge from platform 296-B is about 30 liters per hour, but it flows on an intermittent basis. Alkane concentrations in sewage were as high as 800 ppb but the alkane/isprenoid ratios were not typical of petroleum, and C_{17}, C_{18}, and C_{20} alkanes usually predominated. Also, most samples showed a strong odd-carbon preference, indicating biogenic alkanes. It is likely that alkanes in sewage represent biogenic input from bacteria or feces, for example.

Overflow from the fire-fighting system had an alkane content which was generally in the range of 1-5 ppb. In addition to the petroleum alkanes, some samples contained a group of unresolved hydrocarbons in the region C_{20} to C_{30}. These compounds were particularly notable (44 ppb) in one sample. It is possible that they derived from a rubber hose. One sample of fire-fighting water contained the bacterial metabolite squalene.

Food scraps are dumped into the sea from the quarters platform adjacent to platform 296-B. Platform personnel estimate the quantity to be approximately 1 kg/day. Fish have been observed eating these scraps, and some of this material (particularly corn kernels) was found among stomach contents (Gallaway, 1980). Interestingly, the food preservative butylated hydroxytoluene (BHT) was identified in the alkane fraction of a sheepshead muscle sample (Figure 5). The BHT was particularly prominent in this chromatogram, probably because the alkane content of the sample was relatively low and a high gain was needed to detect the alkanes. With such a high gain, an unresolved "hump" of petroleum hydrocarbons was seen between C_{17} and C_{25}. [The baseline of the chromatogram rose again between C_{25} and C_{31} as column bleed increased, and then fell after the column temperature had reached its maximum.]

Seawater

Initial studies (1976-77) of hydrocarbons in seawater used samples collected at many locations, both adjacent to the

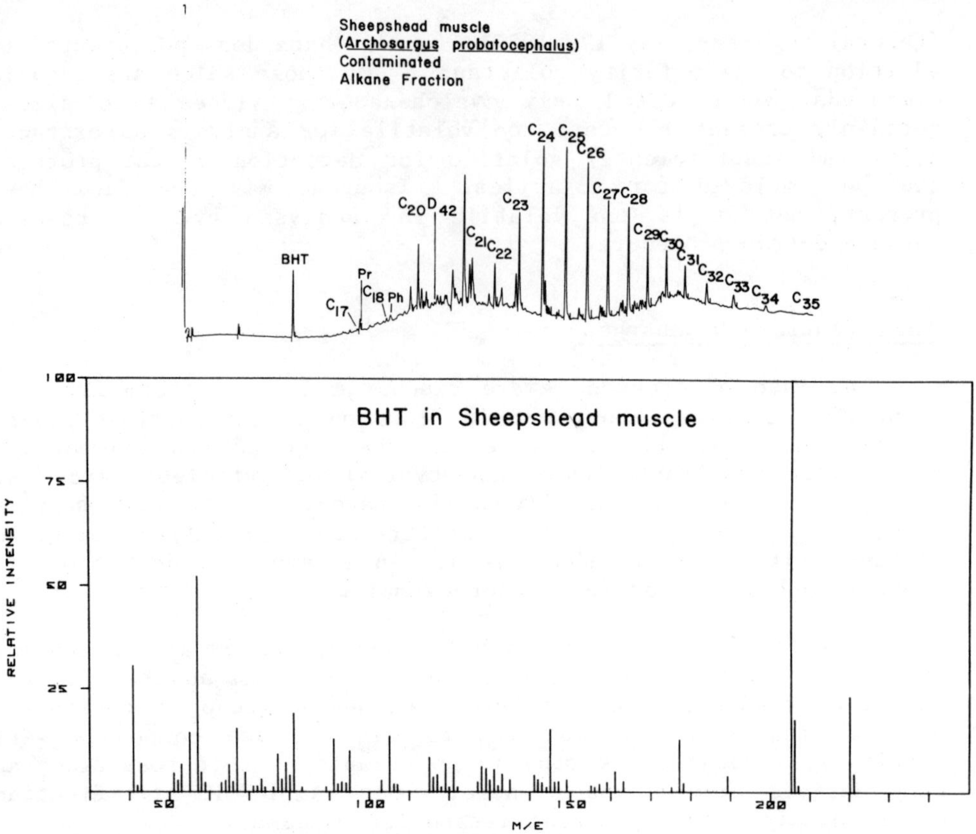

Figure 5. Gas chromatogram of alkane fraction of sheepshead (*A. probatocephalus*) muscle containing BHT (upper), and mass spectrum of this component (lower) collected August 24, 1978.

platforms and up to 10 km from them. Occasional surface seawater samples contained up to 30 ppb of petroleum alkanes and some bottom seawater samples contained up to 25 ppb of biogenic alkanes. No concentration gradients of alkanes were found in surface water samples around the platforms. During the period August 31 - September 1, 1976, however, relatively high concentrations of alkanes were seen down-current (southwest) of the production platforms in bottom water samples (Figure 6). These alkanes were in the range C_{25} to C_{35} and appeared to be biogenic, with a distribution similar to those produced by sulfur-utilizing bacteria (Davis, 1968).

The vessel used for sampling during the first year of the study could not be maneuvered close to the platforms, so subsequent sampling was by divers or from a small inflatable boat.

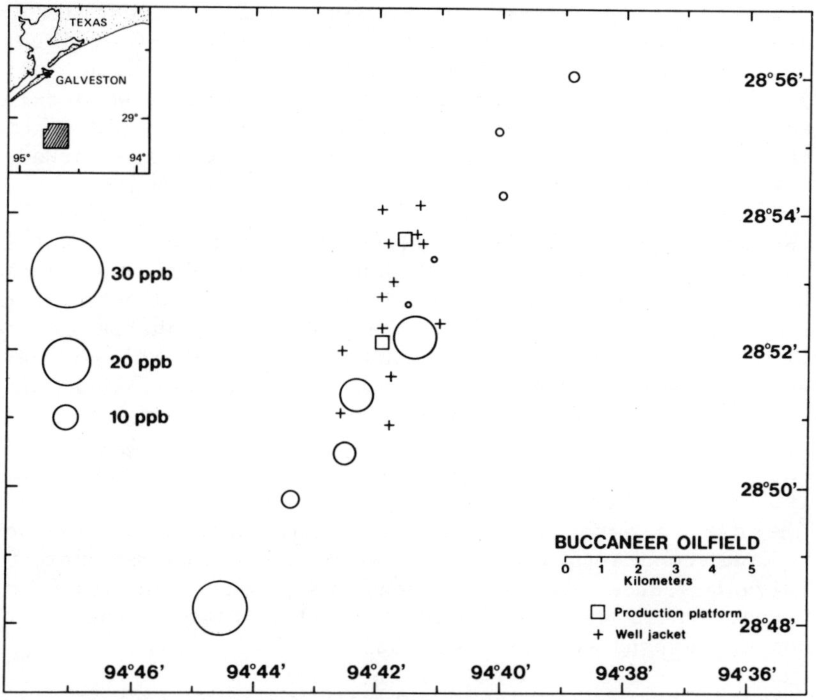

Figure 6. Distribution of biogenic alkanes in bottom water samples during August 31 - September 1, 1976.

Three approaches were used in an attempt to delineate concentration gradients of contaminants in the vicinity of the discharge pipes:

(1) three-dimensional "bullseye" sampling,
(2) sampling from dye-marked plumes, and
(3) tagging effluents with single alkanes.

Three-dimensional "Bullseye" Sampling. Samples were collected from the air/sea interface, at depths of 3, 5, and 8 m from the surface, and at the sediment/water interface. At each depth, a sample was taken directly below the discharge pipe, and an additonal 4-12 samples were taken on transects extending N, S, E, and W of these points. This sampling was repeated during the four seasons of 1978-79. Measurable amounts of petroleum alkanes were usually found directly below the discharge pipe at the air/sea interface and at nearby sampling sites. However, there were no clearly defined concentration gradients. Moreover, the distribution of contaminants was not related to wind or current strength or direction.

Sampling from Dye-marked Plumes. Other investigators used Rhodamine WT dye to investigate the distribution of effluents from the platforms (Armstrong and Hamilton, 1979; Workman and Jones, 1979). Under calm conditons the dye-marked plumes were relatively static and dye penetrated to depths as great as 10 m from the surface. On other occasions, currents swept the dye-marked plumes away from the platforms.

We conducted a similar experiment at a time when the current was flowing at about 3 knots. Rhodamine WT dye was poured into thesea during two consecutive produced water discharge cycles. Care was taken to ensure that the dye and discharge entered the sea at the same point simultaneously. Each of the two resulting dye patches was carried away from the platform, and they were sampled from an inflatable rubber boat.

Alkane and dye concentrations were determined for each sample. While there were clearly defined concentration gradients for the dye in each patch, there was no correlation between dye and alkane concentrations. We were unable to account for any significant source of hydrocarbons other than the produced water discharge, so it was concluded that the distribution of the dye did not correspond to the distribution of alkanes.

Tagging Effluents with Single Alkanes. In a final attempt at identifying concentration gradients in the water column, we decided to use a marker with chemical and physical properties more closely similar to hydrocarbons. It was deemed undesirable to use hydrocarbons labeled with radioisotopes if samples were to be collected by divers. Also, we were reluctant to use large amounts of stable isotopes in an area under intensive scientific scrutiny for fear of invalidating any future study. The only reasonable method of tagging the effluents was to add a relatively large quantity of a single hydrocarbon.

It seemed appropriate to use an alkane for this purpose, since they are non-toxic, odorless, and colorless. The C_{20} alkane n-eicosane was selected since it was in the mid-range of the alkanes in the produced water and since (unlike the odd-carbon alkanes) it is not among the major biogenic hydrocarbons in most organisms.

The total daily discharge of heavy alkanes is approximately 560 g. n-Eicosane usually comprises about 5% of this amount, or about 30 g per day. The produced water discharge generally flows for about three minutes at a time every six minutes. Thus, each three-minute discharge contains about 0.125 g of n-eicosane. We decided to add 1 kg of n-eicosane to the discharge on one occasion during each of six consecutive days to determine its distribution in the environment. This quantity was considerably greater than either the amount in a single discharge or a single day. Also, the amount of n-eicosane used during the six-day period was more than 30 times the amount in the produced water discharge during that time period.

We purchased 6 kg of n-eicosane from a reputable supplier and analyzed it prior to use. It was found to contain only 27% of n-eicosane along with other n-alkanes. It was apparently paraffin wax and was unsuitable for our experiment. In the brief time period (less than 24 hours) before the vessel sailed we were not able to obtain a further supply of 6 kg of n-eicosane, but we were able to purchase 3 kg of n-octadecane, 2 kg of n-eicosane, and 1.5 kg of n-docosane. We modified our protocol to use 0.75 kg of n-docosane on each of the first two days, 1 kg of n-eicosane on each of the next two days, and 1.5 kg of n-octadecane on each of the last two days.

Samples of seawater from the air/sea interface collected below the discharge pipe immediately after each release of hydrocarbon contained the appropriate hydrocarbon in concentrations higher than would be expected if the sample contained only the hydrocarbons from the produced water discharge. No excess of the added hydrocarbon was found in other samples (seawater, surficial sediments, biota) collected during that week.

Our repeated failure to detect hydrocarbon concentration gradients in the water column indicates that these compounds are rapidly diluted and dispersed after discharge.

Surficial Sediments

Our initial studies (1976-77) were on samples collected from locations throughout the study area. When these failed to yield evidence of petroleum contamination, we focused our attention on the immediate vicinity of the production platforms.

Area-wide Studies. During 1976-77, alkane analyses were performed on samples from 45 locations selected for determination of trace metal content and measurement of sediment texture (Anderson et al., 1977). Only a few of these samples contained any evidence at all for petroleum hydrocarbons, and those samples were collected several kilometers from the production platforms.

Platform Studies. In general, we have found concentration gradients for alkanes in surficial sediments in the immediate vicinity of the production platforms; concentrations decreasing with increasing distance from the discharge. Anomalously low concentrations are sometimes found in sediments from below the

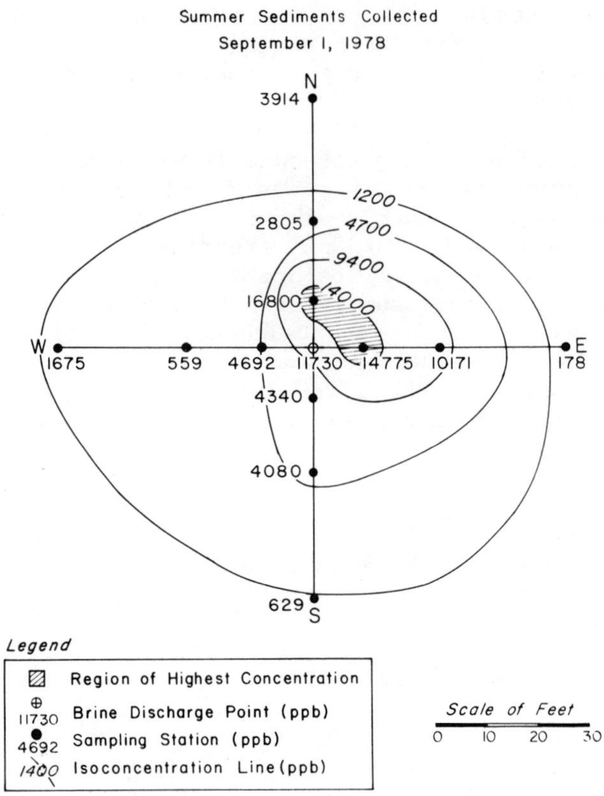

Figure 7. Distribution of alkanes in surficial sediments on September 1, 1978.

structures, presumbly due to the scouring effects of currents round the legs of the structures.

On one occasion when sampling could not be completed in a single day and it extended into a second day, data were obtained which were consistent with prior observations of sediment resuspension in the region (Anderson et al., 1977, 1979). On September 1, 1978 (Figure 7), samples were collected on transects running N, S, E, and W of the production platform. On the following day (Figure 8), samples were collected along intermediate transects. Significantly lower alkane concentrations were found on the second day, indicating that sediment resuspension had taken place.

A particularly notable feature of the hydrocarbons in the surficial sediments from below the production platforms is that they represent relatively fresh oil: i.e., the lighter alkanes were still present in these samples (Figure 9). Weathering would have eliminated these components. The n-heptadecane/pristane and

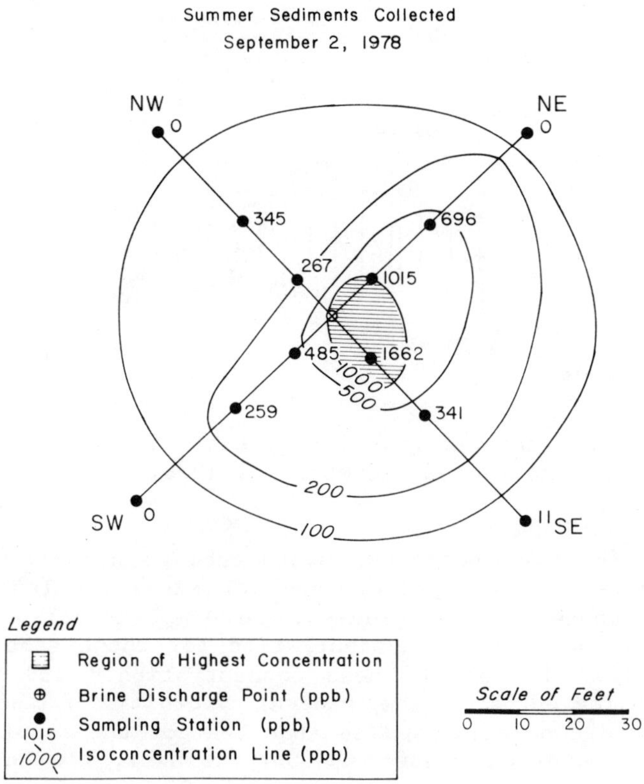

Figure 8. Distribution of alkanes in surficial sediments on September 2, 1978.

n-octadecane/phytane ratios in the surficial sediments are similar to those in the produced water discharge.

<u>Sulfur</u>. Elemental sulfur was found in surficial sediments from locations close to the production platforms, but concentration gradients were not as pronounced as they were for the alkanes. Sulfur was also present in most of the other sediment samples collected elsewhere in the region.

The mean sulfur/alkane ratio was 17.4, considerably lower than in the produced water discharge. The maximum sulfur/alkane ratio observed was only 83.0. As in the case of the produced water discharge, there was no clear correlation between alkane and sulfur content.

Since there are no discharges from the well jackets, samples were taken from one such structure to determine whether there is any seepage of oil along the well pipes which might result in an

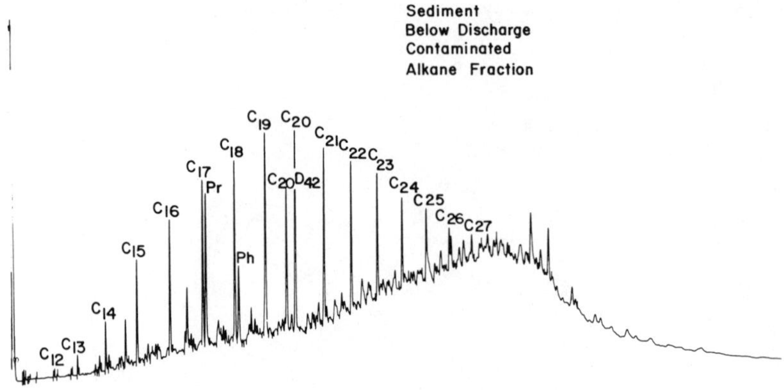

Figure 9. Gas chromatogram of alkane fraction of surficial sediment sample collected December 18, 1978.

additional input of petroleum hydrocarbons into the environment. Sulfur levels were also determined for these samples. There was no alkane concentration gradient in the vicinity of the well jacket. The mean alkane concentration for these samples was 0.12 ppm, less than 10% of the mean concentration (1.39 ppm) for the samples collected from the general vicinity of the production platform, and considerably less than that of the samples collected below the production platform. Surprisingly, the sulfur concentrations near the well jacket (mean 4.10 ppm) are comparable to those near the production platform (mean 3.32 ppm). It seems likely that elemental sulfur is distributed throughout the region of the Buccaneer Field, but we have data only for samples collected in the vicinity of structures.

Sediment Traps

These devices were designed and constructed by Anderson et al. (1979) for sampling particulate matter descending through the water column. They collected fish fecal pellets, bones, and scales, barnacles molts, tissue, and shell debris, bryozoans, algae, metal flakes, and fine sediments.

Petroleum hydrocarbons were usually not found in the contents of these traps. A typical analysis is represented in Figure 10, which provides evidence for the existence of an unusual C_{25} diolefin. We were unable to elucidate the precise structure of this compound, and cannot comment on its significance.

The sediment trap studies provided little evidence for the mode of transport of hydrocarbons through the water column.

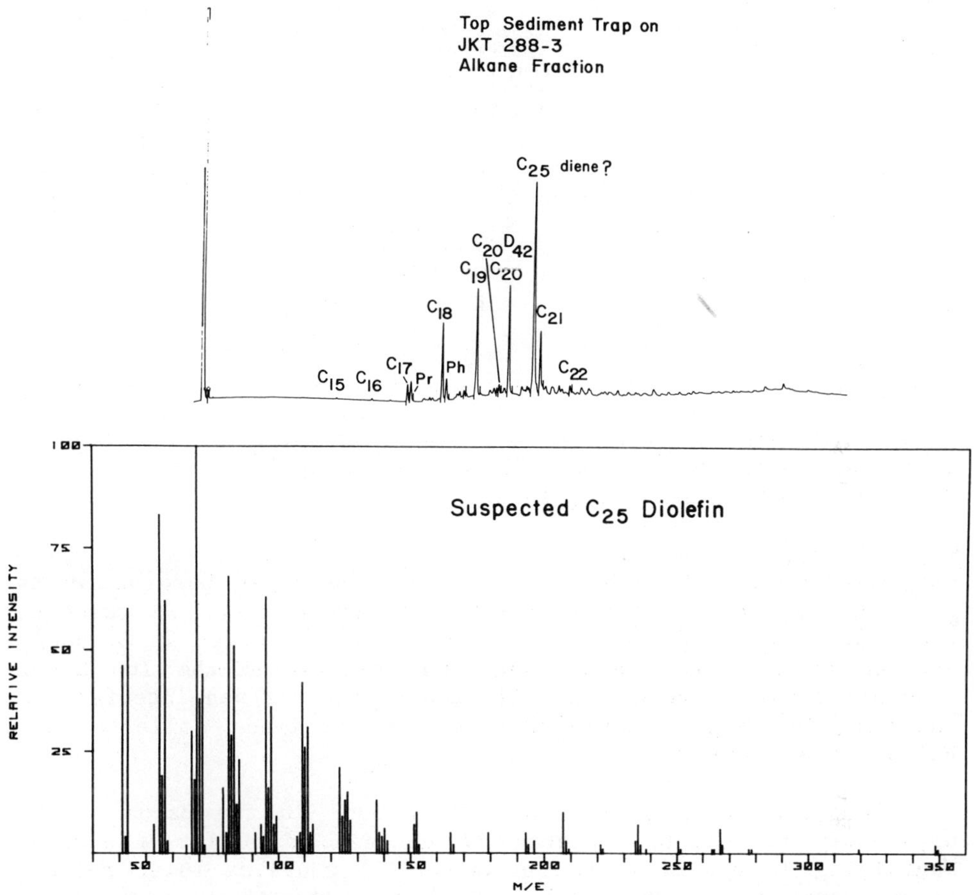

Figure 10. Gas chromatogram of alkane fracton of sediment trap sample (upper) and mass spectrum of an unusual C_{25} diolefin which it contained (lower). Sample collected November 28, 1978.

Platform Fouling Community

The Mediterranean barnacle (*Balanus tintinnabulum*) is the most conspicuous member of the platform fouling community. This population has been fully characterized by Fotheringham (1977), Gallaway (1980), and Gallaway et al. (1979).

Barnacles. During the first two years of the study, flesh from barnacles collected from the production platforms at depths approximately 3 m from the surface was found to contain relatively high concentrations of petroleum alkanes (up to 4 ppm). Those collected from the air/sea interface and at depths greater than 3 m were generally devoid of contaminants. Oil tends to float at

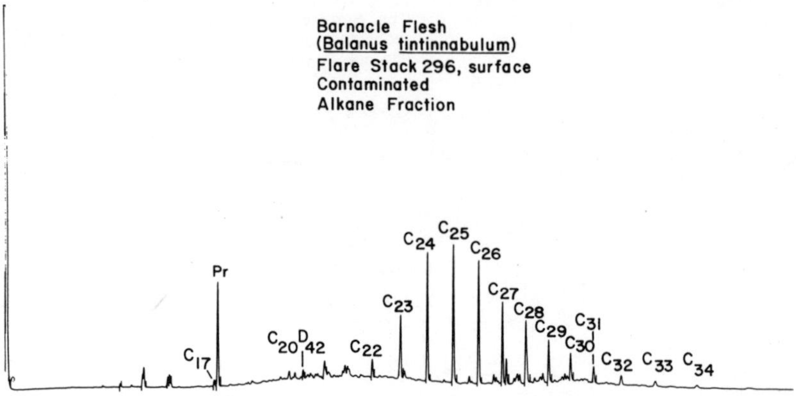

Figure 11. Gas chromatogram of alkane fraction of contaminated barnacles collected from the flare stack associated with platform 296-A, collected August 26, 1978.

the air/sea interface so barnacles in the upper portion of the water column receive the greatest exposure to it. We suspected that more rapid degradation of hydrocarbons occurred in barnacles exposed to the air and/or sun, thus explaining the low alkane concentrations in barnacles collected at the air/sea interface and the somewhat higher concentrations in specimens collected 3 m below the surface.

Degradation of petroleum hydrocarbons in the barnacles was apparently efficient. Figure 11 contains data for a barnacle containing 16 ppm of these compounds; it contains lower relative concentrations of the lighter alkanes than are observed in surficial sediments.

Curiously, almost all of the barnacles collected during the final two years of the study contained very low concentrations of petroleum hydrocarbons. We used similar analytical procedures for barnacles throughout the study, so there is probably no systematic error that could account for this observation, and we have no alternative explanation for the higher concentrations found during the first two years.

Fouling Mat. Some analyses were performed on barnacle shells and their attached organisms, while other analyses were performed on aggregations of the fouling mat. Similar results were obtained from each type of sample. Most samples contained petroleum hydrocarbons, and concentrations were particularly high in those collected just below the air/sea interface. A typical chromatogram for the fouling mat is given in Figure 12.

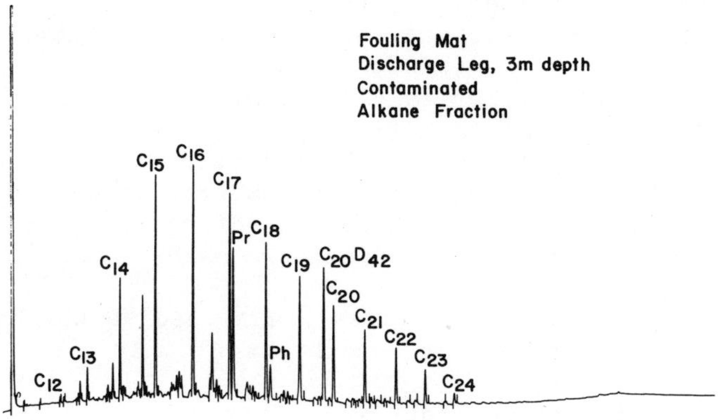

Figure 12. Gas chromatogram of alkane fraction of contaminated fouling mat collected August 26, 1978.

Fish

A large number of fish, representing more than 30 species, was collected for hydrocarbon analysis during the first two years of the project. During subsequent years, when we had more information on the feeding habits and habitats of the different species, we concentrated our efforts on four species which we considered to be particularly significant on the basis of their positions in the food web.

Initial Studies (1976-78). When the project commenced, little was known about the feeding habits or habitats of the fish which could be captured in the Buccaneer Field. Moreover, there was some controversy in the literature concerning the ability (or not) of any marine organisms (including fish) to incorporate petroleum hydrocarbons into their tissues and to concentrate them as they passed through the food web. Consequently, we obtained a wide variety of individual species to determine their content of petroleum hydrocarbons. With the limited resources available, we were able to examine only single specimens of some species and only a few specimens of others.

Species which were found to contain petroleum hydrocarbons were:

Archosargus probatocephalus (sheepshead)
Centropristis philadelphica (rock sea bass)
Chaetodipterus faber (Atlantic spadefish)
Cynoscion arenarius (sand seatrout)
Lutjanus campechanus (red snapper)
Micropogon undulatus (Atlantic croaker)

Pomatomus saltatrix (bluefish)
Porichthys porosissimus (Atlantic midshipman)
Prionotus rubio (blackfin searobin)
Saurida brasiliensis (largescale lizardfish)
Stenotomus caprinus (longspine porgy)
Syacium papillosum (dusky flounder)
Symphurus plagiusa (blackcheek tonguefish)
Synodus foetens (inshore lizardfish)
Urophycis floridanus (southern hake)

Species for which it was not possible to decide whether petroleum hydrocarbons were present or absent were:

Arius felis (sea catfish)
Ogcocephalus parvus (roughback batfish)
Opsanus pardus (leopard toadfish)

Species which were not contaminated were:

Balistes capriscus (gray triggerfish)
Citharichthys spilopterus (bay whiff)
Diplectrum bivittatum (dwarf sand perch)
Echeneis naucrates (sharksucker)
Haemulon aurolineatum (tomtate)
Kyphosus incisor (yellow chum)
Lagodon rhomboides (pinfish)
Menticirrhus americanus (southern kingfish)
Orthopristis chrysoptera (pigfish)
Peprilus burti (Gulf butterfish)
Prionotus ophryus (bandtail searobin)
Scorpaena calcarata (smoothhead scorpionfish)
Serranus subligarus (belted sand bass)

All that we could conclude was that half of the fish that we examined were contaminated, and half were not. The numbers of specimens of each species were too small for us to decide whether any particular species was more susceptible than others to petroleum contamination.

Crested Blenny (*Hypleurochilus geminatus*). Blennies are small fish which feed on and among the platform fouling community. They were also found to be habitat faithful (Gallaway, 1980). Since portions of the fouling community are contaminated with petroleum hydrocarbons (see above), it was of interest to determine whether the blennies became contaminated also.

Virtually every specimen of crested blenny examined contained petroleum alkanes. Concentration maxima were usually in the range C_{20} to C_{31}. A typical analysis is shown in Figure 13. In this specimen, the n-octadecane/phytane ratio is similar to that of the

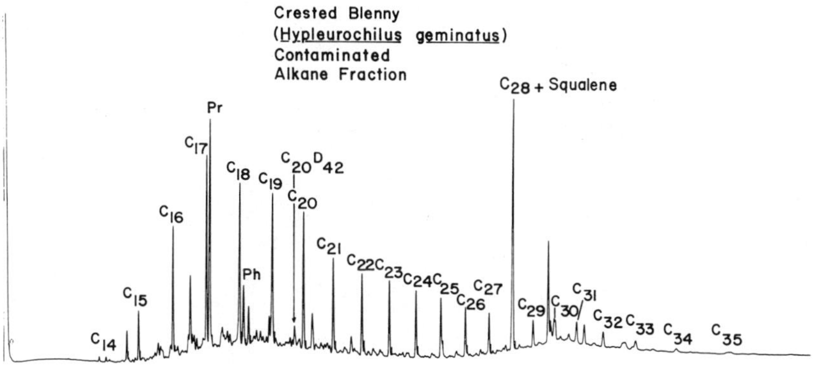

Figure 13. Gas chromatogram of alkane fraction of crested blenny collected March 26, 1979.

produced water, but the n-heptadecane/pristane ratio is distorted by the presence of endogenous pristane of biogenic origin. The mean alkane concentration in this species was 6.8 ppm.

Sheepshead (*Archosargus probatocephalus*). The sheepshead was also found to be habitat faithful. It is well equipped to crush barnacle shells, and consumes their flesh. It also consumes swimming crabs (Gallaway, 1980). Petroleum alkanes were found in about half of the muscle samples from this species and in about a quarter of the liver samples. Typical chromatograms for contaminated muscle and liver samples are given in Figure 14. Concentration maxima were generally at C_{25} or C_{26} for both tissue types. The mean alkane concentrations in muscle and liver tissue were 4.6 and 6.1 ppm, respectively.

Spadefish (*Chaetodipterus faber*). The spadefish is another species which is habitat faithful, but it feeds mainly in the water column (Gallaway, 1980) so one would not expect it to be exposed to petroleum hydrocarbons to the same degree as the crested blenny or sheepshead which feed on the more contaminated sessile organisms. Indeed, the mean alkane concentrations in muscle and liver samples were, respectively, 0.6 and 2.0 ppm.

Red Snapper (*Lutjanus campechanus*). Red snapper apparently migrate to the structures in the Buccaneer Field from a wide region, and become habitat faithful after they arrive. They feed on smaller fish and other organisms (Gallaway, 1980). One might expect these fish to be less contaminated than other species: they were included in the detailed study because they are prized by sportfishermen. About a quarter of the specimens examined contained petroleum alkanes in both their livers and in muscle

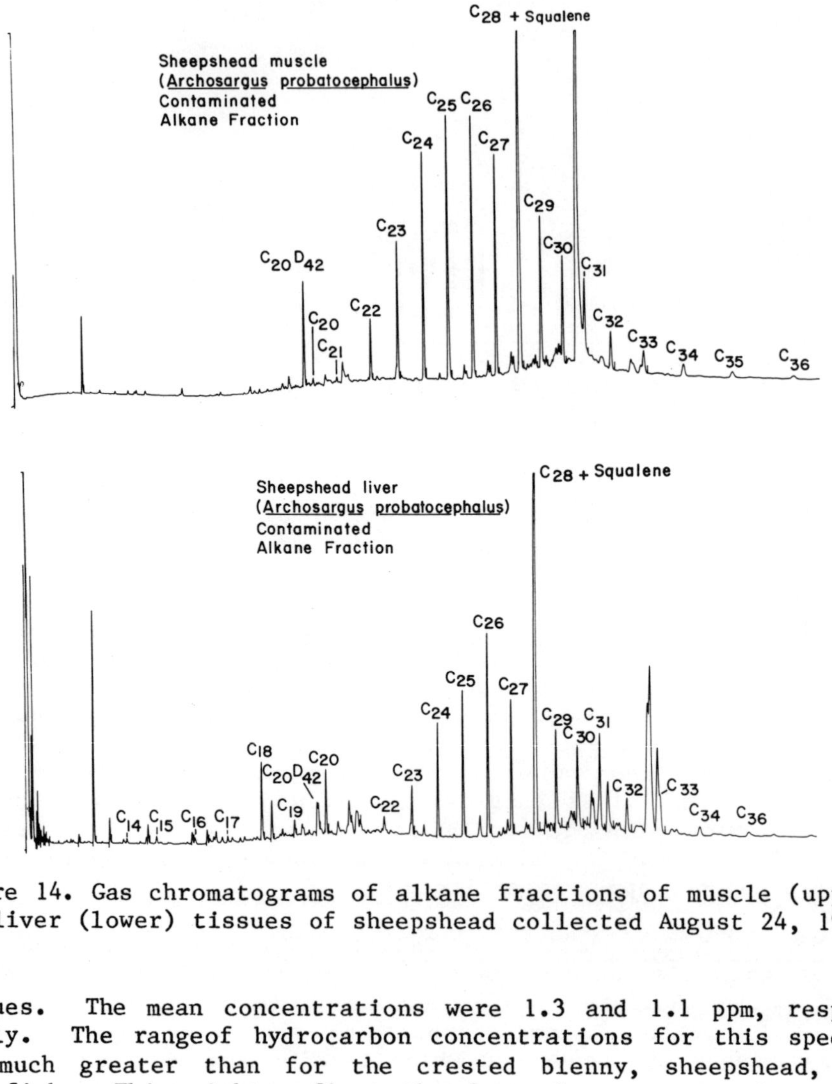

Figure 14. Gas chromatograms of alkane fractions of muscle (upper) and liver (lower) tissues of sheepshead collected August 24, 1978.

tissues. The mean concentrations were 1.3 and 1.1 ppm, respectively. The rangeof hydrocarbon concentrations for this species was much greater than for the crested blenny, sheepshead, and spadefish. This might reflect the fact that some specimens had been resident in the region of the Buccaneer Field longer than others.

Shrimp

Four species of shrimp were examined: *Penaeus aztecus* (brown shrimp), *Penaeus duorarum* (pink shrimp), *Squilla empusa* (mantis shrimp), and *Trachypenaeus similis* (sugar shrimp). Figure 15 shows data for an uncontaminated brown shrimp and one which

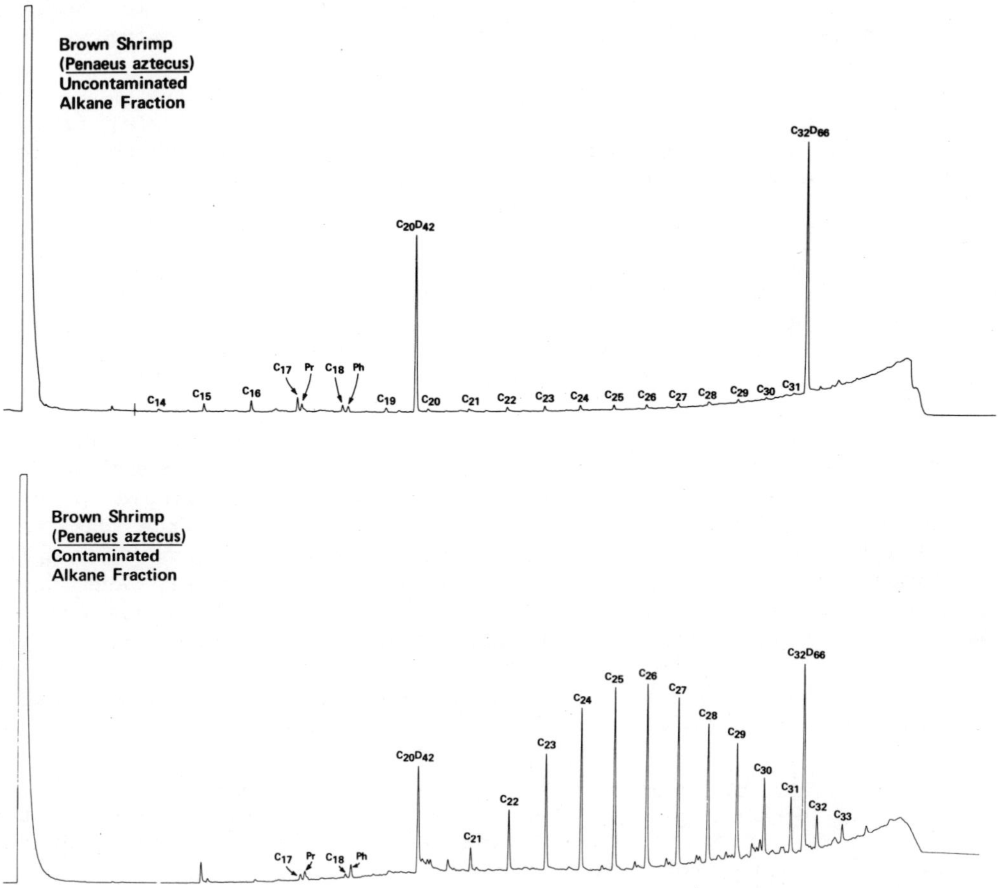

Figure 15. Gas chromatograms of alkane fractions of uncontaminated (upper: August 9, 1976) and contaminated (lower: August 10, 1976) brown shrimp.

contains 2.5 ppm of petroleum alkanes. The latter specimen was the only shrimp encountered which contained any significant quantity of petroleum hydrocarbons, most contained none at all. Shrimp are highly migratory so individual animals would not be expected to spend much time in the region of the Buccaneer Field. This would reduce their probability of ingesting Buccaneer Field contaminants.

Other Biota

Squid and plankton samples collected in the course of other components of the study were examined for their content of petroleum hydrocarbons. In addition, some sessile benthic organisms were collected specifically for us to determine whether they accumulated petroleum hydrocarbons.

Squid. We examined white squid (*Lolliguncula brevis*) collected on four occasions during the first year of the study. None contained petroleum hydrocarbons.

Plankton. Plankton samples were collected both at the air/sea interface and from the water column. No attempt was made to separate and analyze the individual species; a copepod (*Acartia tonsa*), the sergestid shrimp (*Lucifer faxoni*), and a chaetognath (*Sagitta tenuis*) were among the major species obtained (Fotheringham, 1977, 1980; Fotheringham and Brunenmeister, 1975).

The majority of plankton samples from the air/sea interface were contaminated with petroleum alkanes. In two samples, tar balls were readily visible; these have been reported present during microscopic examination of other samples from the same location (Finucane et al., 1979). Most subsurface samples contained only biogenic hydrocarbons: n-C_{15}, n-C_{17}, and pristane. It is possible that the petroleum hydrocarbons observed in surface samples were not contained within the plankton (inside the organisms), but were due to the presence of minute particles of tar. This is still important since biota feeding on plankton would just as likely ingest tar balls as plankton.

Benthic Organisms. Like barnacles, many benthic organisms are sedentary. Examination of species representative of the ecology of the study area provided a more complete picture of the effects of oil production. Forty-eight samples from the classes Bivalvia, Gastropoda, Hydrozoa, and Anthozoa collected from the production platforms, quarters platforms, and well jackets were analyzed. Species examined were:

Alcyonaria (yellow coral)
Arbacia punctulata (sea urchin)
Chloeia viridus (brittle worm)
Fasciolaria hunteria (banded tulip shell)
Hydroida (brown hydroid)
Murex fulvescens (giant eastern murex)
Pagurus floridanus (hermit crab)
Pleuroploea gigantea (Florida horse conch)
Pteria colymbus (winged oyster)

All but one of the Alcyonaria samples contained alkanes in the C_{22} to C_{32} region. While this has been used as an indicator for pollution in other samples, there is evidence that these compounds are biogenic in origin (Koons et al., 1965). Representatives from the class Hydroida had similar profiles, but only one sample from the well jacket location was analyzed so it was not possible to determine whether the hydrocarbons in the production platform samples were the result of oil production activities. Few of the specimens of *Pteria colymbus* collected at the production platform contained petroleum alkanes, while they did contain methylnaphthalenes and benzo[*a*]pyrene. Other species of molluscs (*Mytilus edulis*) have been postulated to metabolize aliphatic and olefinic hydrocarbons but not aromatics (Ehrhardt and Heinemann, 1975). This could also be possible for *P. colymbus*. Specimens of a more mobile mollusc, *Pleuroploea gigantea*, had various hydrocarbon profiles which could be ascribed to petroleum contamination. The number of specimens of other species was not large enough to allow for any valid conclusions.

Lipids

Our interest in lipids dates from the first year of the study. At that time, we had little information concerning the nature of the food web in the Buccaneer Field and we had hoped to trace some of the pathways by identifying the sterols in key organisms. This approach became redundant after direct observations concerning habitats and feeding habits were initiated by Gallaway (1980).

We renewed our interest in lipids during the fourth year of the study after we had found (during a separate study) evidence for a relationship between dietary lipids and ovarian maturation and spawning in penaeid shrimp.

Sterols in the Food Web. Sterols of invertebrates from the region of the Buccaneer Field were examined both as the free sterols and their TMS ethers using packed GC columns, and GC-MS was used for the final identification (Brooks and Middleditch, 1973).

The structures in the Buccaneer Field support a relatively isolated reef ecosystem. If the sterols are unchanged in their composition as they pass through the food web, the steroid profiles of the various organisms would be expected to be similar to those of the plankton. This is generally the case (Table 4). A major exception is the squid, which is capable of biosynthesizing sterols: the endogenous cholesterol predominates over dietary sterols in these specimens. Also, as anticipated (Fagerlund and Idler, 1957), the barnacles contain relatively high concentrations of desmosterol. The sea urchins and polychaete worms, while capable of synthesizing sterols, also contain sterols which are of presumed dietary origin. The shrimp are incapable of the synthesis of sterols *de novo*, although they can accomplish the interconversion of certain sterols. This is apparently what is

TABLE 4. MAJOR STEROLS IN MARINE INVERTEBRATES COLLECTED FROM THE REGION OF THE BUCCANEER FIELD DURING SUMMER, 1976.

Specimen	A[a]	B	C	D	E	F
PLANKTON						
Mixed zooplankton	3.8[b]	13.8	62.4	14.3[c]	3.8[d]	1.9[e]
SQUID						
Loliguncula brevis		2.0	98.0			
SEA URCHIN						
Arbacia punctulata	2.4	6.5	80.6	10.5[c]		
POLYCHAETE WORM						
Chloeia viridis	1.7	7.7	85.5	3.4[c]	1.7[f]	
BARNACLE						
Balanus tintinnabulum	1.2	11.6	60.8	24.9	1.5	
SHRIMP						
Penaeus aztecus		2.0	98.0			
Penaeus duorarum	1.0	1.9	95.2	1.9		
Trachypenaeus similis		2.0	98.0			
Squilla empusa	1.0	2.9	96.1			

[a]A = 22-*trans*-24-norcholesta-5,22-dien-3β-ol, B = 22-dehydrocholesterol, C = cholesterol, D = desmosterol, E = 24-methylcholesta-5,22-dien-3β-ol, F = sitosterol, [b]also contains 22-*trans*-24-norcholest-22-en-3β-ol, [c]also contains 24-methylcholesterol, [d]also contains 24-methylcholestadienol, [e]also contains 24-ethylidenecholesterol, [f]also contains 22-dihydroergosterol.

happening here. These specimens do not reflect the diversity of sterols found at the base of the food web (shrimp are able to convert C_{28} and C_{29} sterols to cholesterol), but the minor sterols are undoubtedly of dietary origin. In these analyses, whole shrimp were examined. However, in other analyses of *P. setiferus* from the Gulf of Mexico and mariculture facilities, we have examined individual tissues. We find that there is a great diversity of sterols in the hepatopancreas, presumably reflecting the dietary composition, and less diversity in the gonads, while the tail muscle contains mainly cholesterol and 22-dehydrocholesterol (Middleditch et al., 1980a).

Maturation and Spawning of Penaeid Shrimp. We have found in the laboratory (Brown et al., 1979, 1980; Lawrence et al., 1980; Middleditch et al., 1979, 1980a) that certain lipids are required for ovarian maturation of penaeid shrimp, and that they are incorporated into eggs that are viable (Ward et al., 1979). In our preliminary studies, we used bloodworms (which are rich in the required lipids) to induce maturation. We have also found that the lipid composition in individual components of a food web is fairly homogeneous. Thus, the sterol profiles of plankton, barnacles, squid, polychaetes, sea urchins, and four species of shrimp from the region of the Buccaneer Field have many features in common (Chang, 1978). Also, the fatty acid profiles of a wide range of invertebrates from West Bay, Galveston are very similar (Middleditch et al., 1980b). We can conclude that the species consumed by the shrimp are less important to maturation than the lipids present in the food web. The primary producers of many of these lipids are the benthic algae, so one would expect to be able to determine whether a particular site is conducive to spawning by examining lipid profiles of the sediments (where the shrimp normally feed). Indeed, our analyses of sediments from shrimp spawning sites located during a study of the Bryan Mound Strategic Petroleum Reserve brine disposal site off Freeport, Texas have revealed the presence of the lipids suspected of being important for ovarian maturation and spawning of penaeid shrimp.

Dietary carotenoids are also needed for ovarian development and the production of viable eggs (Zagalsky et al., 1967). These compounds are also available from sediments (Schwendinger and Erdman, 1963). A colorimetric procedure was used for determining carotenoid concentrations. No attempt was made to identify individual carotenoids.

Sediments from the Buccaneer Field contained sterols, fatty acids, and carotenoids. The sterol fraction was dominated by cholesterol, and there were smaller amounts of 22-dehydrocholesterol and desmosterol. A wide range of fatty acids was encountered, including the polyunsaturated compounds suspected of being vital to ovarian maturation and spawning.

Thus, all of the lipids suspected of being required to stimulate ovarian maturation and spawning are indeed present in the sediments. However, our current state of knowledge is devoid of information concerning the quantities of these lipids that are required in the diet, and concerning other parameters which may be important for ovarian maturation.

DISCUSSION

The conventional method of assessing effluents from offshore production platforms has included measurements of "total oil" by spectrophotometric methods. Our early results showed that it was more appropriate to consider the effluents as a medium containing a number of individual substances; different relative concentrations of these substances could be present in effluent samples containing the same amount of "total oil." During earlier reviews of our research, we suggested that monitoring of individual toxic components of the effluents would be more appropriate than monitoring "total oil." The proposed new Ocean Discharge Criteria (Federal Register, February 12, 1980) include a provision for monitoring effluents for a specific list of toxic pollutants ("priority pollutants").

The intent of these proposed regulations is to protect the environment from exposure to harmful amounts of specific substances with known toxic properties. It appears that the main concern in drafting these regulations was related to limiting the concentrations of toxic substances in the mixing zone below the discharge pipe. The assumption appears to have been made that there is no other location in the region of an oil field which contains significant quantities of contaminants.

We have indeed found that the mixing zone is the only component of the water column in which measurable quantities of petroleum hydrocarbons and sulfur can be seen. Dispersion of these contaminants appears to be rapid and efficient. Observations by Gallaway (1980) indicate that there is only minimal damage to the platform fouling community which can be ascribed to the discharge. If the rate of discharge of produced water from the platforms were to be increased, we are tempted to suggest that little, if any, additional damage to the fouling community would result. If, however, the concentration of the contaminants in the discharge were to be increased beyond a critical level, the factors which control dispersion and dilution in the water column may be insufficient to protect the fouling community from significant damage. Under such circumstances, effects might also be transmitted to other components of the food web. We have insuffiicent information to determine the critical concentration of contaminants at this time, so we are unable to recommend any specific increase in permitted concentrations of contaminants. This appears to be the philosophy of the proposed Ocean Discharge Criteria: limits are imposed upon the concentrations of components of the discharges based upon the state-of-the-art of analytical chemistry, rather than on an assessment of potential harm to the environment.

We feel that some consideration should be given also to the pool of contaminants discovered in the surficial sediments below the production platforms. We venture to suggest that the sediment concentrations of these contaminants depend upon the total quantities of contaminants discharged from the production platforms rather than the concentrations of contaminants in the discharges. Of course, other factors such as water depth, currents, rates of sediment resuspension and deposition, and mode of transport through the water column will also contribute to the concentration of contaminants in the surficial sediments. Therefore, to minimize the size of this pool of contaminants, both the rate of discharge of produced water and its content of toxic substances should be limited. Again, we have insufficient data to suggest critical values for these parameters. We should add that, in the Buccaneer Field, no adverse effects were found which could be ascribed to the presence of contaminants in the surficial sediments.

There are at present no mandated analytical procedures for the analysis of the organic toxic pollutants, although proposed methods based upon gas chromatography, high-performance liquid chromatography, and combined gas chromatography - mass spectrometry (GC-MS) have been described (Federal Register, December 3, 1979). It was indicated in the recently issued Consolidated Permit Regulations that GC-MS would be the method of choice in most instances (Federal Register, May 19, 1980). With the detection limits cited for the proposed GC-MS method, only five organic toxic pollutants would have been reported: acenaphthylene, benzene, ethylbenzene, naphthalene, and toluene. We have in fact found additional organic toxic pollutants at lower concentrations, and we have also found other toxic substances which are not on the formal list of organic toxic pollutants. The priority pollutants were chosen during a consideration of effluents from a wide range of different industries. Most of them will not be found in discharges from offshore production platforms. Conversely, there are toxic substances that we have identified in the effluents which are not on the list of priority pollutants. A more appropriate list of substances to be monitored could be prepared for this industry. This revised list could include alkylated homologs of the aromatic hydrocarbons, since they are similar in physical properties and toxicity to the parent compounds. We should also point out that, of the methods of analysis listed above, only GC-MS provides sufficient sensitivity and specificity for characterizing these complex effluents.

ACKNOWLEDGMENTS

With the exception of some of the samples collected from the platforms, almost all of the samples which we analyzed were collected by other participants in this study. For this assistance, we are particularly indebted to Benny Gallaway, John Anderson, Nick Fotheringham, James Brooks, John Finucane, Ian Workman, and their colleagues (too numerous to mention here).

Laboratory examinations of the samples were conducted with the assistance of Brenda Basile, Evelyn Chang, David West, Mark Spain, Stephen Missler, Rickey Cheung, Alexander Magro, and a number of undergraduate students who passed through our laboratory during the past four years.

Many of the conclusions synthesized from our data were formulated during formal and informal discussions with other participants in this study. It is difficult to recall the sources of individual suggestions, but information on the nature of the Buccaneer food web from Benny Gallaway and his colleagues is worthy of special acknowledgment.

Shell Oil Company allowed us access to their platforms, arranged helicopter transportation on many occasions, and also provided answers to our questions concerning activities in the Field.

This work is a result of research sponsored by the Environmental Protection Agency and Department of Commerce, National Oceanic and Atmospheric Administration, National Marine Fisheries Service, Southeast Fisheries Center, Galveston Laboratory under Contracts 03-6-042-35120, 03-7-042-35127, and 03-78-D08-0044.

REFERENCES

Anderson, J. B., Schwarzer, R. R., and Clark, H. C., 1977, "Sedimentology, geochemistry, and trace metal analysis," in: Jackson, W. B. (ed.), "Environmental Assessment of an Active Oil Field in the Northwestern Gulf of Mexico, 1976-1977." NOAA, pp. 89-174. Available from: NTIS, Springfield, Virginia, Accession No. PB283890.

Anderson, J. B., Schwarzer, R. R., Clark, H. C., Wheeler, R. B., Hokanson, C. L., Anderson, J. E., and Okafor, J., 1979, "Describe the fine sediments and nepheloid layer of the oil field, focusing upon their relationship to heavy metal adsorption/determine levels, pathways, and bioaccumulation of heavy metals in the marine ecosystem in the oil field," in: Jackson, W. B. (ed.), "Environmental Assessment of an Active Oil Field in the Northwestern Gulf of Mexico, 1977-1978. Volume III: Chemical and Physical Investigations." NOAA, pp. 2.3.2/2.4.2-1 to 2.3.2/2.4.2-118. Available from: NTIS, Springfield, Virginia, Accession No. PB80107899.

Armstrong, R. S., and Hamilton, R. C., 1979, "Describe seasonal circulation patterns in the oil field," in: Jackson, W. B. (ed.), "Environmental Assessment of an Active Oil Field in the Northwestern Gulf of Mexico, 1977-1978. Volume III: Chemical and Physical investigations." NOAA, pp. 2.3.9-1 to 2.3.9-250. Available from: NTIS, Springfield, Virginia, Accession No. PB80107899.

Brooks, C. J. W., and Middleditch, B. S., 1973, "Some aspects of mass spectrometry in steriod analysis," in: Heftman, E. (ed.), "Modern Methods of Steroid Analysis." Academic Press, New York.

Brooks, J. M., Estes, E. L., Wiesenburg, D. A., Schwab, C. R., and Abdel-Reheim, H. A., 1980, "Investigations of surficial sediments, suspended particulates, and volatile hydrocarbons at Buccaneer gas and oil field in the northwestern Gulf of Mexico," in: Jackson, W. B., and Wilkens, E. P. (eds.), "Environmental Assessment of Buccaneer Gas and Oil Field in the Northwestern Gulf of Mexico, 1975-1980., Vol. I." NOAA Technical Memorandum NMFS-SEFC-47, 89 pp. Available from NTIS, Springfield, Virginia.

Brown, A., Jr., McVey, J. P., Middleditch, B. S., and Lawrence, A. L., 1979, "Maturation of white shrimp (*Penaeus setiferus*) in captivity," Proc. World Mariculture Soc., 10:435.

Brown, A., Jr., McVey, J. P., Scott, B. M., Williams, T. D., Middleditch, B. S., and Lawrence, A. L., 1980, "The maturation and spawning of *Penaeus stylirostris* under controlled laboratory conditions," Proc. World Mariculture Soc., 11:488.

Chang, E. S., 1978, "Sterols in the Marine Food Web." M.S. Thesis, Univ. of Houston.

Davis, J. B., 1968, "Paraffinic hydrocarbons in the sulfate-reducing bacterium *Desulfovibrio desulfuricans*," Chem. Geol., 3:155.

Ehrhardt, M., and Heinemann, J., 1975, "Hydrocarbons in blue mussels from the Kiel Bight," Environ. Pollut., 9:263.

Fagerlund, U. H. M., and Idler, D. R., 1957, "Marine sterols. IV. 24-Dehydrocholesterol: isolation from a barnacle and synthesis by the Wittig reaction," J. Amer. Chem. Soc., 79:6473.

Finucane, J. H., Collins, L. A., and Barger, L. E., 1979, "Determine the effects of discharges on seasonal abundance, distribution, and composition of ichthyoplankton in the oil field," in: Jackson, W. B. (ed.), "Environmental Assessment of an Active Oil Field in the Northwestern Gulf of Mexico, 1977-1978. Volume II: Data Management and Biological Investigations." NOAA, pp. 2.3.6-1 to 2.3.6-157. Available from: NTIS, Springfield, Virginia, Accession No. PB80165970.

Fotheringham, N., 1977, "Effects of offshore oil field structures on their biotic environment: benthos and plankton," in Jackson, W. B. (ed.), "Environmental Assessment of an Active Oil Field in the Northwestern Gulf of Mexico, 1976-1977." NOAA, pp. 487-549. Available from: NTIS, Springfield, Virginia, Accession No. PB283890.

Fotheringham, N., 1980, "Beachcomber's Guide to Gulf Coast Marine Life." Lone Star Books, Houston, Texas, 124 pp.

Fotheringham, N. and Brunenmeister, S., 1975, "Common Marine Invertebrates of the Northwestern Gulf Coast." Gulf Publishing Co., Houston.

Gallaway, B. J., 1980, "Pelagic, reef and demersal fishes and macrocrustaceans/biofouling communities," in: Jackson, W. B., and Wilkens E. P. (eds.), "Environmental Assessment of the Buccaneer Gas and Oil Field in the Northwestern Gulf of Mexico, 1975-1980, Vol. II." NOAA Technical Memorandum NMFS-SEFC-48, 82 pp. Available from: NTIS, Springfield, Virginia.

Gallaway, B. J., Johnson, M. F., Howard, R. L., Martin, L. R., and Boland, G. S., 1979, "A study of the effects of Buccaneer Oil Field structures and associated effluents on biofouling communities and the Atlantic spadefish (*Chaetodipterus faber*)," in: Jackson, W. B. (ed.). "Environmental Assessment of an Active Oil Field in the Northwestern Gulf of Mexico, 1977-1978. Volume II: Data Management and Biological Investigations." NOAA, pp. 2.3.8-1 to 2.3.8-126. Available from: NTIS, Springfield, Virginia.

Giam, C. S., 1976, "A preliminary environmental assessment of the Buccaneer Oil/Gas Field. IV. Levels of petroleum hydrocarbons in sediments, Buccaneer Oil Field," in: "Pilot Study of the Buccaneer Oil Field (Benthos and Sediments)." Gulf Fisheries Center, Galveston, p. 44.

Koons, C. B., Jamieson, G. W., and Cieresko, L. S., 1965, "Normal alkane distributions in marine organisms: possible significance to petroleum origin," Bull. Am. Assoc. Petrol. Geologists, 49:301.

Lawrence, A. L., Akamine, Y., Middleditch, B. S., Chamberlain, G., and Hutchins, D., 1980, "Maturation and reproduction of *Penaeus setiferus* in captivity," Proc. World. Mariculture Soc., 11:481.

Middleditch, B. S., and Basile, B., 1976, "Deuteriated analogs as internal standards for the quantitation of environmental alkanes by gas chromatography," Analyt. Letters, 9:1031.

Middleditch, B. S., and Basile, B., 1979, "Sulfur-34 as an internal standard for the quantitation of environmental sulfur by combined gas chromatography - mass spectrometry," Analyt. Letters, 12A:777.

Middleditch, B. S, and West, D., 1979, "Bicyclohexyl interference in the quantitation of environmental alkanes by gas chromatography," Analyt. Letters, 12A:1279.

Middleditch, B. S., Missler, S. R., Ward, D. G., McVey, J. P., Brown, A., and Lawrence, A. L., 1979, "Maturation of penaeid shrimp: dietary fatty acids," Proc. World Mariculture Soc., 10:472.

Middleditch, B. S., Missler, S. R., Hines, H. B., McVey, J. P., Brown, A., Ward, D. G., and Lawrence, A. L., 1980a, "Metabolic profiles of penaeid shrimp: dietary lipids and ovarian maturation," J. Chromatog., 195:359.

Middleditch, B. S., Missler, S. R., Hines, H. B., Chang, E. S., McVey, J. P., Brown, A., and Lawrence, A. L., 1980b, "Maturation of penaeid shrimp: lipids in the marine food web," Proc. World Mariculture Soc., 11:463.

Schwendinger, R. B., and Erdman, J. G., 1963, "Carotenoids in sediments as a function of environment," Science, 141:808.

Sweeley, C. C., Elliot, W. H., Fries, I., and Ryhage, R., 1966, "Mass spectrometric determination of unresolved compounds in gas chromatographic effluents," Analyt. Chem., 38:1549.

Ward, D. G., Middleditch, B. S., Missler, S. R., and Lawrence, A. L., 1979, "Fatty acid changes during larval development of *Penaeus setiferus*," Proc. World Mariculture Soc., 10:464.

Workman, I. K., and Jones, C. E., 1979, "Determine effects of oil field discharges on species composition and abundance of pelagic fishes and demersal fishes and macro-crustaceans in the oil field," in: Jackson, W. B. (ed.), "Environmental Assessment of an Active Oil Field in the Northwestern Gulf of Mexico, 1977-1978. Volume II: Data Management and Biological Investigations." NOAA, pp. 2.3.5-1 to 2.3.5-149. Available from: NTIS, Springfield, Virginia.

Zagalsky, P. F., Cheesman, D. F., and Ceccaldi, H. J., 1967, "Studies on carotenoid-containing lipoproteins isolated from the eggs and ovaries of certain marine invertebrates," Comp. Biochem. Physiol., 22:851.

BIOCIDES

Brian S. Middleditch

Department of Biophysical Sciences
University of Houston
Houston, Texas 77004

INTRODUCTION

Sulfur-oxidizing bacteria can produce sulfuric acid, which will corrode metal pipes and containers. Thus, as in the Buccaneer Field, a product stream which contains appreciable quantities of elemental sulfur can become corrosive as a result of bacterial action. One method of minimizing this effect is to reduce the bacterial populations by adding biocides to the product stream. This has been done in the Buccaneer Field throughout the period of our study. Investigation of biocides was a minor component of the research performed by the University of Houston. No definitive results were obtained and time did not permit a thorough discussion of this topic at the Symposium. Since biocides are mentioned in several other Chapters, however, a summary of our knowledge of biocides is presented here.

BIOCIDES EMPLOYED

Biocides used in the Buccaneer Field at the start of the study were K-31 (pentanedialdehyde) and KC-14 (alkyldimethyl benzyl chloride), both supplied by Champion Chemical Company. In addition, a surfactant (Surfatron DQ44) was also employed. The weekly alternation in application of the two biocides was only partially successful in reducing the numbers of *Desulfovibrio* so, in April 1978, Shell began using Magnacide B (Magna Corporation).

Magnacide B is a formulation of acrolein (2-propenal) of approximately 90-94% purity, depending on the source. The acrolein is not manufactured by Magna. Acrolein is a highly volatile, toxic, and reactive substance. Residual acrolein in produced water is scavenged by treatment with sodium bisulfite (Magnatreat OS-L) prior to discharge.

The orginal biocides and surfactant were found to be contaminated with alkanes: K-31, 1.89 ppm; KC-14, 1.60 ppm; Surfatron DQ44, 3.26 ppm. They contributed a negligible amount to the total alkane content of the produced water discharge.

ASSAYS OF BIOCIDES

Spectrophotometric assays for K-31, KC-14, and Surfatron DQ44 were being developed when their use was discontinued. We did not determine what quantities (if any) of these substances were contained in the produced water discharges from the production platforms.

The determination of acrolein in aqueous solution is a difficult analytical task. It is not amenable to the regular purge-and-trap technique commonly used for the determination of low molecular weight substances by GC-MS (Middleditch et al., 1981; Zlatkis and Shanfield, 1979). The preliminary "priority pollutant" protocol for acrolein called for direct aqueous injection into a water-compatible GC column. We were able to obtain greater sensitivity with a purge-and-trap method carried out at elevated temperature - this is also the most recently recommended "priority pollutant" method for acrolein. Even so, the detection limit is only 100 ppb (compared with 10 ppb for most other organic priority pollutants). We were unable to detect acrolein in any produced water samples by this technique. This problem may have been exacerbated by the instability of acrolein in aqueous solution. If some acrolein had been present in these samples at the time of collection, it might have been dissipated by the time that they reached the laboratory.

Information provided by Shell indicated that residual amounts of acrolein in the range 1-3 ppm were measured in produced water immediately after the installation of the injection system, when sufficient time had probably not elapsed for equilibration of acrolein and scavenger concentrations and bacterial counts. Once this had been done no residual was detected until a pump failed and was replaced, and acrolein levels of 2 ppm were reported. After three days, residual acrolein was again below detection limits. The detection limit of the assay procedure employed was apparently 1 ppm.

Rose and Ward (this volume), using a colorimetric assay with a detection limit of 1 ppm, also failed to detect residual acrolein in formation water from platform 296-B.

Given the high reactivity and low half-life of acrolein in the water disposal system and adequate means of neutralizing any residual it is suspected that little, if any, is reaching the ocean. Even if trace amounts should enter the environment they would not be expected to be persistent, but easily and quickly dispersed and inactivated.

No assays were performed to determine the concentrations of sodium bisulfite in the produced water discharges

CONCLUSIONS

We have no data concerning the concentrations of K-31, KC-14, or Surfatron DQ44 in the discharges during the first two years of our study. When acrolein was being used later in the study, we found none of this substance (detection limit, 100 ppb) in produced water samples after they had been returned to the laboratory. It is possible that any acrolein which might have been present in the discharges became hydrolysed prior to analysis.

The presence of any of these substances in the discharges might be cause for concern, but no effects attributable to them could be discerned by those who were studying the bacterial communities, the platform fouling community, or the toxicity of the discharges (see appropriate Chapters in this volume).

REFERENCES

Middleditch, B. S., Missler, S. R., and Hines, H. B., 1981, "Mass Spectrometry of Priority Pollutants." Plenum Press, New York.

Zlatkis, A., and Shanfield, H., 1979, "Concentration techniques for volatile samples," in: Middleditch, B. S. (ed.), "Practical Mass Spectrometry: A Contemporary Introduction." Plenum Press, New York, p. 151.

SEDIMENTOLOGY AND GEOCHEMISTRY OF RECENT SEDIMENTS

John B. Anderson and Richard B. Wheeler[a]

Geology Department
Rice University
Houston, Texas 77001

Rudolph R. Schwarzer

Chemistry Department
Texas Southern University
Houston, Texas 77004

INTRODUCTION

The objective of the sedimentology-geochemistry portion of the Buccaneer Field investigation was to determine whether exploration/production activities have resulted in trace metal contamination of the surrounding environment. Those elements analyzed include Ba, Cd, Co, Cr, Cu, Fe, Mn, Ni, Pb, Sr, and Zn.

In this paper the term "contaminant" refers to trace metal concentrations in sediment samples which are significantly high with respect to suitable control samples. Control samples were collected from areas located some distance from the field and from subsurface deposits obtained by piston cores. Trace metal concentrations from the Bureau of Land Management South Texas Outer Continental Shelf Study (BLM-STOCS: Berryhill, 1976) were also used for comparison with Buccaneer Field data. A detailed discussion of procedures and results has been published elsewhere (Anderson et al., 1977, 1979; Wheeler, 1979; Wheeler et al., 1978, 1981), so this paper provides only a brief description of the methods used and a summary of the results.

[a]Present address: Exxon Production Research Company, Houston, Texas 77005.

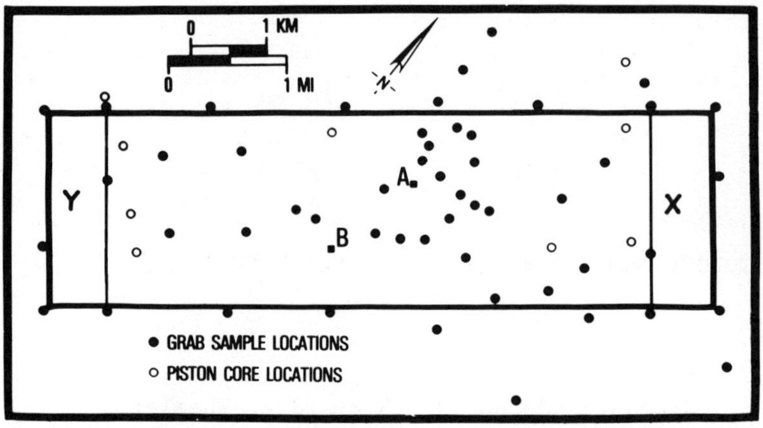

Figure 1. Locations of grab samples and piston cores used in this study. Samples were also collected at 7.5, 23, 53, and 114 m intervals along 4 radial transects around production platforms (labeled A and B) and 1 satellite platform.

METHODS

Bottom samples were taken with a large (20 cm diameter) gravity core. Longer cores were obtained using a piston core and by divers. All samples were stored in plastic containers and refrigerated until analyzed. Sample locations are shown in Figure 1.

Sediment grain size analyses were conducted using an automated settling tube - hydrophotometer system which has been shown to be more accurate than standard sieve and pipette methods (Anderson and Kurtz, 1979). Clay mineral x-ray diffraction analyses were provided by Dr. John Butler, University of Houston, and include semiquantitative determinations of kaolinite, montmorillonite, chlorite, and illite. In addition, heavy mineral compositions of selected sand samples were obtained using standard bromoform separation techniques. Calcium carbonate percentages were determined by a titrimetric technique adapted from Turekian (1956) and total organic carbon data were provided by Behrens (1977).

Trace metal extractions were prepared by partial digestion of a dried sample with sodium hypochlorite and nitric acid. A Perkin-Elmer Model 360 atomic absorption spectrophotometer was used for trace metal analysis, with a nitrous oxide - acetylene flame used for Ba analyses.

Because of the volume of data acquired in this study (>100 samples x 19 variables) was so great, multivariate statistical techniques were used to assist in the identification of sample groups. R-mode clustering (variable vs. variable) was used to identify similarities between trace elements exhibited by their co-occurence and to eliminate highly similar, redundant variables which might bias Q-mode analysis. Q-mode clustering (sample vs. sample) was used to define sedimentary facies on the basis of textural and compositional similarity. Geochemical comparisons are made only between samples which fall within the same clustered facies group.

RESULTS

Sedimentology

Sediments of the Buccaneer Field are comprised of four components. These four components may occur separately as distinct sediment types or occur as mixtures, generally consisting of two or three components in varying proportions. Representative size frequency curves for each component are shown in Figure 2.

Component 1. A coarse sand to gravel-sized calcareous shell hash consisting mostly of mollusk and bryozoan fragments. This shell hash blankets the ridge in the northeastern corner of the study area where it has a minimum thickness of 86 cm. It is also present as a subordinate coarse fraction in finer-grained sediments throughout the area. Similar sediment floors the Freeport Bank to the west of the study area and has been radiometrically dated at 26,900±1800 yr (Curray, 1959).

Component 2. An overcompacted silty clay (Figure 2) containing fossil foraminifera indicative of brackish conditions (Wheeler et al., 1981). This component is interpreted as the Pleistocene Beaumont Formation, which is known to outcrop over broad regions of the Texas Gulf Coast. Radiometric age determinations for this sediment yielded ages of 22,120±1440 yr and 24,180±1170 yr (Behrens, 1977) which coincides with the late Wisconsin low sea level stand.

Component 3. A coarse to very coarse (2.0 ϕ to 0.5 ϕ), well rounded quartz sand (Figure 3). This unit has a heavy mineral assemblage consisting of zircon, tourmaline, rutile, epodite, and brown hornblende, which implies that the Trinity River was its source (Wheeler et al., 1981). It contains fossil foraminifera that today inhabit the turbulent wave zone of the Gulf Coast (Anderson, 1968) and is therefore probably of Pleistocene age. Component 3 occurs mostly as a subordinate component of residual sediments found in scour depressions near the legs of drilling structures.

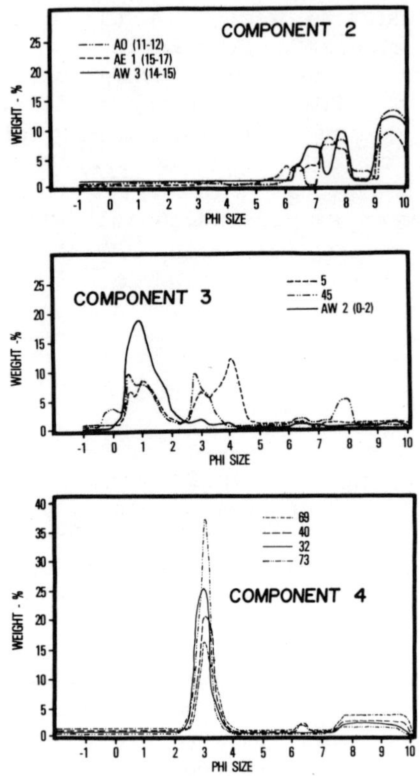

Figure 2. Representative size frequency curves for sediment component 2, 3, and 4. Component 2 is the Beaumont clay. Component 3 ranges from 0.0 ϕ to 2.0 ϕ in size, and is mixed with component 4 in samples 5 and 45. Component 4 consists of well sorted fine (3.5 ϕ) sand. Component 4 is generally mixed with component 2 sediments.

Component 4. The most abundant sediment type in the field area consists of very fine (3.5 ϕ), very angular quartz sand. This component is generally associated with component 2 in varying proportions. The heavy mineral assemblage of this unit consists dominantly of hornblende (avg. 50%), with lesser amounts of epidote and pyroxene (avg. 15% each) and falls within the limits of modern sediments of the region (Van Andel and Poole, 1960). Component 4 also contains foraminifera that are known to be endemic to the inner continental shelf today. Figure 3 is a sediment facies map for the area which includes contours of sand/mud ratios.

The most significant outcome of the sedimentologic investigation is that it shows quite clearly that the Buccaneer Field is

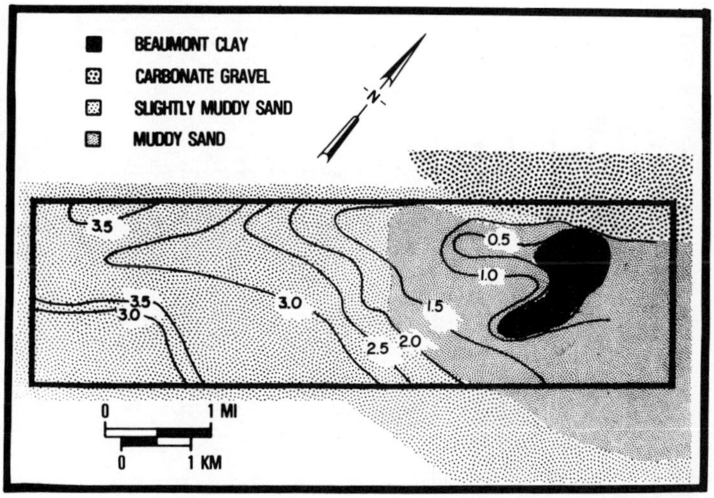

Figure 3. Sediment distribution map for the Buccaneer Field along with contours of the sand/mud ratios of samples. Mud content decreases from northeast to southwest away from the area of exposed Beaumont Formation. These data point to local reworking and southwestward transport of fines in the field.

a highly dynamic sedimentary environment in which relict deposits are actively being exposed, eroded, redistributed by bottom currents, and mixed by organisms. In fact, it is probably fair to characterize the area as a "passive" sedimentary environment in that most fine suspended sediments derived from coastal rivers today bypass the area as they swept seaward. Much of the silt and clay comprising the muddy matrix in muddy sands is derived from the exposed Beaumont Formation. This is implied by the decreasing silt-clay content of surficial sediment from east to west across the study area, away from Beaumont exposures (Figure 3). Erosion of this cohesive sediment is facilitated by bioturbation. Organisms also play a key role as they mix these fines with local sands in distinctly bimodal sediments (Figure 2).

Geochemical Results

Trace metal concentrations were compared for samples with similar texture from platform areas, control areas, subsurface samples, and from the BLM-STOCS area (Table 1). Buccaneer platform samples proved to have higher concentrations of Ba, Cd, Co, Pb, and Sr than any of the control populations. Cr, and to a lesser degree Zn, also have higher average concentrations around platforms than with Buccaneer control areas. On the other hand,

TABLE 1. STATISTICAL SUMMARY GIVING MEAN ($\overline{x}$) AND COEFFICIENT OF VARIATION (v = standard deviation/$\overline{x}$) FOR SAMPLE POPULATIONS. BUCCANEER PLATFORM, STUDY AREA, AND BLM-STOCS SAMPLES ARE ALL FROM GROUP III OF THE Q-MODE CLUSTER ANALYSIS. SUBSURFACE AND BEAUMONT SUMMARIES GIVEN FOR COMPARISON. NUMBER OF SAMPLES ANALYZED GIVEN IN PARENTHESES AFTER NAME OF SAMPLE POPULATION.

Element	Buccaneer Platform (4)		Buccaneer Control (22)		BLM-STOCS Control (6)		Buccaneer Subsurface (8)		Beaumont Formation(10)	
	$\overline{x}$	v	$\overline{x}$	v	$\overline{x}$	v	$\overline{x}$	v	$\overline{x}$	v
Ba	403.	(0.61)	151.	(0.80)	168.	(0.63)	38.	(0.17)	44.	(0.31)
Cd	1.1	(0.05)	0.8	(0.16)	0.1	(0.34)	0.2	(1.38)	<0.5	(1.41)
Co	8.8	(0.13)	8.1	(0.16)	n.d.[a]		6.6	(0.15)	13.3	(0.23)
Cr	13.3	(0.87)	9.7	(0.17)	24.2	(0.29)	18.8	(0.27)	22.9	(0.64)
Cu	4.3	(0.08)	4.7	(0.24)	7.4	(0.23)	6.9	(0.32)	15.4	(0.41)
Fe	0.69	(0.18)	0.85	(0.25)	2.22	(0.15)	1.19	(0.37)	2.00	(0.35)
Mn	142.	(0.10)	164.	(0.22)	338.	(0.09)	164.	(0.07)	197.	(0.36)
Ni	14.7	(0.05)	17.3	(0.09)	20.2	(0.25)	5.8	(0.50)	21.5	(0.55)
Pb	10.4	(0.25)	9.4	(0.18)	9.0	(0.20)	8.6	(0.19)	19.4	(0.46)
Sr	60.7	(0.65)	25.1	(0.92)	n.d.[a]		10.1	(0.63)	10.3	(0.95)
Zn	29.6	(0.14)	29.0	(0.23)	28.8	(1.16)	36.1	(0.26)	59.3	(0.39)

[a]n.d. = not determined

the Beaumont Formation has higher average concentrations of all metals, except Ba, Cd, and Sr, than platform samples and is possibly the source of Co, Cr, Pb, and Zn in platform samples.

A number of samples collected near platforms were found to contain a distinctively well-crystallized expandable clay mineral, probably a bentonitic clay, which is atypical of marine clays (Butler, personal communication). Only one control sample contained these clays. Platform samples containing this well-crystallized clay average 450 ppm Ba while those without average 248 ppm Ba. The source of this excess Ba could be drilling muds that were dumped onto the seabed during the drilling operation. Alternatively, high Ba concentrations in platform samples could be attributed to the abundance of calcium carbonate, mostly in the form of aragonitic shell debris, on and beneath structures. As alkali metals, both Ba and Sr have much the same chemistry as calcium (Turekian, 1955) and both may be concentrated in the calcareous shells of organisms.

The source of Cd in platform samples is not certain. One suspected source is that of corroded metal debris and storage batteries which litter the seafloor beneath platforms.

CONCLUSIONS

The Buccaneer Gas and Oil Field represents a dynamic sedimentary environment in which relict sediments are being reworked into equilibrium with modern conditions. As a result, sediment-bound trace contaminants are not likely to be deposited in the field. Rather the finer clay mineral and organic material on which trace metals are likely passengers is being swept seaward and dispersed. Thus, there is little that can be concluded about contamination of bottom sediments in the field.

For the most part, sediments within the field are not enriched in trace metals relative to control samples (Buccaneer Field control areas, subsurface samples, and BLM-STOCS samples). Those samples that are contaminated were taken within 100 m of structures. A possible source of Ba is drilling muds, though enrichment of Ba in platform samples could also be related to the high $CaCO_3$ (mostly in the form of aragonite) concentration in sediments taken around structures. In any case, Ba in the form of barite ($BaSO_4$) poses little threat to the marine environment.

Platform samples are also enriched in Cd relative to control samples. The source of this Cd is undetermined, but corrosion of the structures and metal debris on the seafloor are suspected.

The overall conclusion from this study must be that what little potential impact was identified in the Buccaneer Field is being mitigated by natural processes. This does not mean that bottom sediment contamination does not result from drilling and production operations in other areas where sedimentation of fines occurs. This study does serve to emphasize the value of sedimentologic work in evaluating geochemical data in any aquatic environment.

ACKNOWLEDGEMENT

This work is a result of research sponsored by the Environmental Protection Agency and the Department of Commerce, National Oceanic and Atmospheric Administration, National Marine Fisheries Service, Southeast Fisheries Center, Galveston Laboratory under Contract No. 03-6-042-35116

REFERENCES

Anderson, J. B., 1968, "Ecology of foraminifera from Mississippi Sound and surrounding waters," J. Alabama Acad. Sci., 1:1.

Anderson, J. B., and Kurtz, D. D., 1979, "'RUASA': an automated rapid sediment analyser," J. Sed. Petrol., 49:625.

Anderson, J. B., Schwarzer, R. R., and Clark, H. C., 1977, "Sedimentology, geochemistry, and trace metal analysis," in: Jackson, W. B. (ed.), "Environmental Assessment of an Active Oil Field in the Northwestern Gulf of Mexico, 1976-1977." NOAA, pp. 89-174. Available from: NTIS, Springfield, Virginia, Accession No. PB283890.

Anderson, J. B., Schwarzer, R. R., Clark, H. C., Wheeler, R. B., Hokanson, C. L., Anderson, J. E., and Okafor, J., 1979, "Describe the fine sediments and nepheloid layer of the oil field, focusing upon their relationship to heavy metal adsorption/ determine levels, pathways, and bioaccumulation of heavy metals in the marine ecosystem in the oil field," in: Jackson, W. B. (ed.), "Environmental Assessment of an Active Oil Field in the Northwestern Gulf of Mexico, 1977-1978. Volume III: Chemical and Physical Investigations." NOAA, pp. 2.3.2/2.4.2-1 to 2.3.2/2.4.2-118. Available from: NTIS, Springfield, Virginia, Accession No. PB80107899.

Behrens, E. W., 1977, "Total organic carbon and carbon isotopes of sediments," in: Jackson, W. B. (ed.), "Environmental Assessment of an Active Oil Field in the Northwestern Gulf of Mexico, 1976-1977." NOAA, pp. 737-759. Available from: NTIS, Springfield, Virginia, Accession No. PB283890.

Berryhill, H. L. (ed.), 1975, "Assessment of the South Texas Outer Continental Shelf: Geologic Investigations, October 25, 1974 - December 1, 1975." Bureau of Land Management, New Orleans, Louisiana.

Curray, J. R., 1959, "Sediments and History of the Holocene Transgression: Continental Shelf, Northwest Gulf of Mexico." Ph.D. Thesis, Scripps Inst. Oceanog., La Jolla, California.

Turekian, K. K., 1955, "Paleoecological significance of the strontium-calcium ratio in fossils and sediments," Geol. Soc. Bull., 66:155.

Turekian, K. K., 1956, "Rapid technique for determination of carbonate content of deep-sea cores," Amer. Assoc. Petrol. Geologists Bull., 40:2507.

Van Andel, T. H., and Poole, D. M., 1960, "Sources of recent sediments in the northern Gulf of Mexico," J. Sed. Petrol., 30:91.

Wheeler, R. B., 1979, "Environmental Trace Element Geochemistry of Sediments of the Buccaneer Offshore Oil and Gas Field: Factors Controlling Concentrations of Trace Elements in Sediments." M.S. Thesis, Rice University, Houston, Texas.

Wheeler, R. B., Schwarzer, R. R., and Anderson, J. B., 1978, "Assessment of environmental impact of offshore production in the Buccaneer oil field: sedimentologic and geochemical results," Proc. Offshore Technol. Conf. (Houston, Texas), 3082:285.

Wheeler, R. B., Anderson, J. B., Schwarzer, R. R., and Hokanson, C. L., 1981, "Sedimentary processes and trace metal contaminants in the Buccaneer oil/gas field, northwestern Gulf of Mexico," Environ. Geol., in press.

SURFICIAL SEDIMENTS AND SUSPENDED PARTICULATE MATTER

James M. Brooks, Denis A. Wiesenburg, Claude R. Schwab

Department of Oceanography
Texas A&M University
College Station, Texas 77843

Ernie L. Estes

Department of Marine Sciences
Texas A&M University
Galveston, Texas 77550

Robert F. Shokes

Science Applications, Inc.
1200 Prospect Street
La Jolla, California 92038

INTRODUCTION

Although extensive offshore oil and gas development has occurred in the outer continental shelf (OCS) areas of Texas and Louisiana, few multidisciplinary programs have been undertaken to study the environmental consequences of such development. One of the few programs to study the effects of Gulf of Mexico petroleum development is the "Environmental Assessment of the Buccaneer Gas and Oil Field in the Northwestern Gulf of Mexico". This program, initiated in 1975, was funded by the Environmental Protection Agency through interagency agreement with the National Oceanic and Atmospheric Administration and managed by the National Marine Fisheries Service, Southeast Fisheries Center, Galveston Laboratory in Galveston, Texas. Objectives of the project were (1) to identify and document the types and extent of biological, chemical and physical alterations of the marine ecosystem associated with Buccaneer Field, (2) to determine specific pollutants, their quantity and effects, and (3) to develop the capability to describe and predict fate and effects of Buccaneer Field contaminants. After a survey study initiated in 1975, four successive one-year programs were undertaken to study various aspects of biology, chemistry, physics, and geology at the Buccaneer Field. This paper reports surficial sediment and suspended particulate studies undertaken in 1978-1979 (third year) and 1979-1980 (fourth year).

Suspended sediments play an important part in transport of contaminants in an ecosystem. Platform contaminants introduced into the water column can be absorbed or adsorbed by suspended sediments (Meyers, 1976; Meyers and Quinn, 1973). Results of trace metal (Anderson et al., 1977, 1979; Tillery, 1980) and hydrocarbon (Middleditch and West, 1980; Middleditch et al., 1977, 1979) studies around the Buccaneer Field indicate these contaminants are concentrated in surficial sediments near the production platforms. Clays are largely responsible for scavenging contaminants from seawater (Wheeler, 1977) because of their large surface/area volume ratios and their electrochemical properties. Different clay minerals and grain sizes have different sorption capacities which can influence the distribution of contaminants. The suspended particulates may also be responsible for removing Buccaneer Field contaminants from the area. Contaminants introduced into a strongly stratified water column can be absorbed onto particles and transported out of the Buccaneer Field before settling through the thermocline and contributing contaminants to surficial sediments. The movement, sedimentation, and bioturbation of surficial sediments can also control the accumulation of Buccaneer Field contaminants.

Suspended particulates found within the water column originate from terrigenous runoff, resuspension of bottom sediments, and productivity. The type and quantity of sediments discharged into the Gulf of Mexico from rivers and estuaries are controlled by precipitation and drainage rates. Other important parameters include wind speed and direction, and thermohaline wedging within coastal water bodies (Anderson and Broeckstra, 1977). Once introduced into the Gulf, the transport of suspended sediment is dependent upon circulation dynamics within the region and the density stratification of the water column. Pycnoclines in the water column act as a barrier against transport of suspended particulates to the sediments. The increased eddy diffusivity across a pycnocline acts to slow the rate of settling of suspended particulates, often producing particulate maxima at these boundaries. Resuspension of bottom sediments is dependent on current speeds, sediment type, and turbulent mixing.

The Buccaneer Gas and Oil Field is located approximately 50 km south-southeast of Galveston sea buoy (Figure 1). The field was selected because: (1) the field had been in production for

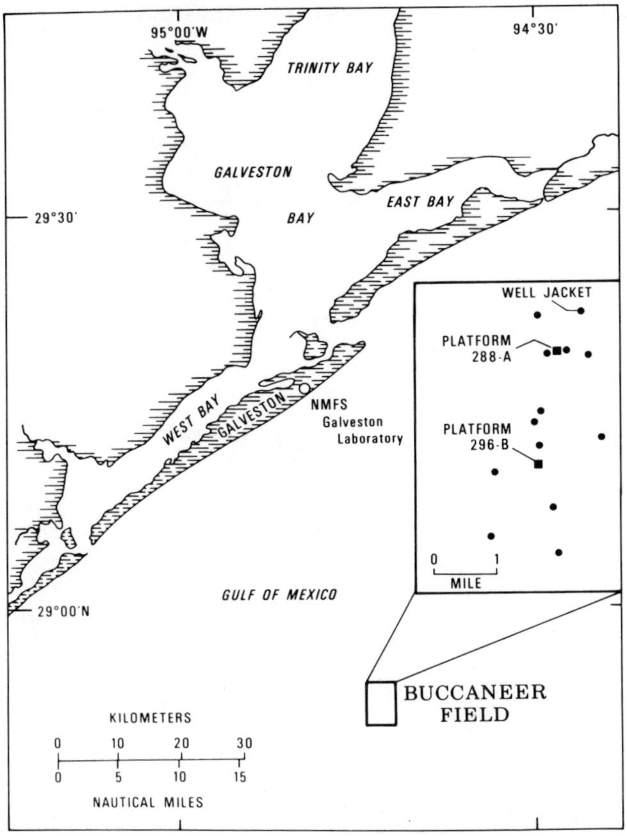

Figure 1. Location of Buccaneer Gas and Oil Field.

about 15 years allowing for full development of oil-field-associated marine communities; (2) it is isolated from other fields; (3) it produces both oil and gas; (4) its location simplified logistics; and (5) the Texas offshore area had not been fully developed for gas and oil production. The field was developed by the Shell Oil Company in four blocks of land that were leased in 1960 and 1968. In the development of the field, 17 structures have been built; two are production platforms, two are quarters platforms, and 13 are satellite structures surrounding well jackets. This study involved investigations of the two production platforms and several well jackets and flare stacks.

METHODS

Water column sampling was performed aboard the GUS III and TANYA and JOE. Surficial sediment samples were collected by divers

TABLE 1. WATER COLUMN AND SURFICIAL SEDIMENT SAMPLING PERIODS.

Season	Dates
Water Column	
Summer	17-19 August 1978
Fall	29-31 October 1978
Winter	18-20 February 1979
Spring	28-30 May 1979
Summer	21-22 August 1979
Winter	7-8 January 1980
Sediments	
Summer	31 August 1978
Fall	1 September 1978
Winter	18-19 December 1978
Spring	17 May 1979
Summer	14 August 1979
Winter	8 December 1979

from LGL Ecological Research Associates and by box corings. Table 1 gives the dates of the sampling periods.

Water Column

Water column samples were taken to determine total suspended matter, dissolved and particulate organic carbon, ATP, chlorophyll a, phaeophytin, calcium carbonate, silicate, phosphate, nitrate, nitrite, clay mineralogy, and particle size. Water column samples were taken using large Niskin samplers (12-l or 30-l volume).

For total suspended matter measurements, known volumes of seawater were filtered through preweighed 47-mm, 0.40-micron Nuclepore filters. After filtration the filters were rinsed with distilled, deionized water to remove all salt and then stored in dust-free containers in a desiccator until re-weighing (see Brooks et al., 1980). The continuous supended material distribution was determined by a Martek Model XMS "In Situ" Transmissometer lowering at each station. This instrument provided a continuous profile of the suspended matter distribution from the surface to near-bottom at each station.

Dissolved and particulate organic carbon (DOC, POC) were determined following the method of Fredericks and Sackett (1970). For chlorophyll and phaeophytin, 300 ml of seawater were filtered through GF/C glass filters (4.5 cm diameter). The filters were desiccated, wrapped in aluminum foil, and frozen. Chlorophyll and its degradation product phaeophytin were measured using a Turner Fluorometer (Model 111) following the method as described by Yentsch and Menzel (1963) and Holm-Hansen et al. (1965). Samples for ATP analysis were filtered onto 0.40-micron, 47-mm Nuclepore filters. Immediately after filtration the filters were placed in vials with 5 ml of boiling TRIS-buffer for 5 minutes. The vials with filters were labeled and frozen at -20°C until analysis. The ATP was determined by the technique of Holm-Hansen and Booth (1966) using a JRB Inc. ATP Photometer.

Nutrient sampling involved placing unfiltered aliquots of seawater into 300-ml Whirl-Pak plastic containers and freezing until analysis. The methods of Strickland and Parsons (1972) were used in these determinations, with specific procedures as given by Technicon Instruments Corporation of Tarytown, New York Industrial Methods No. 100-70W (NO_2 and NO_3), 161-71-W/B (NO_2), 155-71W (PO_4), and 105-71W (SiO_2). Clay mineralogy was determined after the methods of Huang (1975, 1976). Particle size determinations of suspended siliciclastic material in the water column were determined by optical microscopy.

Surficial Sediments

Surficial sediments were collected by divers using one-liter wide-mouth plastic jars, also by gravity or box coring. Samples were split in the laboratory for organic carbon, calcium carbonate, carbon isotope, clay mineralogy, and particle size determinations. Separate cores were taken for Pb-210 dating.

Organic carbon samples were freeze-dried and homogenzied in an organic carbon-free ball mill. Accurately weighed one-gram aliquots of each homogenized sample were placed in ceramic or metal boats and treated with dilute hydrochloric acid to eliminate inorganic carbon. The samples were combusted at >800°C and the CO_2 measured manometrically. Calcium carbonate was determined manometrically in a Craig-type combustion system (see Brooks et al., 1980). Clay mineralogy was determined after the method of Huang (1975). Particle size samples were analyzed for particle texture using the sieve and settling tube methodology of Folk (1974). Stable carbon isotopes on the organic fraction of the sediment were determined after the method of Sackett et al. (1970). Radio analyses of one-cm sediment layers for Pb-210 specific activity (dpm g^{-1}) were conducted throughout the upper 20 to 50 cm of cores. Analytical procedures followed those of Shokes

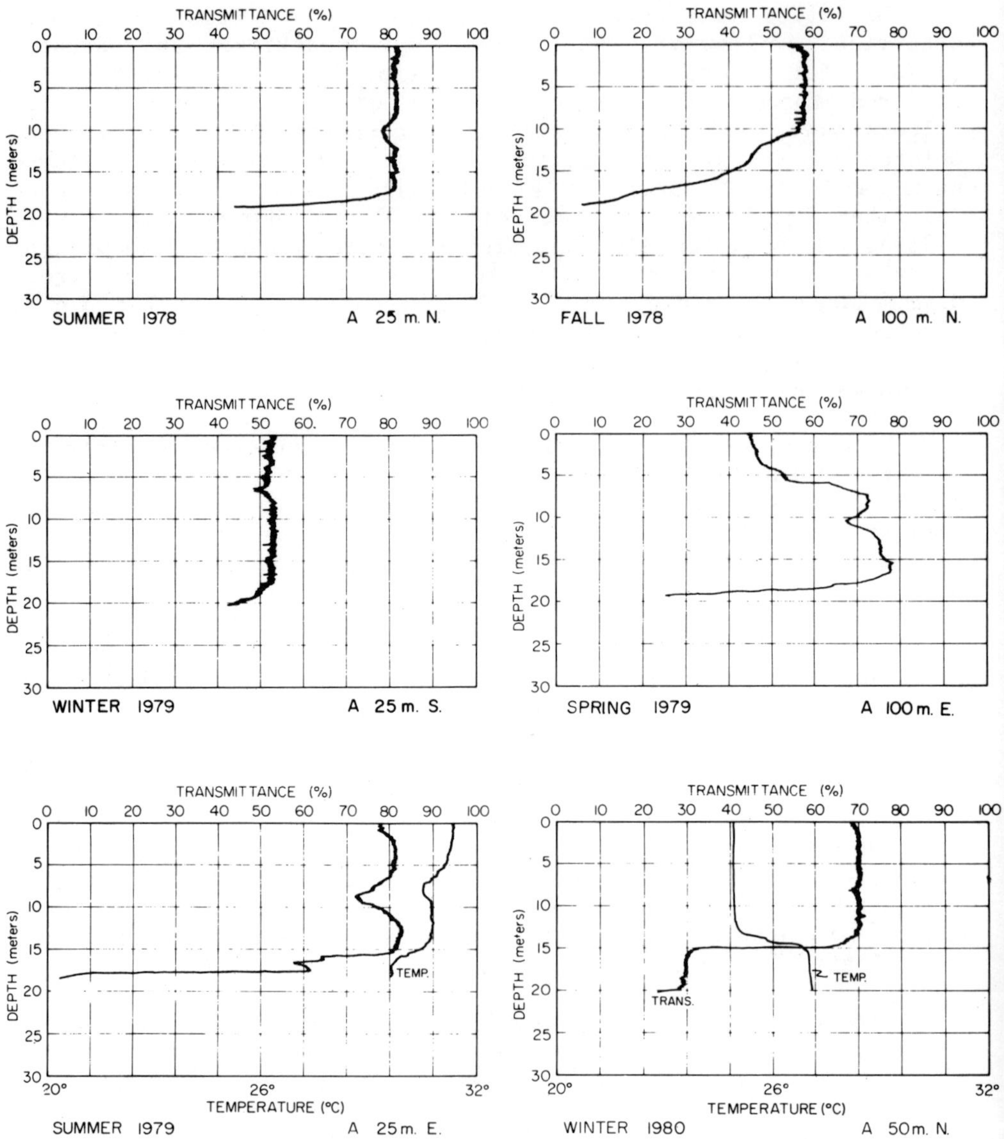

Figure 2. Representative transmissometry profiles obtained during six seasonal samplings at platform 288-A. Temperature profiles also displayed in summer 1979 and winter 1980.

(1976) using low level beta spectroscopy after a strong acid leach of the sedimentary materials and a chromatographic separation of the Pb-210 from interfering isotopes. The combined precision and analytical error on the scale length of these determinations allow

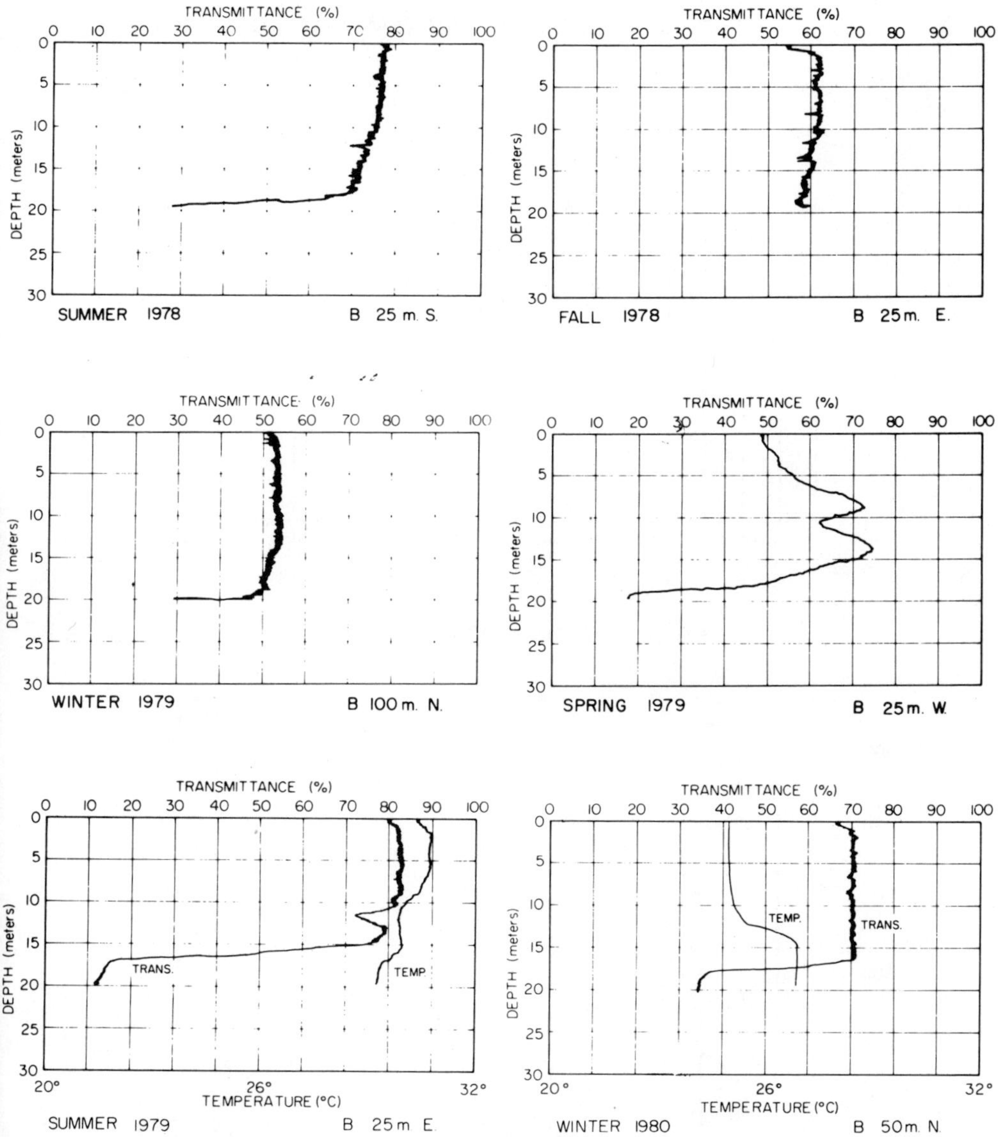

Figure 3. Representative transmissometry profiles obtained during six seasonal samplings at platform 296-B. Temperature profiles also displayed in summer 1979 and winter 1980.

measurement of sedimentation rates within ±0.3 mm yr^{-1} in the past 45 to 50 years.

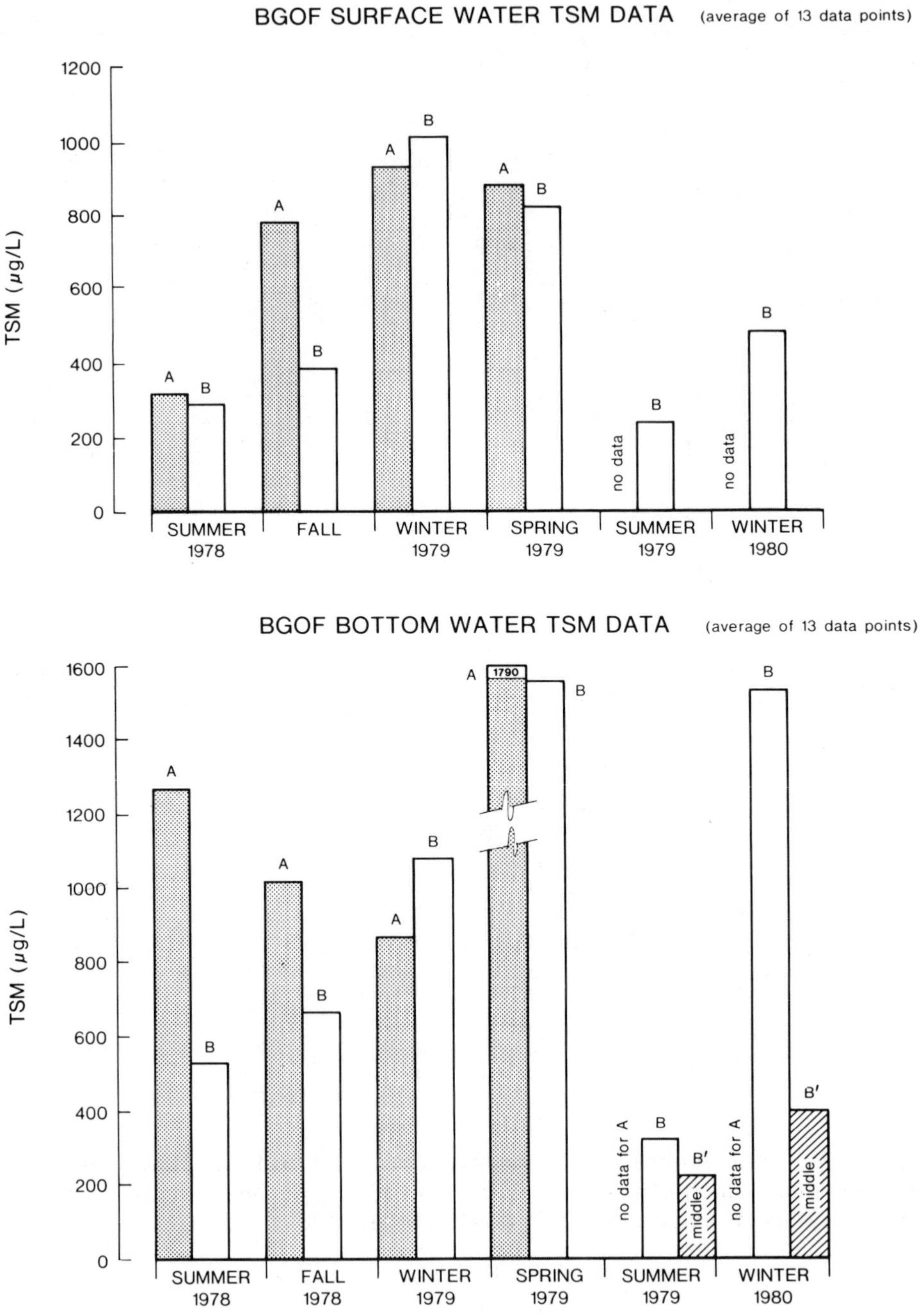

Figure 4. Histogram of total suspended matter concentrations during six seasonal samplings at platforms 288-A (A) and 296-B (B) in surface and near-bottom waters. Concentrations represent an average of 13 or more stations around each platform.

TABLE 2. SUMMARY OF TOTAL SUSPENDED PARTICULATE DISTRIBUTIONS (μg/l).

Season	Platform A		Platform B		
	Surface	Bottom	Surface	Mid	Bottom
Summer 1978[a]	321 ± 159	1266 ± 1006	288 ± 138	-	530 ± 366
Fall 1978[a]	781 ± 281	1015 ± 394	388 ± 119	-	661 ± 231
Winter 1979[a]	929 ± 267	865 ± 131	1008 ± 221	-	1075 ± 239
Spring 1979[a]	874 ± 310	1790 ± 738	814 ± 187	-	1554 ± 260
Summer 1979[b]	-	-	233 ± 61	219 ± 61	323 ± 275
Winter 1980[b]	-	-	481 ± 128	396 ± 129	1529 ± 3214

[a]Each number represents mean or standard deviation of 13 stations near each platform.
[b]Each number represents mean or standard deviation of 20 station near each platform.

RESULTS AND DISCUSSION

Suspended Particulates

Little was known about the composition of suspended particulates within the Buccaneer Field prior to this study. Manheim et al. (1972) presented a fairly extensive survey of suspended matter in the northern Gulf of Mexico and found TSM levels greater than 1 mg/l were restricted to within 100 km offshore in Texas waters. The mineral composition of surface suspensates from the Texas region were found to consist of montmorillonite-kaolinite mixtures with montmorillonite being most abundant. Anderson et al. (1977) measured several water samples in the Buccaneer Field for TSM and turbidity, finding concentrations ranging from <1.0 to 17 mg/l. Danek and Tomlinson (1980) found TSM levels were generally less than 5 mg/l and frequently less than 1.0 mg/l during his current and hydrography study in 1978-1979.

Total Suspended Particulates - Transmissometry. Total suspended matter distributions in surface and near-bottom waters reflected closely the profiles obtained by transmissometry. Figures 2 and 3 show representative profiles during six seasonal samplings at platforms 288-A and 296-B, respectively. There were few variations in transmittance profiles at stations taken around the platforms, indicating no observable effect of the structures on TSM levels. These profiles are typical of the Gulf shelf areas near the Buccaneer Field. Variations between platforms were only observed in the fall 1978 and summer 1979 samplings (see Figures 2 and 3). The near-bottom nepheloid layer at platform 288-A which was not observed at platform 296-B can be attributed to closer proximity to shore. The difference in the depth of the mid-depth nepheloid layer in summer 1979 between the two platforms results from different thermal structures. No anomalies are present that can be related to either platform presence or brine discharge.

Figure 4 and Table 2 summarize TSM data for the six samplings. Lowest TSM concentrations were obtained in surface waters during the summer samplings, averaging between 0.2 and 0.3 mg/l. During summer months, the Gulf shelf water is well stratified with a shallow thermocline. The low summer TSM levels result from this seasonal stability which inhibits upward flux of suspended sediment. The productivity of the Gulf surface water is also diminished during this period, due to nutrient depletion in the surface layer. The low summer TSM levels are consequently reflected in low biomass values at the surface.

TSM increased dramatically in near-bottom waters in the summer samplings, probably from resuspension of sediments by bottom currents. This near-bottom nepheloid layer was very sharp and located within two to three meters of the bottom during both years. A much smaller mid-depth nepheloid layer was also observed at many of the summer stations. This mid-depth layer may be associated with primary productivity and/or accumulation of particulates at the pycnocline. The temperature profiles obtained in summer 1979 indicate the mid-depth nepheloid layer was associated with a sharp decrease in temperature (Figure 3).

Near-bottom TSM measurements (Table 2) were similar during both summer and fall samplings; however, a much broader nepheloid layer was observed in the fall. Fall transmission records indicate that the bottom nepheloid layer extended from the bottom (20 m) to a depth of about 10 m (cf. 2-3 m in the summer). This feature reflects stronger currents during this sampling period along with decreased stability of the water column, both of which would produce increased turbulent mixing. A small surface TSM maximum was observed at many of the fall stations.

Highest surface TSM concentrations and lowest transmission levels were recorded during the winter 1979 sampling. The Gulf shelf water usually overturns and thoroughly mixes after the first winter (or fall) storm lowers surface temperatures and provides sufficient wind energy to drive the mixing process. During December, deeper shelf Gulf water can have a mixed layer depth of 80-100 m (Wiesenburg and Schink, 1978). Our data indicate that the water column in the Buccaneer Field was well mixed during the winter 1979 sampling. The TSM data reflect the water column structure during this sampling. There was little to no structure in transmission records. Temperature profiles (not shown) also showed a uniform surface-to-bottom structure. Nearly uniform surface and near-bottom TSM levels were observed. No significant near-bottom nepheloid layer existed during this season. The water column was homogeneous and the TSM data showed no other discernable trends.

During the winter 1980 sampling, overturn of the water column had not occurred. Temperature profiles (Figures 2 and 3) indicated that the bottom ~5 meters of the water column was 1.5°C warmer than the overlying water column. Mixing of cooled surface waters had only penetrated the top 12 m of the water column. The near-bottom warm layer contained a strong nepheloid layer from resuspension of bottom sediments by currents.

The transmission records during spring 1979 were more complex. Three distinct nepheloid layers were observed in each profile. A surface layer, containing approximately 0.8 mg/l of TSM, existed in the upper 5 m of the water column. This maximum results from spring runoff. Anderson et al. (1977) report that maximum river discharge on the Sabine, Trinity, and Brazos Rivers generally occurs from March to June. Maximum suspended sediment discharges occur in May. Transmission levels increased from approximately 40% to 75% in the 5 to 8 m depth interval indicating decreased turbidity. A small but distinct mid-depth nepheloid layer was found around 12 m. A bottom nepheloid layer was observed in the bottom two to three meters. The highest TSM levels (~1.6 mg/l) observed during this study were at near-bottom depths during the spring.

The spring TSM distributions reflect a complex interaction of processes. The shelf water is changing from a vertically mixed system to a characteristic seasonal thermocline situation. The mid-depth nepheloid layer is probably associated with the onset of this seasonal thermocline, since sinking particles tend to congregate at density interfaces. The dramatic bottom nepheloid layer is characteristic of this area in all seasons, except winter when it is less pronounced. There is little difference in TSM composition in bottom water between fall and spring samplings (Figure 4). Also, there is no exceptional increase in organic TSM in the spring, which rules out a productivity factor causing the high TSM at the bottom. We must conclude that the bottom nepheloid layer in the spring, as in the fall, results from increased bottom turbulent mixing due to bottom currents.

In general, the TSM distributions observed at the Buccaneer Field stations are typical of the Gulf shelf region. No noticable differences between up-current and down-current stations were observed; at least no significant differences and none that could be attributed to the presence of the production platforms. Danek and Tomlinson (1980), using an average TSM for August of 0.5 mg/l and 1.2 mg/l at surface and near-bottom, respectively, estimated a flux of suspended sediment at about 1.92 g/sec per square meter of water surface. Using the same procedure and TSM values for February of 0.92 mg/l (surface) and 0.9 mg/l (bottom) results in an average suspended sediment transport of about 2.86 g/sec per square meter of water surface. Consequently, an average flux of material of about 2 g/sec per square meter of water surface can be expected at the Buccaneer Field; however, this value can vary considerably under storm conditions when the TSM concentration and the current meter measurements near platform 296-B indicated the currents averaged nearly 60 cm/sec, and TSM values varied from 2.0 mg/l to 6.4 mg/l near the bottom. This resulted in a suspended material flux of about 40 g/sec per square meter of water surface (Danek and Tomlinson, 1980).

Organic Carbon. Dissolved organic carbon (DOC) levels in surface and near-bottom samples showed little vertical or horizontal variation (Table 3). DOC concentrations, throughout the year, averaged 1.2 ± 0.2 mg C/l. Winter 1979 values, however, were observed to be significantly higher than in other seasons. These winter levels were associated with higher particulate,

TABLE 3. SUMMARY OF DISSOLVED ORGANIC CARBON CONCENTRATIONS (mg C/1).

Season	Platform A Surface	Platform A Bottom	Platform B Surface	Platform B Mid	Platform B Bottom
Summer 1978[a]	1.33 ± 0.24	1.23 ± 0.16	1.38 ± 0.11	-	1.34 ± 0.14
Fall 1978[a]	1.20 ± 0.17	1.14 ± 0.20	1.14 ± 0.11	-	1.15 ± 0.08
Winter 1979[a]	1.71 ± 0.09	1.84 ± 0.19	1.80 ± 0.31	-	1.82 ± 0.23
Spring 1979[a]	1.15 ± 0.12	0.97 ± 0.10	1.30 ± 0.16	-	1.01 ± 0.09
Summer 1979[b]	-	-	1.94 ± 0.58	1.65 ± 0.20	1.78 ± 0.18
Winter 1980[b]	-	-	1.61 ± 0.37	1.38 ± 0.15	1.31 ± 0.15

[a]Each number represents mean or standard deviation of 13 stations near each platform.
[b]Each number represents mean or standard deviation of 20 station near each platform.

chlorophyll, and ATP levels. The high DOC levels result from one of two processes: resuspension of bottom sediments containing high interstitial DOC concentrations, or excretion of dissolved organics by cells associated with the much higher productivity during this period. The latter process is more probable since during this season the phytoplankton contribution to the TSM is almost 50% (Brooks et al., 1980). The phytoplankton were also distributed similarly to the DOC on a vertical basis. At platform 288-A, the bottom DOC was 8% higher than at the surface while the phytoplankton proportion of the TSM increased from 41% at the surface to 49% at the bottom. This relationship may be fortuitous, but there does seem to be an obvious correlation with higher phytoplankton productivity and higher DOC levels in the winter. There were also high DOC levels in the summer 1979 sampling which also correlated with higher phytoplankton levels.

Particulate organic carbon (POC) levels showed significant seasonal changes. Summer and fall POC concentrations ranged from 0.60 to 0.94 mg/l (Table 4) with only slight vertical gradation. This observation is in direct contrast to the high TSM levels in the bottom waters during these seasons. The relatively uniform POC distributions compared to the variable TSM levels indicate that the high TSM levels in near-bottom waters result from resuspension of bottom sediments containing smaller percentages of organic matter than surface TSM. Thus, the bottom nepheloid layers (discussed in the previous section) are confirmed not to be of organic origin.

During winter 1979, high POC values are associated with an unstable water column, high TSM and increased biological activity. Winter 1980 POC levels were more characteristic of summer and fall values. There is little vertical gradation in POC, except in the spring. Spring POC concentrations average 50% higher in surface than near-bottom water. These high levels also reflect the large surface TSM layer observed during this period, a result of fresh water runoff (see previous section). In general, POC levels appeared typical of this region of the Gulf continental shelf.

<u>Biological Indicators</u>. Tables 5 and 6 present average surface and near-bottom chlorophyll <u>a</u> and ATP concentrations during the seasonal samplings near platforms 288-A and 296-B. Tables 7 and 8 show the particulate organic composition at these platforms. Since carbon constitutes from 40 to 50% by weight of most marine organic materials, we used (for purposes of our calculations) a factor of 2.25 for conversion of particulate carbon into particulate organic matter (POM). For conversion of chlorophyll <u>a</u> and ATP into organic carbon, Holm-Hansen and Paerl (1972) among others have used factors of 100 and 250, respectively. Although these multipliers may vary somewhat under

TABLE 4. SUMMARY OF PARTICULATE ORGANIC CARBON CONCENTRATIONS (μg C/1).

Season	Platform A Surface	Platform A Bottom	Platform B Surface	Platform B Mid	Platform B Bottom
Summer 1978[a]	65 ± 8	86 ± 21	63 ± 14	-	75 ± 9
Fall 1978[a]	66 ± 8	72 ± 11	65 ± 10	-	67 ± 11
Winter 1979[a]	187 ± 17	191 ± 11	156 ± 32	-	161 ± 25
Spring 1979[a]	166 ± 42	106 ± 44	150 ± 24	-	96 ± 11
Summer 1979[b]	-	-	80 ± 12	78 ± 14	94 ± 20
Winter 1980[b]	-	-	92 ± 24	79 ± 17	92 ± 21

[a]Each number represents mean or standard deviation of 13 stations near each platform.
[b]Each number represents mean or standard deviation of 20 station near each platform.

TABLE 5. SUMMARY OF CHLOROPHYLL A DISTRIBUTIONS (mg/m^3).

Season	Platform A Surface	Platform A Bottom	Platform B Surface	Platform B Mid	Platform B Bottom
Summer 1978[a]	0.068 ± 0.022	0.195 ± 0.189	0.059 ± 0.016	-	0.125 ± 0.038
Fall 1978[a]	0.114 ± 0.028	0.129 ± 0.026	0.097 ± 0.026	-	0.119 ± 0.022
Winter 1979[a]	1.729 ± 0.208	1.872 ± 0.307	1.883 ± 0.469	-	1.883 ± 0.469
Spring 1979[a]	0.491 ± 0.214	0.557 ± 0.068	0.389 ± 0.152	-	0.499 ± 0.131
Summer 1979[b]	-	-	0.498 ± 0.186	0.36 ± 0.105	0.358 ± 0.125
Winter 1980[b]	-	-	0.573 ± 0.092	0.413 ± 0.138	0.449 ± 0.104

[a]Each number represents mean or standard deviation of 13 stations near each platform.
[b]Each number represents mean or standard deviation of 20 station near each platform.

TABLE 6. SUMMARY OF ADENOSINE TRIPHOSPHATE DISTRIBUTIONS (ng/1).

Season	Platform A Surface	Platform A Bottom	Platform B Surface	Platform B Mid	Platform B Bottom
Summer 1978[a]	349 ± 201	493 ± 239	552 ± 336	-	806 ± 368
Fall 1978[a]	288 ± 126	339 ± 154	302 ± 151	-	301 ± 122
Winter 1979[a]	644 ± 212	791 ± 263	749 ± 196	-	793 ± 194
Spring 1979[a]	537 ± 155	553 ± 238	532 ± 363	-	380 ± 224
Summer 1979[b]	-	-	301 ± 140	375 ± 185	416 ± 248
Winter 1980[b]	-	-	487 ± 473	334 ± 216	354 ± 203

[a]Each number represents mean or standard deviation of 13 stations near each platform.
[b]Each number represents mean or standard deviation of 20 station near each platform.

TABLE 7. PARTICULATE ORGANIC COMPOSITION AT PLATFORM 288-A (μg/1).

Season	Depth	TSM[b]	POM[c]	Biomass[c,d]	Phytoplankton[c,e]
Summer[a]	SFC	321	146	196	14
	BT	1266	194	277	43
Fall[a]	SFC	781	148	162	25
	BT	1015	162	189	27
Winter[a]	SFC	929	420	373	387
	BT	865	429	445	420
Spring[a]	SFC	874	373	301	110
	BT	1790	238	310	126

[a]Compositions represent averages of 13 stations near platform 288-A.
[b]See Table 1.
[c]Carbon concentrations (Table 4) multiplied by 2.25 to convert to particulate organic matter.
[d]ATP concentrations (Table 6) times 250 to convert to biomass C.
[e]Chlorophyll concentrations (Table 5) times 100 to convert to phytoplankton C.

TABLE 8. PARTICULATE ORGANIC COMPOSITION AT PLATFORM 296-B (μg/1).

Season	Depth	TSM[c]	POM[d]	Biomass[d,e]	Phytoplankton[d,f]
Summer 1978[a]	SFC	288	141	310	14
	BT	530	168	454	27
Fall 1978[a]	SFC	338	146	171	23
	BT	661	150	169	27
Winter 1979[a]	SFC	1008	351	421	423
	BT	1075	362	445	423
Spring 1979[a]	SFC	814	447	299	90
	BT	1554	216	214	113
Summer 1979[b]	SFC	233	180	169	112
	MID	219	175	211	81
	BT	323	210	234	81
Winter 1980[b]	SFC	481	207	274	129
	MID	396	178	188	93
	BT	1529	207	199	101

[a]Compositions represent averages of 13 stations near platform 296-B.
[b]Compositions represent averages of 20 stations near platform 296-B.
[c]See Table 1.
[d]Carbon concentrations (Table 4) multiplied by 2.25 to convert to particulate organic matter.
[e]ATP concentrations (Table 6) times 250 to convert to biomass C.
[f]Chlorophyll concentrations (Table 5) times 100 to convert to phytoplankton C.

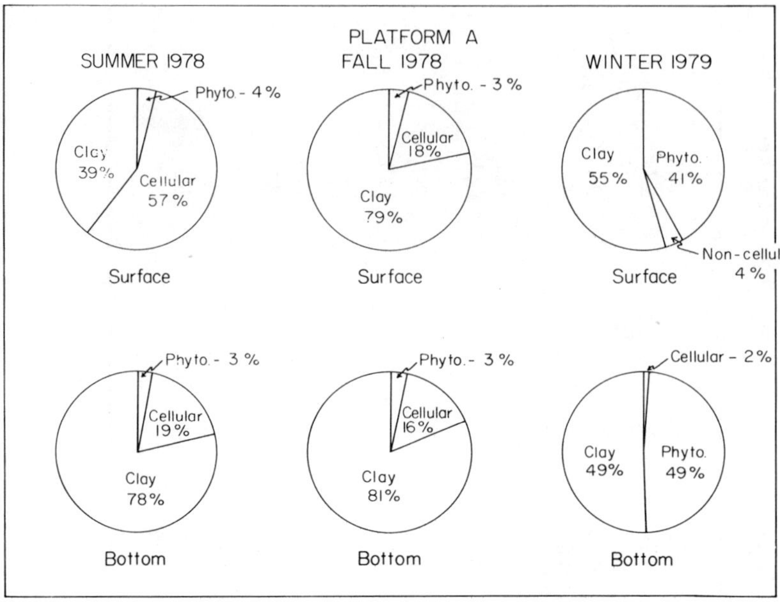

Figure 5. Composition of suspended particulates at stations near platform 288-A during seasonal samplings. Phyto. = phytoplankton biomass estimated from chlorophyll a. Cellular = cellular biomass estimated from ATP minus phytoplankton biomass.

different environmental conditions, they are useful in examining the composition of the suspended particulates. We have used them to calculate POM values from the biological data. Chlorophyll a is used to estimate that fraction of the organic carbon which derives from the primary producers. ATP is an estimate of the living cellular material present, whether from phytoplankton, zooplankton, and/or bacteria. These parameters were used to determine the suspended particulate compositions. The equations used for the calculations are given below:

$$\text{POM} = \text{POC} \times 2.25 \quad (1)$$

$$\text{Biomass} = (\text{ATP} \times 250) \times 2.25 \quad (2)$$

$$\text{Phyto.} = (\text{Chl } \underline{a} \times 100) \times 2.25 \quad (3)$$

These equations were used to convert all data from organic carbon concentrations to POM data so the POM component of the TSM could be deciphered. Figures 5 through 8 are pie diagrams which depict the percentage distributions of constituents making up the TSM. Cellular material is equivalent to total biomass (equation 2), while non-cellular material is the difference between POM and

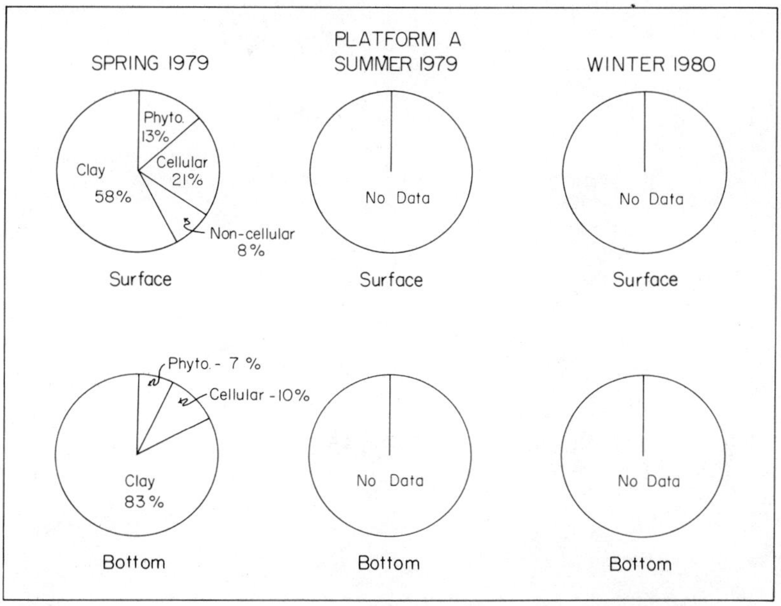

Figure 6. Composition of suspended particulates at stations near Platform 288-A during seasonal samplings. Phyto. = phytoplankton biomass estimated from chlorophyll a. Cellular = cellular biomass estimated from ATP minus phytoplankton biomass.

biomass, if any. The phytoplankton constituent of the TSM is part of the overall biomass data, but can be estimated separately by measuring chlorophyll.

Tables 7 and 8 indicate that very little particulate organic matter is non-cellular material. In most cases, cellular biomass (determined by ATP concentrations) is greater than particulate organic matter determined by POC. An exception, however, is surface water during the spring, when as much as 25% of the POC was non-cellular organics. This is consistent with the surface nepheloid layer which is characterized by high TSM and POC levels during this period. The high levels of non-cellular organics in this layer probably reflect a fresh water outflow from Galveston Bay. Such an outflow would input land plant debris as non-cellular, non-living POC. There also were measurable concentrations of non-cellular organics during the winter 1979 and summer 1979 cruises in surface waters. This could represent excretion of cellular material from the large phytoplankton communities as a result of the very high productivity and resulting inefficient assimilation.

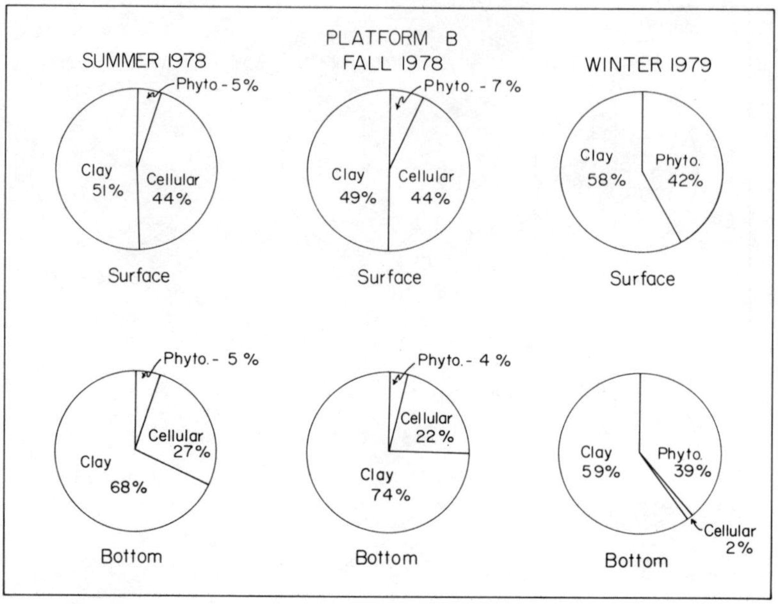

Figure 7. Composition of suspended particulates at stations near platform 296-B during seasonal samplings. Phyto. = phytoplankton biomass estimated from chlorophyll a. Cellular = cellular biomass estimated from ATP minus phytoplankton biomass.

Figures 5 through 8 indicate that, during the summer and fall 1979, phytoplankton constituted a very small percentage of the total suspended particulates. Lowest chlorophyll concentrations were also observed during summer 1978 in surface waters. These low values reflect stratification of the water column during the summer months, which results in low productivity. Productivity was considerably higher in bottom waters during summer 1978. There were only small differences between surface and bottom chlorophyll levels during the fall. Winter 1979 chlorophyll concentrations were more than an order of magnitude higher than the previous two seasons. Little vertical gradation in phytoplankton biomass was observed during the winter; a result of strong vertical mixing of the water column. Figures 5 and 7 show that, during the winter 1979 sampling, virtually all cellular material consisted of phytoplankton, both in surface and bottom waters. Phytoplankton constituted 41 and 49% of the particulate material in surface and bottom waters, respectively. Although there was a large standing crop in the water column during spring 1979, phytoplankton only constituted 13 to 7% of the TSM, respectively. The spring phytoplankton population was high, but percentages are lower because of the high TSM levels. The standing crop of phytoplankton in summer 1979 was only slightly

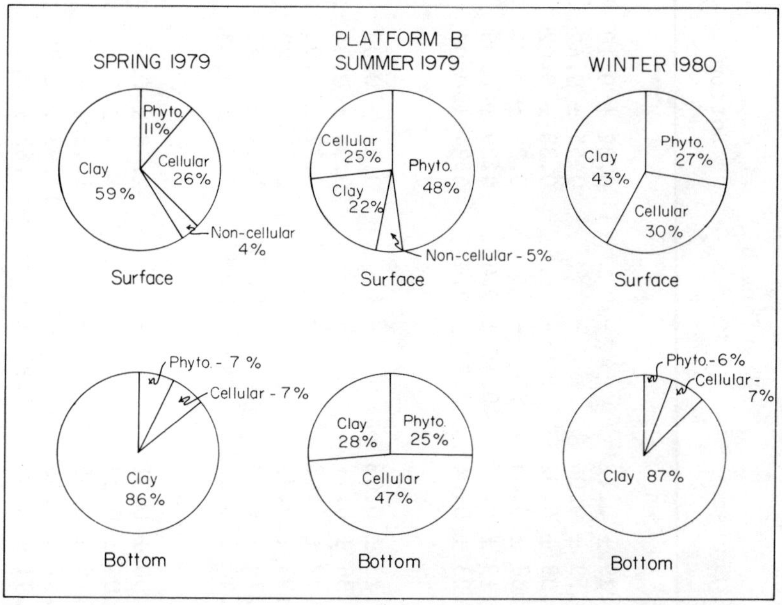

Figure 8. Composition of suspended particulates at stations near platform 296-B during seasonal samplings. Phyto. = phytoplankton biomass estimated from chlorophyll a. Cellular = cellular biomass estimated from ATP minus phytoplankton biomass.

higher than in the spring, but because of the TSM levels, phytoplankton constituted 48 and 25% of the particulates in the surface and near-bottom waters, respectively. Phytoplankton levels were high in the winter 1980, but do not represent the bloom conditions observed the previous winter.

ATP concentrations are a measure of total cellular material. Table 6 illustrates that, although less pronounced, ATP concentrations show the same general trends as chlorophyll concentrations. This was expected since bacterial and zooplankton biomass would be associated with the plankton. The cellular material (bacteria, zooplankton, and phytoplankton), generally constituted between 20 and 35% of the total suspended particulates. However, during the winter and summer 1979 when the phytoplankton standing crop was high, cellular material consituted as much as 75% of suspended particulates.

Calcium carbonate levels were low during all seasons and represent an insignificant fraction of the TSM. Only during the fall were sufficient volumes of water filtered to determine carbonate concentrations. All $CaCO_3$ concentrations were below 0.4% of the total suspended particulates. The carbonate observed

TABLE 9. SUMMARY OF NUTRIENT CONCENTRATIONS (μM)[a].

Season	Platform A Surface	Platform A Bottom	Platform B Surface	Platform B Mid	Platform B Bottom
Phosphate					
Summer 1978	0.31 ± 0.03	0.32 ± 0.05	0.29 ± 0.12	-	0.32 ± 0.10
Fall 1978	0.52 ± 0.05	0.55 ± 0.05	0.59 ± 0.06	-	0.55 ± 0.42
Winter 1979	0.39 ± 0.02	0.40 ± 0.02	0.39 ± 0.02	-	0.38 ± 0.02
Spring 1979	0.30 ± 0.05	0.36 ± 0.04	0.38 ± 0.04	-	0.39 ± 0.04
Summer 1979	-	-	0.19 ± 0.04	0.19 ± 0.03	0.29 ± 0.06
Winter 1980	-	-	0.28 ± 0.04	0.27 ± 0.04	0.34 ± 0.07
Nitrate					
Summer 1978	0.31 ± 0.05	0.31 ± 0.05	0.23 ± 0.03	-	0.25 ± 0.02
Fall 1978	0.38 ± 0.06	0.40 ± 0.09	0.36 ± 0.11	-	0.34 ± 0.04
Winter 1979	0.34 ± 0.03	0.34 ± 0.03	0.32 ± 0.03	-	0.32 ± 0.03
Spring 1979	2.73 ± 0.32	1.74 ± 0.05	3.03 ± 0.85	-	1.59 ± 0.20
Summer 1979	-	-	0.79 ± 0.14	0.68 ± 0.13	1.44 ± 0.43
Winter 1980	-	-	1.40 ± 0.07	1.48 ± 0.09	2.20 ± 0.53
Silicate					
Summer 1978	2.53 ± 0.07	3.26 ± 0.52	2.00 ± 0.31	-	2.58 ± 0.41
Fall 1978	5.23 ± 0.20	5.35 ± 0.16	5.04 ± 0.32	-	4.94 ± 0.55
Winter 1979	1.25 ± 0.05	1.25 ± 0.05	1.15 ± 0.05	-	1.15 ± 0.05
Spring 1979	4.75 ± 2.54	13.06 ± 0.35	4.57 ± 0.45	-	11.99 ± 2.98
Summer 1979	-	-	1.91 ± 0.29	1.91 ± 0.31	4.62 ± 2.14
Winter 1980	-	-	0.92 ± 0.14	1.60 ± 0.33	3.76 ± 2.08

[a]Each number represents mean or standard deviation of 13 stations near each platform.

TABLE 10. CLAY MINERAL COMPOSITION IN WATER COLUMN[a].

Season	Smectite (%)	Illite (%)	Kaolinite (%)	Quartz (%)
Summer 1978	62	4	7	27
Fall 1978	36	5	10	49
Winter 1979	3	39	38	20
Spring 1979	16	55	7	21

[a]Numbers represent average of at least 16 samples from each season taken at surface and near-bottom depths near platforms 288-A and 296-B. Values are percent of total clay.

was probably from organisms which have carbonate exoskeletons. The low values observed are thought to be typical of this region of the Gulf.

Nutrients. The general nutrient levels (Table 9) in the Buccaneer Field measured during this study are typical of inner shelf Gulf coastal areas (Sackett and Brooks, 1978). There are no discernible relationships to either the platforms or direction from the platforms. The surface-to-bottom variations that were observed result from inflow of fresh water at the surface and regeneration in the bottom water.

Clay Mineralogy. Table 10 summarizes the seasonal changes in clay mineralogy at the Buccaneer Field (Brooks et al., 1980). The summer samples were the highest in smectite. The winter samples were highest in both illite and kaolinite and lowest in smectite concentration. The spring samples showed illite concentration to be the highest of all. In all the samples, the quartz concentration remains fairly constant to slightly higher towards the bottom of the water column with the highest concentration in the fall.

Particle Size. Particle size data were tested statistically to determine whether any significant temporal or spacial trends existed. No consistent trends were noted either temporally or spacially. The lack of correlation may be due to the impossibility of synchronous sampling of a moving water mass, or in fact a wide variability in suspended particulate material present. Table 11 list the mean and standard deviation for suspended siliciclastic particle determinations, and illustrates the variability involved.

TABLE 11. SUMMARY OF SUSPENDED SEDIMENTS - SILICICLASTIC FRACTION.

Season		Median[a]	Mean[b]	St. Dev.[c]	Skewness[d]	Kurtosis[e]
PLATFORM 288-A						
Fall, 1978	Surf.	8.21 ± 2.76	8.05 ± 2.70	0.76 ± 0.30	-0.34 ± 0.24	0.89 ± 0.32
	Bottom	6.70 ± 4.31	6.52 ± 4.19	0.63 ± 0.44	-0.33 ± 0.28	0.74 ± 0.49
Winter, 1978	Surf.	8.58 ± 0.28	8.61 ± 0.23	0.81 ± 0.19	+0.00 ± 0.16	0.99 ± 0.27
	Bottom	8.67 ± 0.36	8.69 ± 0.20	0.77 ± 0.08	-0.03 ± 0.26	0.95 ± 0.18
Spring, 1979	Surf.	9.26 ± 0.16	9.09 ± 0.13	0.66 ± 0.09	-0.21 ± 0.13	0.84 ± 0.14
	Bottom	8.64 ± 0.29	8.69 ± 0.21	0.68 ± 0.04	+0.05 ± 0.10	0.97 ± 0.20
PLATFORM 296-B						
Fall, 1978	Surf.	6.31 ± 4.15	6.30 ± 4.14	0.59 ± 0.39	-0.09 ± 0.19	0.58 ± 0.39
	Bottom	3.46 ± 4.40	3.41 ± 4.33	0.36 ± 0.50	-0.09 ± 0.15	0.38 ± 0.05
Winter, 1978	Surf.	8.27 ± 2.64	8.16 ± 2.59	0.73 ± 0.27	-0.29 ± 0.26	1.00 ± 0.60
	Bottom	8.84 ± 0.56	8.73 ± 0.47	0.83 ± 0.15	-0.26 ± 0.27	1.23 ± 0.46
Spring, 1979	Surf.	9.01 ± 0.21	8.93 ± 0.13	0.72 ± 0.07	-0.21 ± 0.18	0.91 ± 0.20
	Bottom	8.46 ± 0.19	8.56 ± 0.17	0.75 ± 0.12	+0.10 ± 0.12	1.09 ± 0.24
Summer, 1979	Surf.	9.29 ± 0.22	9.09 ± 0.14	0.72 ± 0.08	-0.04 ± 0.18	0.91 ± 0.16
	Mid.	9.20 ± 0.24	9.05 ± 0.16	0.73 ± 0.07	-0.36 ± 0.18	0.96 ± 0.18
	Bottom	9.13 ± 0.33	9.00 ± 0.24	0.73 ± 0.08	-0.32 ± 0.17	0.89 ± 0.15
Winter, 1980	Surf.	8.82 ± 2.03	8.67 ± 1.99	0.58 ± 0.17	-0.42 ± 0.15	0.95 ± 0.36
	Mid.	8.71 ± 2.12	8.55 ± 2.07	0.63 ± 0.17	-0.36 ± 0.26	0.94 ± 0.32
	Bottom	9.04 ± 0.38	8.92 ± 0.24	0.68 ± 0.29	-0.31 ± 0.25	0.88 ± 0.14

[a]Median particle size, in Phi units
[b]Graphic mean particle size, in Phi units
[c]Inclusive graphic standard deviation, in Phi units (Sorting)
[d]Inclusive graphic skewness
[e]Inclusive graphic kurtosis

Surficial Sediments

The preliminary sediment sampling of the Buccaneer Field was performed by Harper, Scrudato, and Giam in November 1975. Based on 34 sampling stations, the surficial sediments were broadly classified as poorly sorted, fine-grained silty sands containing low concentrations of clay-sized material. It was noted that the sediments became progressively finer-grained toward the northeast within the field, and that the fine-grained sand content in the Buccaneer Field ranged from 70 to 80%.

In 1976, Anderson, Schwarzer, and Clark undertook a more comprehensive investigation of a ~24 km^2 area around the Buccaneer Field which involved both geophysical and sedimentological aspects. Surficial sediment was obtained from 93 and 80 sampling stations in 1976-1977 and 1977-1978, respectively, within the Buccaneer Field using both grab samplers and short cores. The cores exhibited extreme bioturbation and homogenization, and the agitation of the sediment was believed to contribute to the loss of fine-grained material, as it was continually reexposed to bottom current activity. Behrens (1977) using these same cores measured total organic carbon and carbon isotopes.

During the third (1978-1979) and fourth (1979-1980) years of the program, we analyzed surficial sediments along transects aligned north, east, south, and west of platforms 288-A and 296-B with sampling locations at 10, 25, 50, and 100 m from the brine discharge pipe. Well jacket and flare stack samples were also collected. Grain size distribution was measured six times seasonally over these two years. Percent organic carbon, percent calcium carbonate, and their respective $\delta^{13}C$ values, and clay mineralogy were determined twice yearly (winter and summer samplings). Pb-210 distributions were determined from six cores within the field.

Particle Size Figures 9 through 13 illustrate the particle size distribution in the immediate vicinity of the platforms. Since samples were taken on transects radiating from the discharge leg located on the west side of each platform, the 10 and 25-meter east stations are actually directly beneath the platforms. During summer 1978, the sediments at platform 288-A were poorly to extremely poorly sorted, with sorting being poorest to the northeast. Sedimentary material beneath the platform was relatively coarse grained, and exhibited a decrease in mean grain-size with distance from the platform. There was some evidence of removal of fine-grained material by current winnowing, as evidenced by sorting and skewness. Fall 1978 data showed a major increase in particle size except to the south of the platform, indicating an intervening period of high current flow and associated winnowing of sediments, particularly to the east.

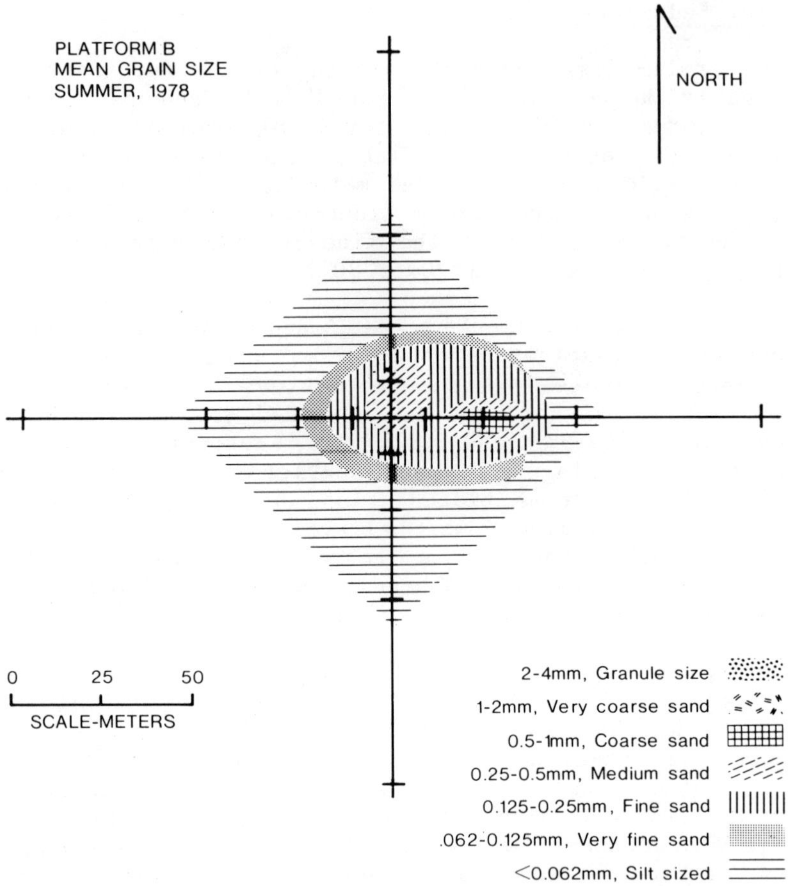

Figure 9. Mean grain size distribution at platform 296-B during summer 1978.

Sorting values also indicated some deposition of finer-grained material to the north. Winter 1978-1979 data indicated mild erosion of the sediment along the east and west sampling traverses, with deposition to the south and at 10 m east. Data from the spring 1979 cruise indicated widespread deposition of fine-grained material near the platform with mild winnowing occurring only beneath the platform and directly to the east (see Brooks et al., 1980 for data). Summer 1979 data indicated an extreme increase in mean grain size to the immediate north and northwest of platform 288-A, with a corresponding increase in sorting. This indicates effective winnowing of the surficial sediments by current activity had occurred in these areas between spring and summer samplings.

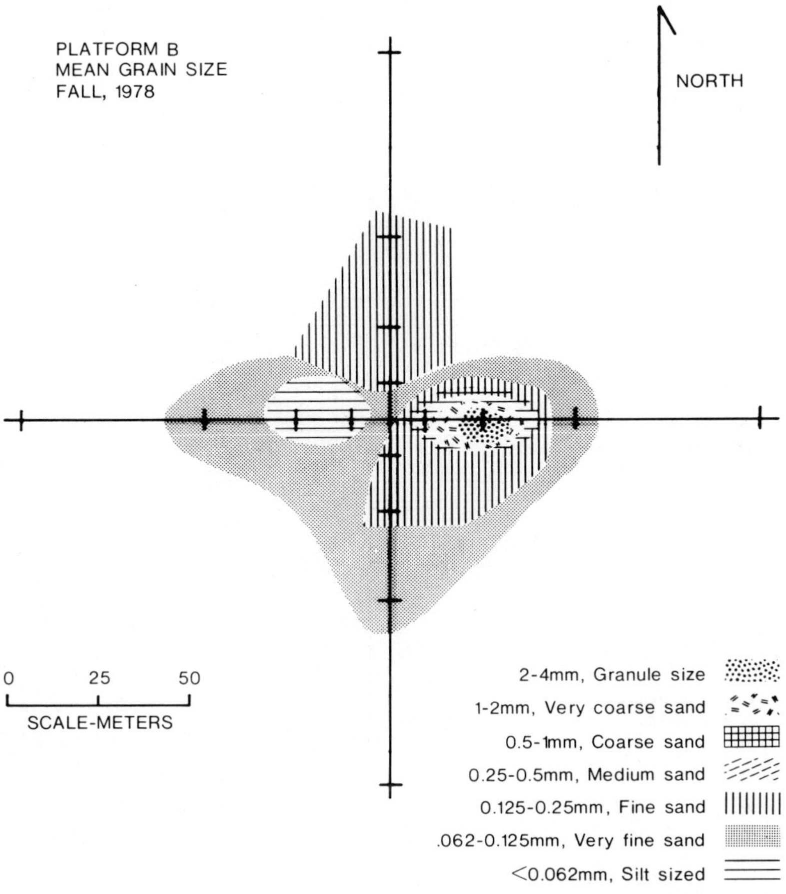

Figure 10. Mean grain size distribution at platform 296-B during fall 1978.

At platform 296-B, the sampling during summer 1978 indicated the surficial sediments were coarser-grained to the east and directly beneath the platform than on the surrounding seabed (Figure 9). The sediments were very poorly sorted beneath the platform, but became more highly sorted with increasing distance from the platform. Fall 1978 collections indicated particle size had decreased to the west and north, but increased to the east (Figure 10). Sorting increased to the east and west, but decreased to the north and south. The data indicated a period of deposition of fine-grained material, particularly to the west of the platform, with winnowing of sediments being particularly effective to the immediate east, and less so to the north and south.

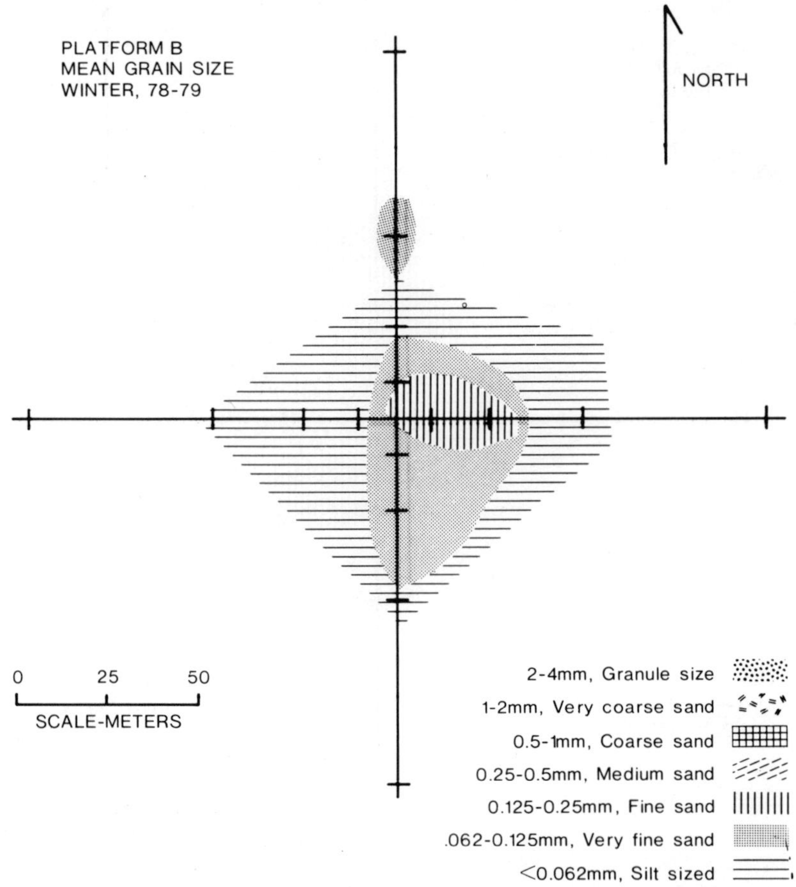

Figure 11. Mean grain size distribution at platform 296-B during winter 1979.

Data from the winter cruise in 1978-1979 indicated a general period of deposition of fine-grained material occurred throughout the area (Figure 11), suggesting a period of low current velocities. Spring 1979 collection (Figure 12) suggests a period of winnowing of surficial sediments and indicates the presence of strong current activity beneath the platform. Deposition of fine-grained material was confined to the south transect, suggesting current flow in a general north to south direction, with eddy effects producing erosion beneath the platform. Summer 1979 data (Figure 13) indicated deposition of fine-grained material to the north and south, with winnowing effective beneath and to the northeast of the platform. Winter 1979-1980 data indicated continued winnowing of sediments beneath the platform and deposition of fine-grained material to the northeast.

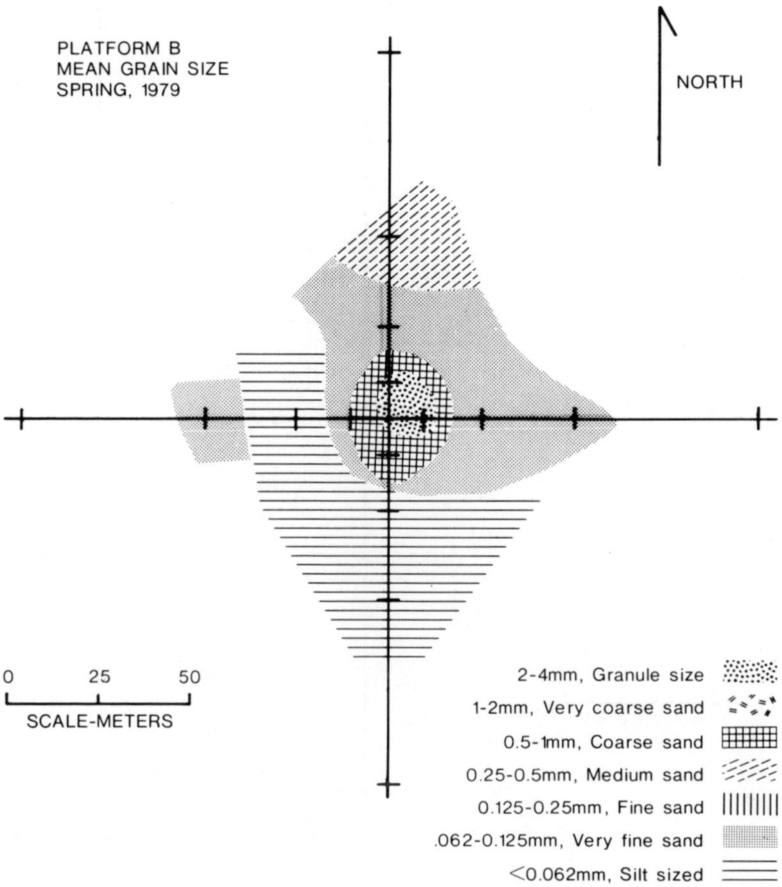

Figure 12. Mean grain size distribution at platform 296-B during spring 1979.

The dynamic nature of the sediments flux at both platforms is illustrated in Table 12, which specifies the processes active between each seasonal samplings.

<u>Organic Carbon</u>. Behrens (1977) determined the organic carbon content of surficial sediments in the general area of the Buccaneer Field. Total organic carbon (TOC) content of the sediments in the area ranged from less than 0.1% to near 1.0%. Figures 14 and 15 show organic carbon measurements on a closely spaced grid around platforms 288-A and 296-B. There were decreasing organic carbon gradients away from the platforms. TOC varied between 0.05 and 1.23% in the immediate vicinity of the platforms. The high levels beneath the structures probably result from high productivity associated with the platform. Concen-

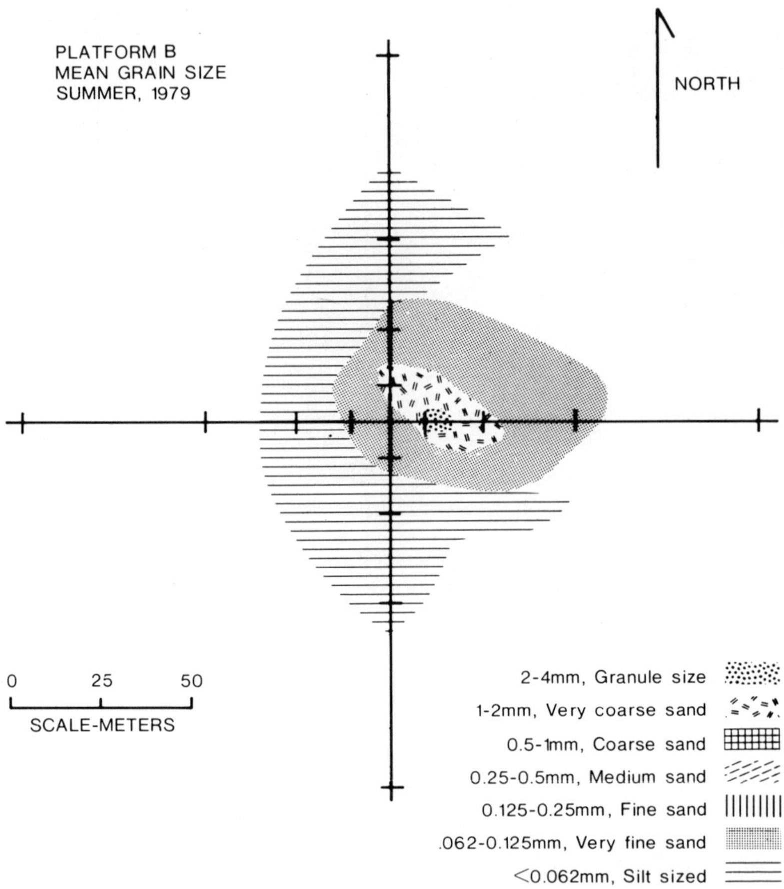

Figure 13. Mean grain size distribution at platform 296-B during summer 1979.

trations of organic platform contaminates are not sufficient to cause a measurable increase in sedimentary organic carbon (Middleditch and West, 1980). A seasonal pattern in TOC levels beneath the platforms was apparent during most samplings. Summer samplings showed a fairly predominant northeast flux of organic carbon which may be a reflection of prevailing east-northeast bottom currents during the summer months (Armstrong and Hamilton, 1979). In this case, sediments below and surrounding the platform which are relatively high in organic carbon contents were swept away and redeposited in the direction of current flow. Similar contours for the winter data indicated a southerly flux of organic carbon which agreed well with the 180° current change occurring during and lasting through the winter months (Armstrong and Hamilton, 1979). The $\delta^{13}C$ values in the 217 measurements by

TABLE 12. SURFICIAL SEDIMENT VARIATIONS THROUGH TIME AT PLATFORMS 288-A AND 296-B.

	Platform 288-A
Summer to Fall 1978	Active current activity - winnowing effect
Fall to Winter 1978	Mild winnowing by currents
Winter to Spring 1979	Deposition of fine-grained material
Spring to Summer 1979	Winnowing of sediments by currents
Summer to Winter 1979	No data collected
	Platform 296-B
Summer to Fall 1978	Winnowing to east; deposition to west
Fall to Winter 1978	Deposition of fine-grained material
Winter to Spring 1979	Deposition to south and east, winnowing beneath platform
Spring to Summer 1979	Deposition beneath platform and to north and south; erosion occurring to east
Summer to Winter 1979	Extreme winnowing beneath platform, deposition of fine-grained material to northeast

Behrens (1977) ranged from -19 to -23°/oo, with the bulk of the samples between -20.0 and -21.3°/oo. The most negative samples consisted of Beaumont sediment with a mean $\delta^{13}C$ of -22.7. These values are consistent with determinations reported by Brooks et al. (1980). The $\delta^{13}C$ values versus organic carbon (%) were plotted for both summer 1978 and winter 1978-1979 samples collected around platform 288-A (Figure 16). The summer data points all lie within a relatively narrow range of both $\delta^{13}C$ and percent organic carbon and reflect normal open marine pelagic values (Sackett, 1964). Winter data, however, are markedly lighter in isotopic composition and show a general increase in percent organic carbon. One possible explanation for this observation is that the sediments might be receiving a greater terrestrial contribution from coastal bays and estuaries during the winter months. The $\delta^{13}C$ of terrestrial organic carbon is generally about -27°/oo whereas marine values are closer to -20°/oo (Wickman, 1952). An increase in the terrestrial component would tend to lower the observed $\delta^{13}C$ values.

Calcium Carbonate. $CaCO_3$ concentrations were anomalously high beneath and near the platforms. Concentrations ranged from 1.8 to over 70%, with concentrations directly beneath the platforms ranging from 20 to 74%. High $CaCO_3$ concentrations

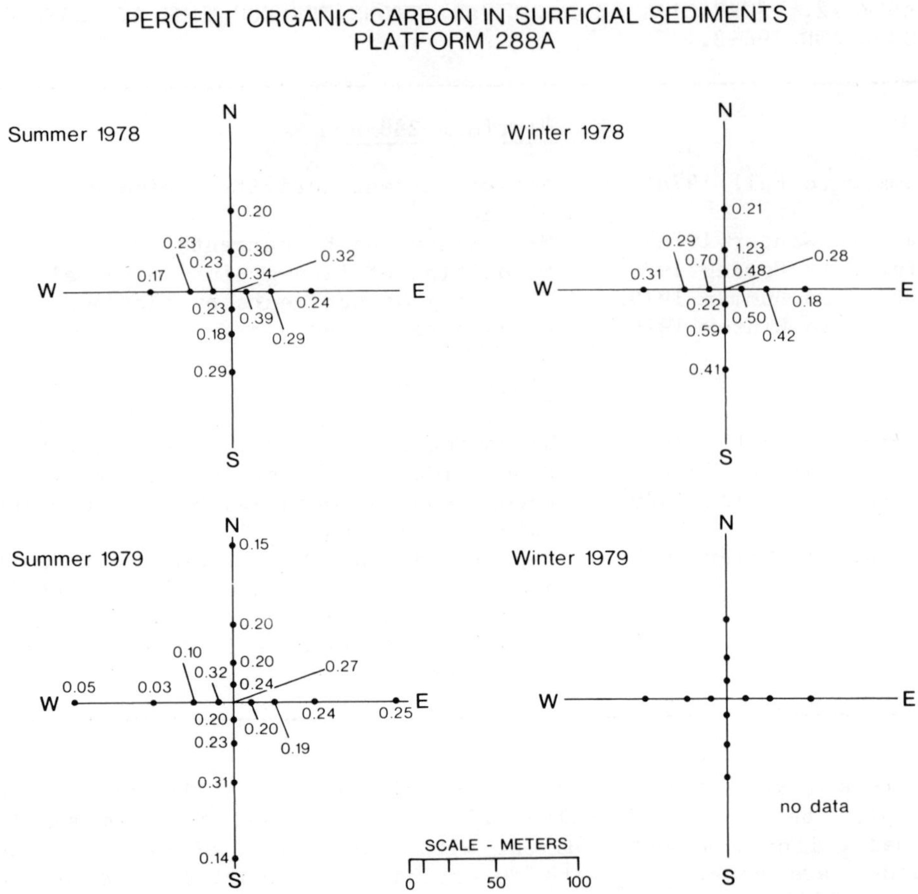

Figure 14. Percent organic carbon around platform 288-A.

associated with the platforms are indicative of the high productivity in the surface waters near the structures and contributions from barnacles and other calcium-containing organisms on the platform itself. Concentrations showed a decreasing gradient with distance from the platforms. As with organic carbon, the effect of seasonal current patterns was generally observed in the grids beneath the structures with concentration gradients being skewed to the north and east in summer and to the south in winter. Since most of the surficial sediment $CaCO_3$ particles consisted of shell hash with a large grain size, $CaCO_3$ patterns paralleled the changing sedimentological patterns discussed for grain size (Figures 17 and 18).

Sedimentation Rates. Sedimentation rates were determined in the Buccaneer Field by radiocarbon (Behrens, 1977) and Pb-210

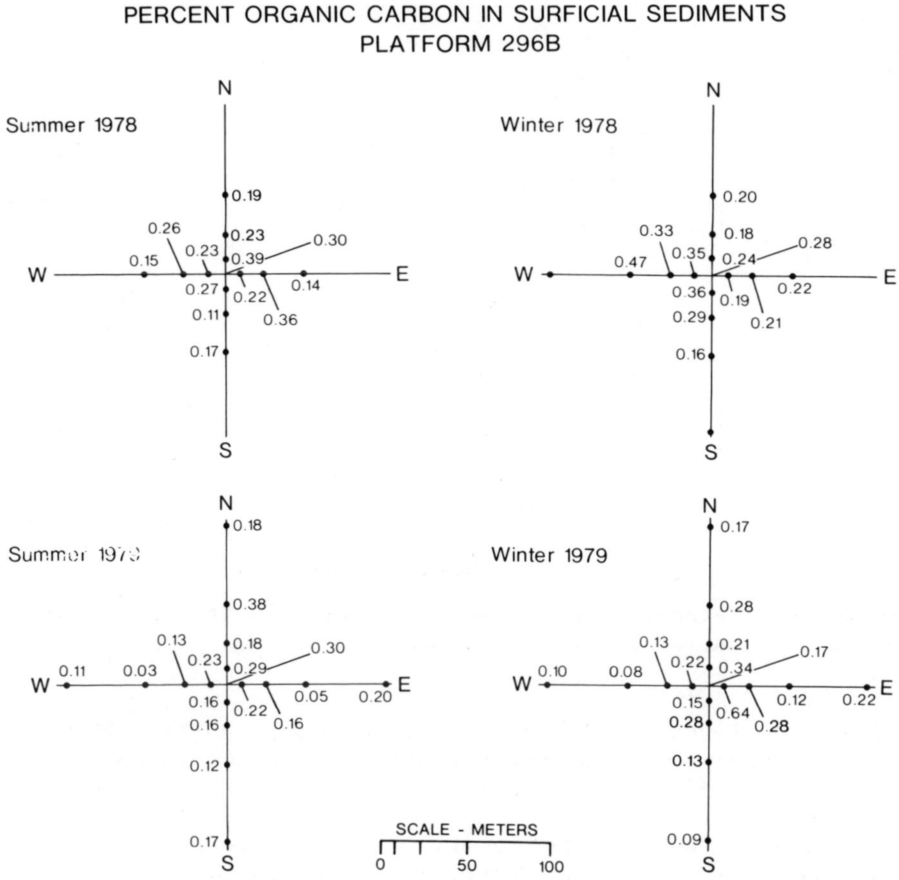

Figure 15. Percent organic carbon around platform 296-B.

dating techniques (Brooks et al., 1980). The radiocarbon dates reported by Behrens (1977) were obtained on the disseminated total organic carbon in the sediments. He estimated the sedimentation rate in the Buccaneer Field by radiocarbon dates and also by the depth of the Beaumont Formation. Cores taken 15, 152, and 1800 m east of platform 288-A had 1 m, 24 cm, and 7 cm, respectively, eroded from the sedimentary column. Observations of the depth of the Beaumont Formation in cores suggested that somewhat more erosion - up to 1.5 m - occurs locally.

The location of cores taken for Pb-210 dating are shown in Figure 19. The relatively low, and uniform, surface sediment Pb-210 activity among the six stations (see Figure 20) indicates a similarly uniform sedimentary composition and origin. The bulk of the materials are terrigenous and are moderately coarse-grained.

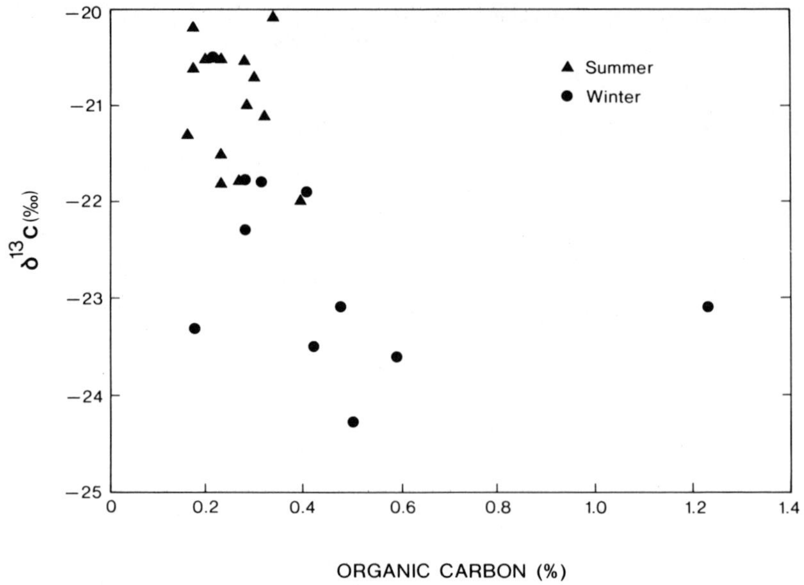

Figure 16. Percent organic carbon versus $\delta^{13}C$ from platform 288-A during both summer and winter samplings.

Although Stations 2 and 3 have slightly higher surface values, the probable cause is an apparently greater degree of mixing in the top 1 or 2 cm at Stations 1, 5, and 6.

There is a definite trend in apparent sedimentation rates among the six stations. Stations 2, 3, and 4 on the northern (shoreward) side of the platform are characterized by lower (1.5 to 2.1 mm/yr) effective sedimentation rates than are Stations 5 and 6 (3.5 and 3.7 mm/yr, respectively) on the seaward side of the platform. Station 1 sediments were apparently mixed to a degree that Pb-210 chronology was not possible. However, based on the radiocarbon and particle size measurements, these Pb-210 sedimentation rates do not appear real. The absence of a surface mixed layer resulting from bioturbation may support other data which indicate the sediments are scoured and not deposited in the Buccaneer Field. The lower apparent sedimentation rates in the upper part of the field may indicate greater erosion of sediment.

CONCLUSIONS

Typical of most inner shelf regions, the quantity and composition of the suspended particulates in the Buccaneer Field show large spatial and temporal variations. TSM varied from ~0.3 mg/l to several mg/l. Near-bottom nepheloid layers were present

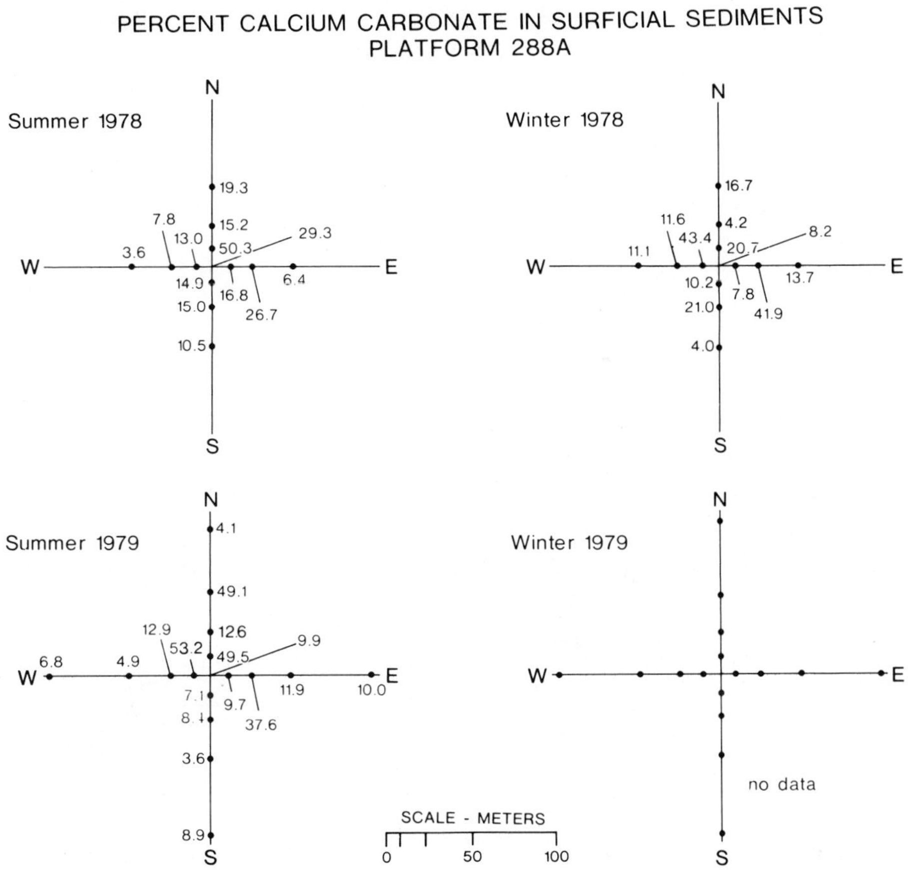

Figure 17. Percent calcium carbonate around platform 288-A.

in the Buccaneer Field during all samplings, except winter 1979 when TSM was high throughout the water column because of complete vertical mixing. The bottom nepheloid layer varied from a few meters to 10 meters in thickness. It was usually associated with a uniform termperature layer beneath a thermocline and resulted from resuspension of bottom sediments. Mid-depth nepheloid layers observed during some samplings also were associated with density surfaces in the water column. Surface nepheloid layers were observed in the spring as a result of fresh water runoff from coastal rivers, bays, and estuaries. Danek and Tomlinson (1980) estimated an average flux of material through the Buccaneer Field at about 2 g/sec. Seasonal changes in the suspended particulate content of the water column generally reflected changing current and mixing patterns.

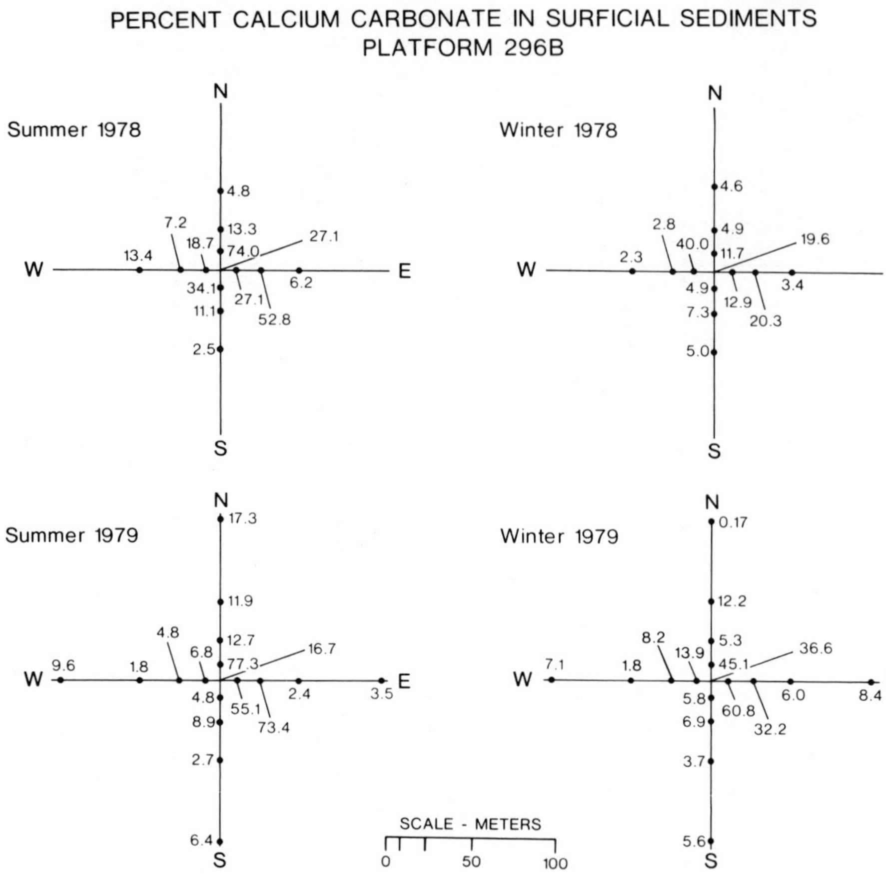

Figure 18. Percent calcium carbonate around platform 296-B.

The composition of the suspended particulates varied considerably over the sampling periods. Clay was the dominant component of the particulates during all seasons, except summer 1979, when low TSM levels resulting from strong stratification of the water column and low turbulent mixing were accompanied by a large phytoplankton standing crop. The organic fraction of the suspended particulates consisted almost exclusively of cellular material (phytoplankton, zooplankton, and/or bacteria). Only when there was a large fresh water lens at the surface or a very high standing crop of phytoplankton were significant concentrations of particulate non-cellular carbon observed. In most seasons, the cellular material generally constituted 20 to 30% of the suspended particulates. In summer and winter 1979, due to a dramatic increase in phytoplankton productivity, organic suspended particulates constituted as high as 78% of the TSM. Almost all the POC during the winter 1979 sampling was phytoplankton. During

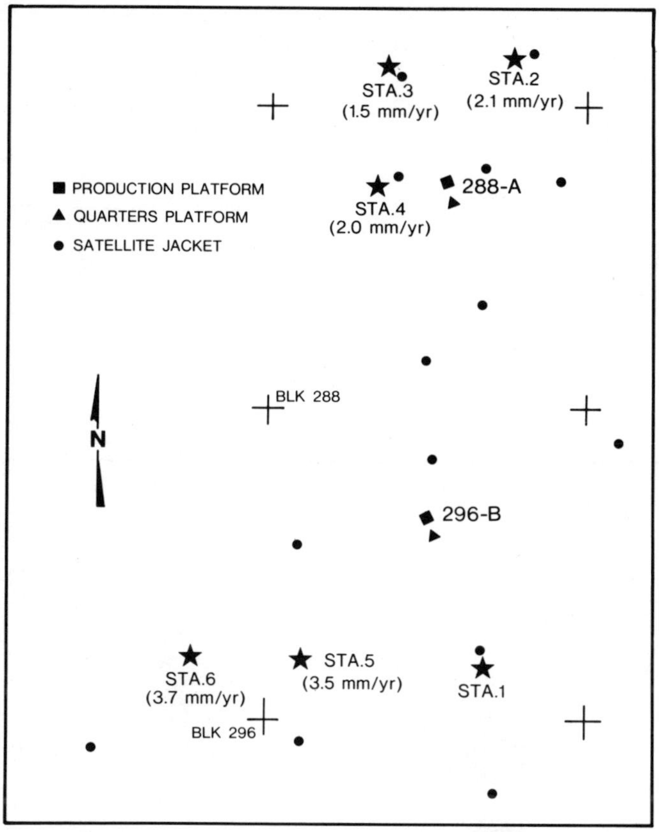

Figure 19. Locations of cores for Pb-210 dating. Sedimentation rates are in parentheses.

other seasons, phytoplankton constituted only a minor fraction of the cellular material. Higher phytoplankton standing crops were associated with an increase in the dissolved organic concentrations in the water column by as much as 50%. Most other DOC concentrations throughout the year were around 1.2 ± 0.2 mg C/l.

Suspended particulate data indicated that the water column was stratified during all samplings except winter. During the summer and fall samplings, a strong pycnocline resulting from temperature stratification apparently existed in the field, effectively inhibiting transport of surface introduced contaminants to the bottom. Only during the winter was the water column thoroughly mixed, due to strong turbulent activity. The stratification of the water column during the majority of the year appears to act as a barrier against introduction of platform contaminants to the sediments near the platforms. Platform-derived contaminants in the water must be rapidly removed from the field.

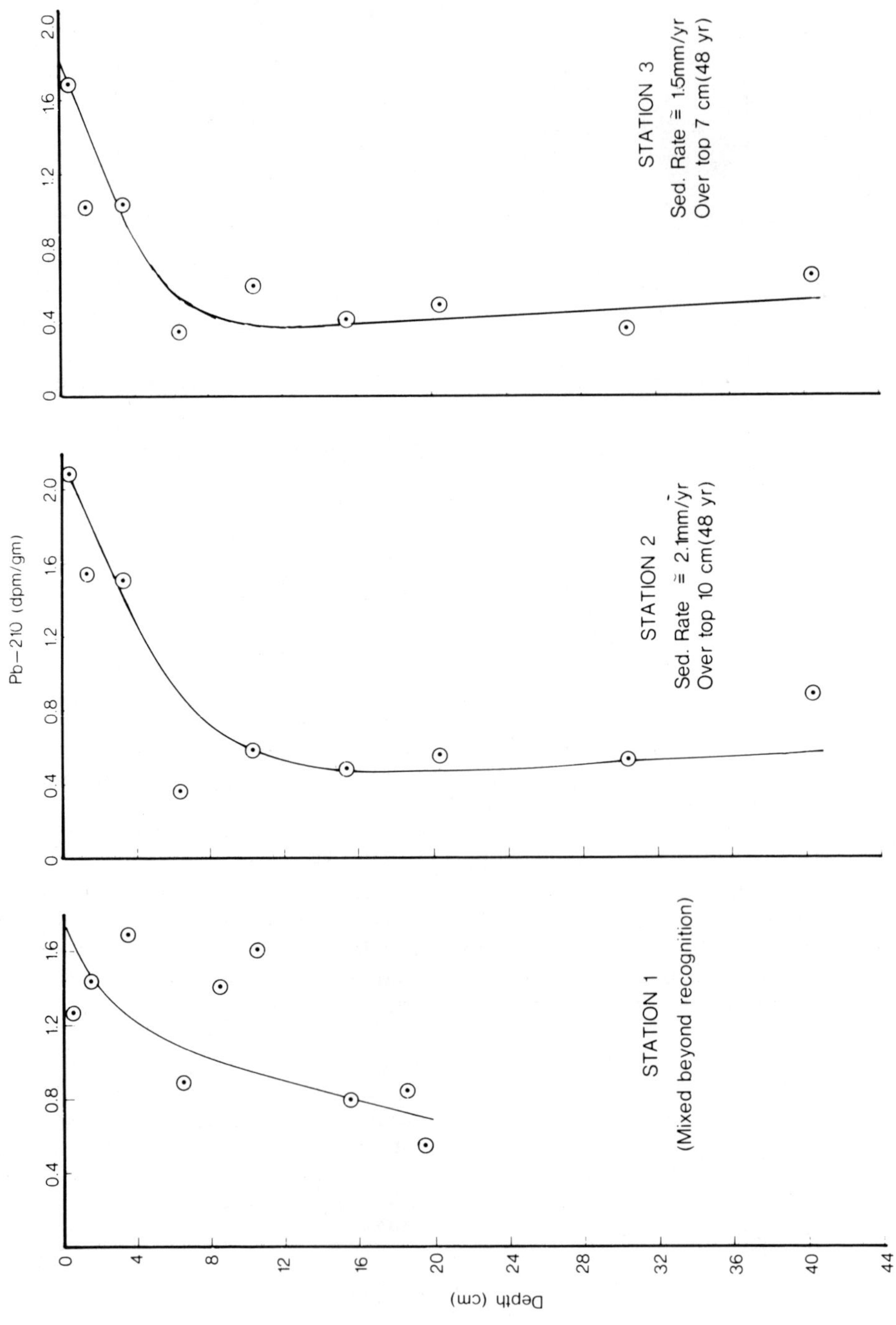
STATION 3
Sed. Rate ≅ 1.5mm/yr
Over top 7 cm(48 yr)
STATION 2
Sed. Rate ≅ 2.1mm/yr
Over top 10 cm(48 yr)
STATION 1
(Mixed beyond recognition)
Pb−210 (dpm/gm)
Depth (cm)
0
0.4
0.8
1.2
1.6
2.0
0
4
8
12
16
20
24
28
32
36
40
44

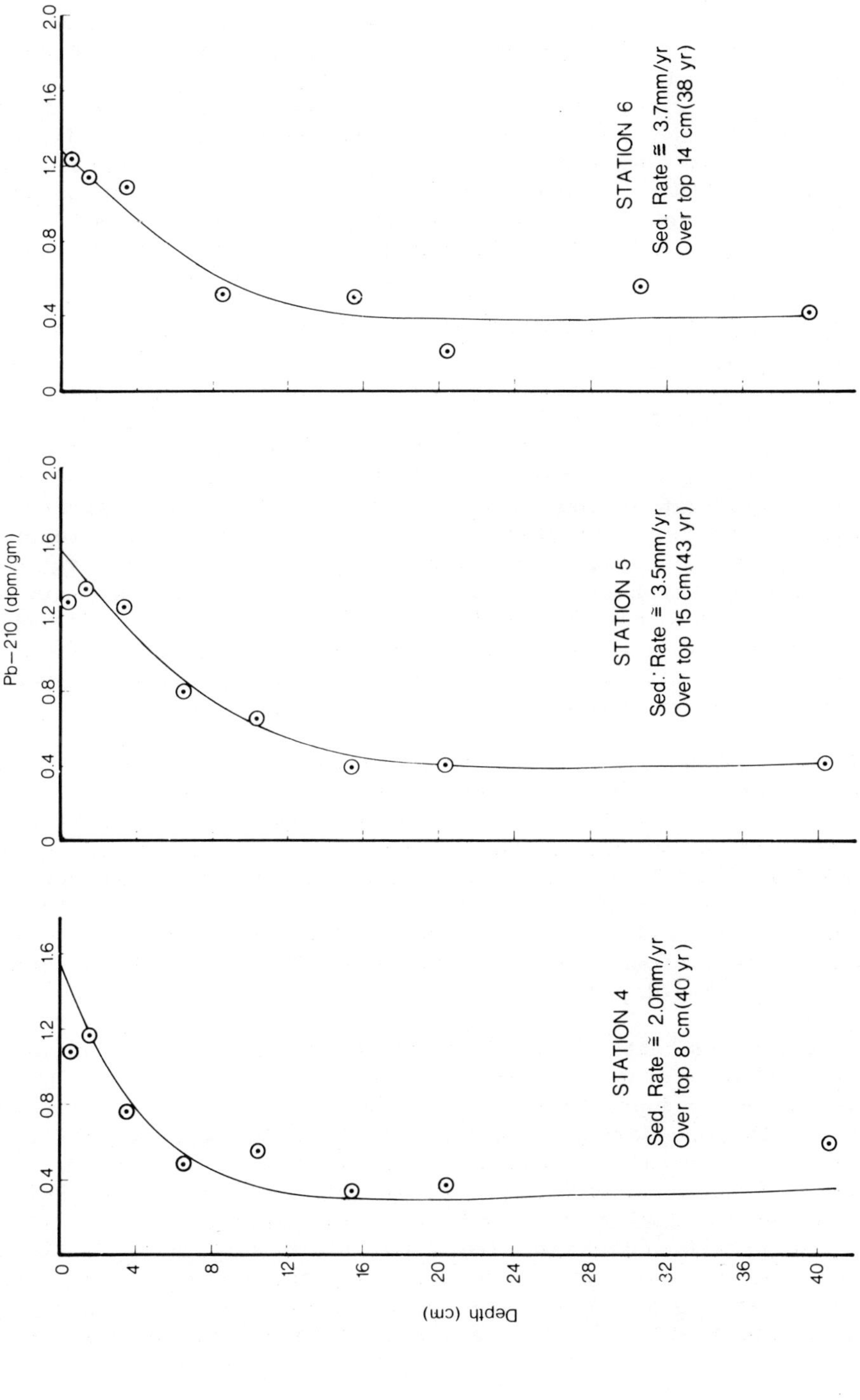

Figure 20. Pb-210 profiles in Buccaneer Field (see Figure 19 for location of stations).

The Buccaneer Field production platforms did not measurably alter the bulk composition of suspended particulates of biological activity (as measured by chlorophyll and ATP) in their immediate vicinity. These results can be explained by considering the small volumes discharged by the platforms (i.e., from the brine, oil spillage, etc.) compared against the large volume of water transported through the field. Pollutants introduced into the water column from the platforms could be rapidly transported out of the system either because of hydrographic conditions or perhaps by attachment to suspended particulates. Once out of the field, rapid dilution would obscure their presence and diminish their effect.

Bottom sediments in the Buccaneer Field area consist of relict calcareous shell hash, silts and clays of the Beaumont Formation, and sands which are dominated by a strong 3.5 ϕ fraction (Anderson et al., 1977). The Buccaneer Field is a highly dynamic and passive sedimentary environment in which relict sediments are presently being redistributed across the seafloor. Sedimentology observations in the study area indicate considerable variation in textural properties which reflect scour and redeposition by bottom currents and reworking of ancient sediments. The sandy nature of the bottom sediments in the Buccaneer Field attest to the high energy bottom conditions which prevail in the field. These conditions are probably sufficient to distribute most contaminated material deposited beneath the platforms over such a large area as to be undetectable.

There was a decreasing concentration of both organic and inorganic carbon in the sediment away from the platform. The gradient reflects high productivity associated with platforms. The distribution of the carbon in surficial sediments was quite variable and showed temporal changes reflecting seasonal current patterns. Sedimentation rates in the field determined by Pb-210 and radioactive dating suggest that erosion of sediments is the dominant process. Over one meter of sediment may have been eroded near Buccaneer Field platforms.

Contaminants introduced to the sediments in the Buccaneer Field may also be rapidly removed from the platform vicinity by suspension and redeposition. Surficial sediment data indicated that there is considerable movement of the fine grain material in the area. Direction of sediment transport appeared to be controlled by seasonal current patterns. The near-bottom nepheloid layer, observed during all seasons, also indicated that fine-grained surficial sediments within the field were in a continual state of resuspension and reworking. Only contaminants associated with very coarse-grained material would be expected to permanently remain in the field. These observations were confirmed by trace metal and hydrocarbon studies in the field which show seasonal contamination patters.

ACKNOWLEDGEMENTS

The assistance of Rick Weber, Hussein Abdel-Reheim, Charlie Coleman, and Keith Klatt during the sampling cruises is gratefully acknowledged. Laboratory analyses of organic carbon were conducted with the assistance of Ron Pflaum and Roger Burke. Charlie Coleman spent the many hours required to analyze the sedimentological parameters. Dr. Wen Huang of the Department of Geology at Texas A&M University made the clay mineralogy determinations. Valuable discussions with other participants in the study were instrumental in understanding the implications of our results. Discussions with Brian S. Middleditch and Benny J. Gallaway are especially acknowledged. We also wish to thank Charles W. Caillouet, William B. Jackson, and others at the National Marine Fisheries Service for their editiorial comments on our reporting.

This work is a result of research sponsored by the Environmental Protection Agency and Department of Commerce, National Oceanic and Atmospheric Administration, National Marine Fisheries Service, Southeast Fisheries Center, Galveston Laboratory under Contracts Nos. 03-78-D08-0043 and NA79-GA-C-00037.

REFERENCES

Anderson, J. B., and Broekstra, B., 1977, "Thermal wedging in Lake Macatawa, Michigan: long-term impact of channelization," J. Great Lakes Res., 3:159

Anderson, J. B., Schwarzer, R. R., and Clark, H. C., 1977, "Sedimentology, geochemistry, and trace metal analysis," in: Jackson, W. B. (ed.), "Environmental Assessment of an Active Oil Field in the Northwestern Gulf of Mexico, 1976-1977." NOAA, pp. 89-174. Available from: NTIS, Springfield, Virginia, Accession No. PB283890.

Anderson, J. B., Schwarzer, R. R., Clark, H. C., Wheeler, R. B., Hokanson, C. L., Anderson, J. E., and Okafor, J., 1979, "Describe the fine sediments and nepheloid layer of the oil field, focusing upon their relationship to heavy metal adsorption/determine levels, pathways, and bioaccumulation of heavy metals in the marine ecosystem in the oil field," in: Jackson, W. B. (ed.), "Environmental Assessment of an Active Oil Field in the Northwestern Gulf of Mexico, 1977-1978. Volume III: Chemical and Physical Investigations." NOAA, pp. 2.3.2/2.4.2-1 to 2.3.2/2.4.2-118. Available from: NTIS, Springfield, Virginia, Accession No. PB80107899.

Armstrong, R. S., and Hamilton, R. C., 1979, "Describe seasonal circulation patterns in the oil field," in: Jackson, W. B. (ed.), "Environmental Assessment of an Active Oil Field in the Northwestern Gulf of Mexico, 1977-1978. Volume III: Chemical and Physical investigations." NOAA, pp. 2.3.9-1 to 2.3.9-250. Available from: NTIS, Springfield, Virginia, Accession No. PB80107899.

Behrens, E. W., 1977, "Total organic carbon and carbon isotopes of sediments," in: Jackson, W. B. (ed.), "Environmental Assessment of an Active Oil Field in the Northwestern Gulf of Mexico, 1976-1977." NOAA, pp. 737-759. Available from: NTIS, Springfield, Virginia, Accession No. PB283890.

Brooks, J. M., Estes, E. L., and Huang, W. H., 1980, "Investigations of surficial sediments and suspended particulates at Buccaneer field," in: Jackson, W. B., and Wilkens E. P. (eds.), "Enivironmental Assessment of the Buccaneer Gas and Oil Field in the Northwestern Gulf of Mexico, 1975-1980, Vol. II." NOAA Technical Memorandum NMFS-SEFC-36, 261 pp. Available from: NTIS, Springfield, Virginia.

Danek, L. J., and Tomlinson, M. S., 1980, "Currents and hydrography of the Buccaneer Field and adjacent waters," in: Jackson, W. B., and Wilkens E. P. (eds.), "Environmental Assessment of the Buccaneer Gas and Oil Field in the Northwestern Gulf of Mexico, 1975-1980, Vol. VI." NOAA Technical Memorandum NMFS-SEFC-40, 33 pp. Available from: NTIS, Springfield, Virginia.

Folk, R. L., 1974, "Petrology of Sedimentary Rocks." Hemphill Publishing Co., Austin, Texas.

Fredericks, A. D., and Sackett, W. M., 1970, "Organic carbon in the Gulf of Mexico," J. Geophys. Res., 75:2199.

Holm-Hansen, O., and Booth, C. R., 1966, "The measurement of ATP in the ocean and its ecological significance," Limnol. Oceanogr., 11:510.

Holm-Hansen, O., and Paerl, H. W., 1972, "The applicability of ATP determination for estimation of microbial biomass and metabolic activity," Mem. Ist. Ital. Idrobiol., 29:1.

Holm-Hansen, O., Lorenzen, C. J., Holmn, R. W., and Strickland, J. D. H., 1965, "Fluorometric determination of chlorophyll," J. Conseil Perm. Intern. Exploration Mer, 30:1.

Huang, W. H., 1975, "Use of Local Minerals in the Treatment of Radioactive Waste." Technical Report Series No. 136, International Atomic Energy Agency, Vienna.

Huang, W. H., 1976, "Genesis of kaolinite in swamp environment." International Geological Congress, Sydney, Australia.

Manheim, F. T., Hathaway, J. C., and Uchupi, E., 1972, "Suspended matter in surface waters of the northern Gulf of Mexico," Limnol. Oceanogr., 17:17.

Meyers, P. A., 1976, "Association of hydrocarbons and mineral particles in saline solution," Nature, 244:23.

Meyers, P. A., and Quinn, J. G., 1973, "Factors affecting the association of fatty acids with mineral particles in seawater," Geochim. Cosmochim. Acta, 37:1745.

Middleditch, B. S, and West, D., 1980, "Hydrocarbons, biocides, and sulfur," in Jackson, W. B. and Wilkens, E. P. (eds.), "Environmental Assessment of the Buccaneer Gas and Oil Field in the Northwestern Gulf of Mexico, 1978-1979, Vol 8." NOAA Technical Memorandum NMFS-SEFC-41, 119 pp. Available from: NTIS, Springfield, Virginia.

Middleditch, B. S., Basile, B., and Chang, E. S., 1977, "Hydrocarbon levels and pathways in the major components of a marine ecosystem," in: Jackson, W. B. (ed.), "Environmental Assessment of an Active Oil Field in the Northwestern Gulf of Mexico, 1976-1977." NOAA, pp. 633-735. Available from: NTIS, Springfield, Virginia, Accession No. PB283890.

Middleditch, B. S., Basile, B., and Missler, S. R., 1979, "Determine levels, pathways, and bioaccumulation of selected discharge constituents (non-metals) in the marine ecosystem in the oil field/hydrocarbon modeling," in: Jackson, W. B. (ed.), "Environmental Assessment of an Active Oil Field in the Northwestern Gulf of Mexico, 1977-1978. Volume III: Chemical and Physical Investigations." NOAA, pp. 2.4.1-1 to 2.4.1-302. Available from: NTIS, Springfield, Virginia, Accession No. PB80107899.

Sackett, W. M., 1964, "The depositional history of isotopic organic carbon composition of marine sediments," J. Mar. Geol., 2:173.

Sackett, W. M., and Brooks, J. M., 1978, "Low-molecular-weight hydrocarbon and hydrographic project," in: "Environmental Studies, South Texas Outer Continental Shelf, Biology and Chemistry." Bureau of Land Management, New Orleans, Louisiana.

Sackett, W. M., Nakaparksin, S, and Dalrymple, D., 1970, "Carbon isotope effects in methane production by thermal cracking," in: Hobson, G. D., and Speers, G. C. (eds.), "Advances in Organic Geochemistry 1966." Pergamon Press, Oxford, England, p. 37.

Shokes, R. F., 1976, "Rate-Dependent Distributions of Lead-210 and Interstitial Sulfate in Sediments of the Mississippi River Delta." Ph.D. dissertation, Texas A&M University, College Station, Texas.

Strickland, J. D. H., and Parsons, T. R., 1972, "A Practical Handbook of Seawater Analysis." Fisheries Research Board of Canada, Ottawa, Canada.

Tillery, J. B., 1980, "Trace metals," in: Jackson, W. B., and Wilkens E. P. (eds.), "Environmental Assessment of the Buccaneer Gas and Oil Field in the Northwestern Gulf of Mexico, 1975-1980, Vol. VI." NOAA Technical Memorandum NMFS-SEFC-42, 39 pp. Available from: NTIS, Springfield, Virginia.

Wheeler, R. B., 1977, "The Fate of Petroleum in the Marine Environment." Exxon Production Research Co. in-house report, Houston, Texas.

Wickman, R. E., 1952, "Variations in the relative abundance of the carbon isotopes in plants," Geochim. Cosmochim. Acta, 2:243.

Wiesenburg, D. A., and Schink, D. R., 1978, "Use of the Texas A&M Deep Towed Pumping System in the Gulf of Mexico Aboard the Research Vessel GYRE during Cruise 77-G-14, 3-7 December 1977." Technical Report No. 78-4-T, Texas A&M University, College Station, Texas.

Yentsch, C. S., and Menzel, D. W., 1963, "A method for the determination of phytoplankton chlorophyll and phaeophytin by fluorescence," Deep-Sea Res., 10:221.

TOTAL ORGANIC CARBON AND CARBON ISOTOPES OF SEDIMENTS

E. William Behrens

University of Texas Marine Science Institute
700 The Strand
Galveston, Texas 77550

INTRODUCTION

This research project addresses the question: do the isotopes of carbon in the organic matter of the sediments within and around the Buccaneer Field indicate that any alteration of the sedimentary organic matter has resulted from drilling and producing operations? Isotopes are a potentially useful tool to use in this inquiry, because various possible sources of carbon are characterized by distinctive stable isotope ratios and/or radiocarbon contents. Crude oil is commonly isotopically lighter than organic matter in recent marine sediments by more than 5 per mil $\delta^{13}C$; and, of course, any carbon brought up from an oil or gas field would be devoid of radiocarbon.

As seen in Figure 1, the Buccaneer Field lies in a swale 20 to 20.5 m (66 to 67 ft) deep between two ridges which shoal to roughly 15 m and trend northeast - southwest parallel to the shoreline. The ridges are characterized by sands with patches or strips of shell gravels. The swale has muddier sands with mud content increasing considerably in the silled depression to the east and northeast of the field. The shelly sediments tend to be regularly bedded, the muddy sands highly mixed by bioturbation, and the muddiest sediments more or less regularly laminated with little burrowing disturbance. For more information on sediment distribution, see Anderson et al. (this volume).

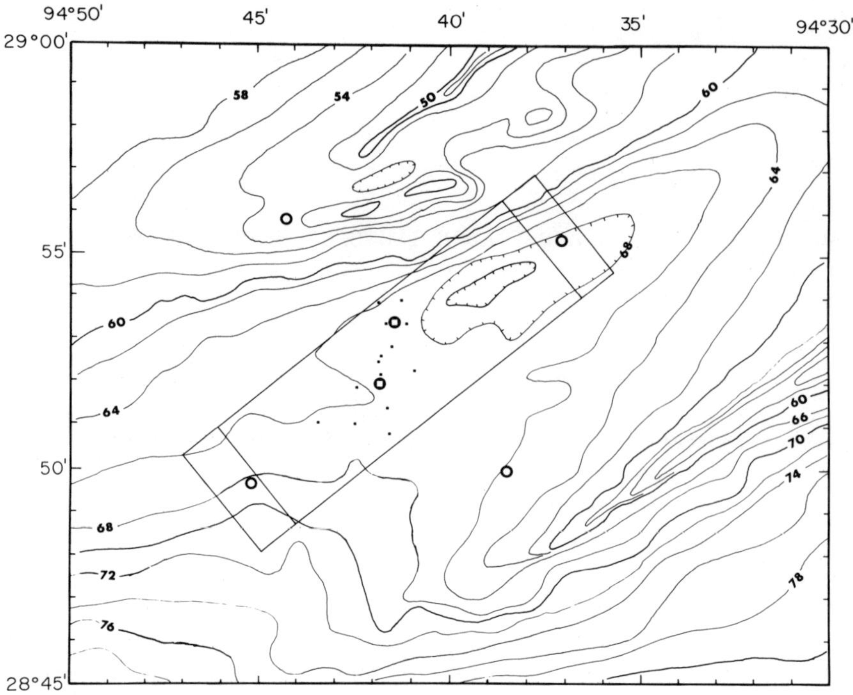

Figure 1. Buccaneer Field and surrounding area. Rectangular outline is of more intensively studied area shown in Figures 3 and 5. Small, solid squares are well jackets; larger, open squares are production platforms (more northerly one is 288-A). Open circles are coring sites in control areas. Contours are in feet below mean low water.

MATERIALS AND METHODS

Samples of the surficial marine sediments were removed from short cores (up to 20 cm) and samples of older sediments from longer piston cores (approximately 1 m long). The locations of the surficial samples are shown in Figures 3 and 5. The piston cores were taken on transects north, east, south, and west from platform 288-A (the more northerly of the two shown in Figure 1) at distances of 15 m (15 ft), 53 m (175 ft), 76 m (250 ft), 152 m (500 ft), and some at 460 m (0.25 n.mi.), 910 m (0.5 n.mi.), and 1.8 km (1 n.mi.). Three cores were taken at most stations. Four piston cores were also taken in each of four control areas located northwest (inshore), northeast, southeast (offshore), and southwest and at distances of about 4 km from the field.

The selected sediment samples were acidified in HCl to remove carbonate, diluted and decanted to remove excess salts, dried at less than 50°C, and ground to a powder. The powdered samples were then combusted with oxygen in a LECO radiofrequency furnace to obtain CO_2 for total organic carbon measurement (manometrically) and for isotopic analyses. The method is described by Calder (1969) and Parker et al. (1972). The stable isotope parameter is

$$\delta^{13}C = 1000 \times \frac{^{13}C/^{12}C \text{ sample} - {^{13}C/^{12}C} \text{ standard}}{^{13}C/^{12}C \text{ standard}}$$

where the standard of reference is PDB. Mean standard deviation of several sets of replicate analyses was 0.175.

Radiocarbon dating requires much more material than the other types of analyses, so samples from short intervals (4 to 5 cm) were taken from equivalent depths in several of the replicate cores at each station for this purpose. Results are reported as uncorrected radiocarbon years before present (B.P.).

PRELIMINARY RESULTS AND PREDICTIVE MODEL

In order to develop a predictive model of the carbon isotope effects that might result from the oil field production activities, the specific difference between the produced crude and background organic carbon stable isotopes was determined. $\delta^{13}C$ for the crude was measured at -30.8. The Texas and Louisiana continental shelves were sampled extensively by Plucker (1970) and Gearing (1975), and the data summarized by Gearing et al. (1977). Surficial sediments of the Texas-Louisiana shelf have an organic carbon mean $\delta^{13}C$ of -20.6 (standard deviation = 0.6) while that of 188 samples within the study area is -20.6 (standard deviation = 0.6). Within the broader region, Texas shelf samples southwest of the study area are a little more positive (mean $\delta^{13}C$ = -20.2, standard deviation = 0.3) while shelf sediments to the east of the study area have a mean value of -21.3 (standard deviation = 0.2).

These sediment values result primarily from the isotopic fractionation occurring during photosynthesis by marine phytoplankton. They may also be influenced by river and estuarine inputs of terrestrial and intracoastal carbon sources. Another possible input is from marine erosion of older deposits. The Beaumont formation underlies the Holocene sediment throughout this area (Nelson and Bray, 1970). Beaumont sediments are distinctively clayey in texture and are either medium to dark gray or strongly stained with brown to yellow bands and laminae of iron oxides. $\delta^{13}C$ of organic carbon from the Beaumont averages -22.7. This is similar to a Texas coastal fresh water lake sediment with a $\delta^{13}C$ value of -22.4 and probably reflects the exposure and/or accumulation of Beaumont sediments to/in a

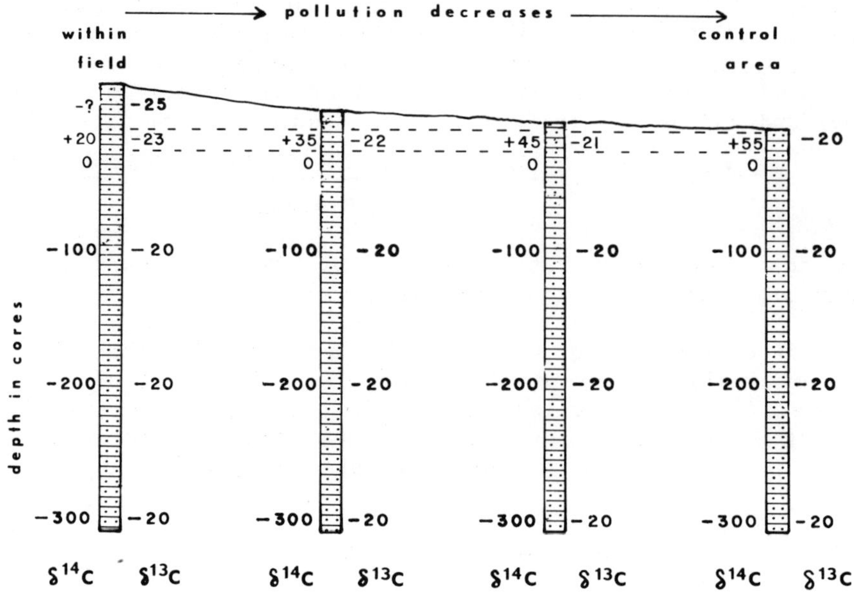

Figure 2. Predictive model for carbon isotope distribution. Assumes a source of isotopically lighter carbon with no ^{14}C within the oil field. Values are meant to be relative only.

terrestrial environment during the last Pleistocene low stand of sea level. This is supported by two radiocarbon dates for Beaumont samples: 22,120 ± 1440 and 24,810 ± 1170 B.P.

In addition to the above information, a contribution of carbon with no ^{14}C from the oil field to the surrounding sediments might also be detectable in very modern surficial sediment containing the distinctive positive $\delta^{14}C$ anomaly which resulted from the addition of an excessive amount of ^{14}C created by nuclear device testing. This anomaly began to be detected in the marine environment in the early 1950's and has a $\delta^{14}C$ value of about +50 to +60 (e.g., Land et al., 1979).

Figure 2 incorporates several predictions which can be modeled from the above information assuming a contribution of oil field hydrocarbons to the surrounding marine sediments. $\delta^{13}C$ values would become more negative from an uncontaminated control area to the hydrocarbon source. If the positive ^{14}C anomaly were present in the surficial sediments, it would diminish towards the source. If it were absent, surficial sediments would appear older toward the source, and possibly age reversals (apparently older sediments overlying younger sediments) would appear in cores from near the source. Relative to true ages, erroneously old ^{14}C dates

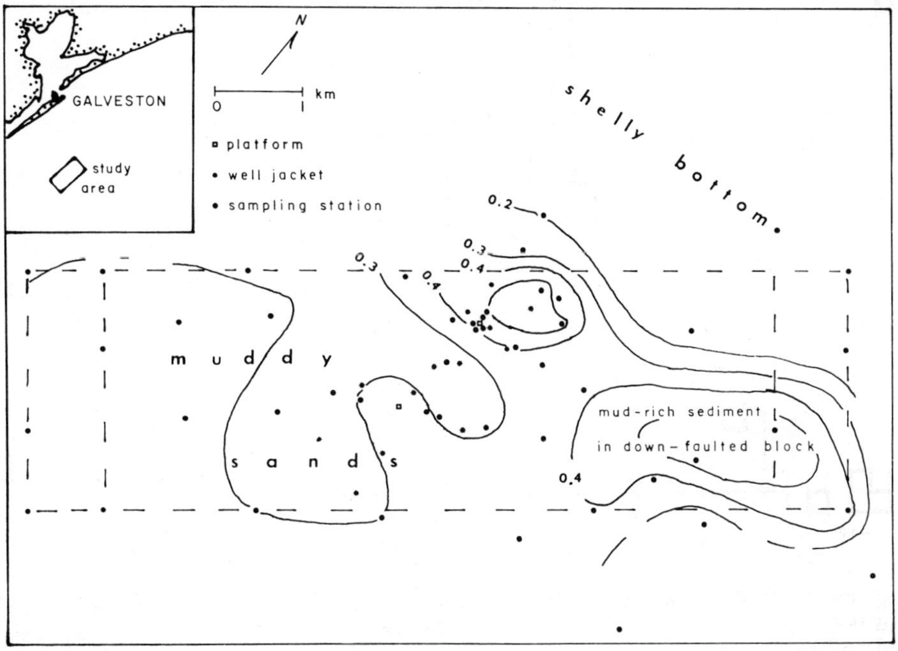

Figure 3. Areal distribution of total organic carbon content. Contours are in tenths of a percent.

in the upper parts of cores would depress the apparent rate of sedimentation in the uppermost interval dated but increase it from there down-core until uncontaminated dates were obtained.

RESULTS

Total Organic Carbon

Total organic carbon (TOC) content of the sediments in the study area is now ranging from less than 0.1% to just under 1%. The mean of 220 analyses is 0.33%. Most of the samples with over 0.4% TOC are either from the muddier control area to the northeast of the field (mean of 35 samples = 0.51%) or from the Beaumont Formation (mean of 8 samples = 0.51%). Consequently the mean value for samples within the field and in comparable control areas is only 0.29% and individual values rarely occur outside the range of 0.15 to 0.42%. The areal distribution of surficial sediment TOC values (Figure 3) shows the concentration of high values in the muddiest sediments to the east-northeast of the field and an

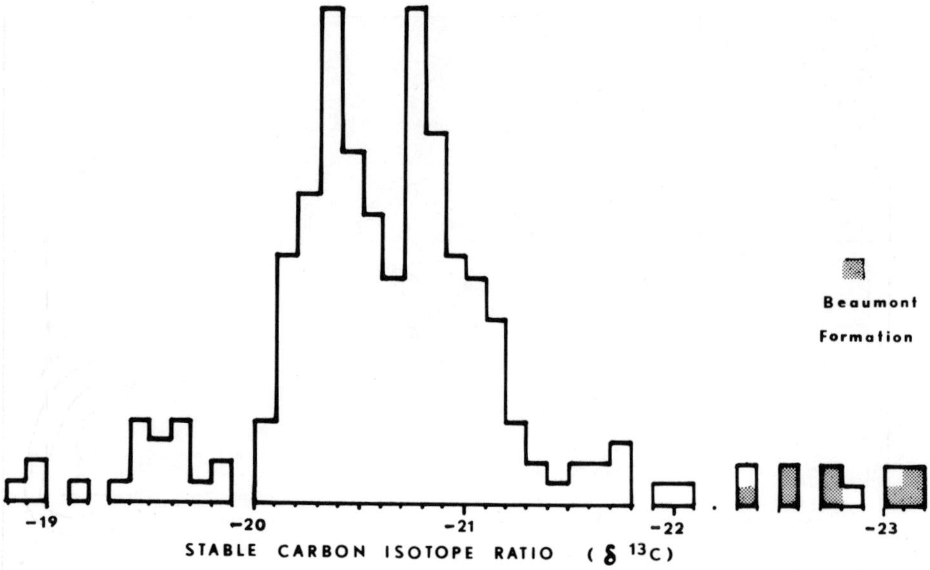

Figure 4. Histogram of $\delta^{13}C$ values. Modes are at -20.3 to -20.4 and -20.7 to -20.8. Mean for Texas continental shelf sediments west of study area is -20.2; mean for Texas-Louisiana shelf sediments east of the study area is -21.3. Mean for both shelf areas equals mean for this data set (= -20.6).

otherwise irregular pattern that bears no apparent relationship to the production platforms.

Stable Carbon Isotopes

The frequency distribution of 217 $\delta^{13}C$ measurements is shown in Figure 4. The most negative tail of this distribution contains eight samples of Beaumont sediment ($\delta^{13}C$ = -22.7). The tail on the left side of the distribution consists of 19 samples with $\delta^{13}C$ more positive than -20.0.

The $\delta^{13}C$ values of the bulk of the samples fall between -20.0 and -21.3 and have a distinctively bimodal distribution. The areal distribution of the two modes is shown in Figure 5. Although there are two major concentrations of the more negative mode near or in the field, several smaller occurrences of it suggest it is distributed well beyond the limits of the field.

In three dimensions no strong tendency in $\delta^{13}C$ distribution is evident except that positive values predominate in the lowest

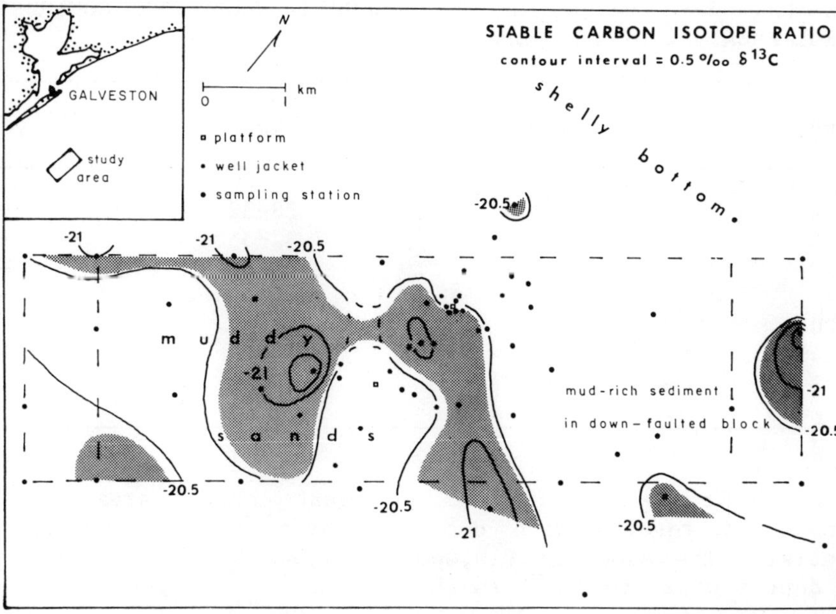

Figure 5. Areal distribution of $\delta^{13}C$. Contours are at 0.5 intervals of $\delta^{13}C$. Shaded area is within the more negative mode of Figure 4.

part of the Holocene near the Beaumont contact (mean of 14 samples = −20.12). Data from one core suggest that the apparent erratic distribution of $\delta^{13}C$ values may be due to sampling cores discontinuously. Core 56 from the southwest control area was sampled continuously at 4 cm intervals and shows a smooth transition from the positive mode at the surface to a negative peak at about 30 cm depth and then back to the center of the negative mode below 50 cm (Table 1).

Radiocarbon

Most radiocarbon dates were obtained on the disseminated total organic carbon in the sediments. Eight of eleven shell dates taken for comparison show a range of only 1100 years and an average 4340 B.P. Most of these were from the inshore control area to the northwest of the field. Only a few shell dates that are compatible with TOC dates in the same cores were used for analysis of sedimentation rates. Radiocarbon data are given in Table 2.

TABLE 1. $\delta^{13}C$ VALUES OF CORE SAMPLES FROM THE SOUTHWESTERN CONTROL AREA VS. DEPTH (cm).

Depth	$\delta^{13}C$	Depth	$\delta^{13}C$
0- 4	-20.4	28-32	-21.2
4- 8	-20.5	32-36	-21.1
8-12	-20.6	40-44	-20.8
12-16	-20.7	44-48	-20.6
16-20	-20.9	48-52	-20.7
20-24	-21.1	52-56	-20.7
24-28	-21.2	56-60	-21.6

Dates from a core in the northeast control area also give inconsistent results with depth. Three of four samples have essentially the same age (10,860 to 11,540 B.P.) although coming from depths of 25 to 75 cm in the cores.

The oldest Holocene sediments lying just above the Beaumont contact date at about 12,000 B.P. The average rate of deposition for Holocene sediments based on 28 pairs of dates is 0.076 mm/year (0.76 cm/century). However, for the most part, the rates decline systematically upward in the cores. Sedimentation rates are also less within the field than in the control area to the southeast (offshore).

DISCUSSION

Due to differences in sediment characteristics, the areas to the northwest (inshore) and northeast should not be used as controls for comparison with conditions within the Buccaneer Field. Organic carbon comparisons strongly support this, because TOC content is anomalously low in the shellier and anomalously high in the muddier sediments occurring in those areas. Both areas have inconsistent radiocarbon age sequences within cores. The offshore (southeastern) area was concluded to be the best control area for comparison to conditions within the field.

From radiocarbon ages and calculated rates of deposition, the thickness of sediment containing the predicted positive $\delta^{14}C$ anomaly in the control area (offshore) would be less than 0.75 mm, and thus it is not detectable. There is no evidence that this material was even deposited because, if it had been, it should have affected the most recent date by making it appear younger than it really is. If this had occurred, the uppermost part of

TABLE 2. RADIOCARBON AGES BEFORE PRESENT (B.P.).

Within oil field				Control areas			
Station[a]	Interval sampled (cm)	^{14}C age	Material	Station[b]	Interval sampled (cm)	^{14}C age	Material
50E	0- 5	6,210± 650	TOC	NW	0- 1	4,240±100	shell
50E	5- 10	7,420± 670	TOC	NW	0- 1	4,560± 70	shell
50E	10- 15	7,470± 430	TOC	NW	14-16	4,460± 70	shell
50E	40- 50	12,330±4,080	TOC	NW	15-17	4,610± 70	shell
				NW	33-35	4,130±100	TOC
500E	0- 5	2,510± 170	TOC	NW	48-52	3,850±440	TOC
500E	0- 5	4,000± 120	shell	NW	48-52	3,680±200	shell
500E	5- 10	5,450± 290	TOC				
500E	5- 10	4,700± 180	shell	NE	8-12	3,890±150	TOC
500E	10- 15	6,340± 220	TOC	NE	23-27	11,540±460	TOC
500E	28- 32	8,760± 710	TOC	NE	48-52	10,860±380	TOC
500E	50- 60	11,490±8,035	TOC	NE	73-77	11,120±440	TOC
6000E	0- 5	1,520± 280	TOC	SE	0- 4	680±240	TOC
6000E	5- 10	2,490± 160	TOC	SE	4- 8	1,250±170	shell
6000E	10- 15	4,460± 230	shell	SE	4- 8	1,420±220	TOC
6000E	10- 15	5,080± 520	TOC	SE	8-12	1,780±330	TOC
6000E	27- 33	9,370± 600	TOC	SE	12-16	2,140±130	shell
				SE	37-43	3,350±350	TOC
3000N	91- 99	8,940± 510	TOC	SE	37-43	2,230±170	shell
3000N	99-107	9,500± 740	TOC				

[a]Number represents distance in feet from platform 288-A; letter indicates direction (E = east, N = north). [b]NW = northwest (inshore), NE = northeast, SE = southeast (offshore).

TABLE 3. RATES OF SEDIMENTATION.

3A. Control site

Interval[a] (cm)	Material[b]	Mid-point[c] (cm)	Interpolated age (B.P.)	Sed. rate (cm/100 yr)
0- 2	TOC	1	340	0.29
0- 6	Sh	3	625	0.48
2- 6	TOC-Sh	4	965	0.70
2- 6	TOC-TOC	4	1,050	0.54
2-10	TOC-TOC	6	1,230	0.73
6-10	Sh-TOC	8	1,515	0.75
6-10	TOC-TOC	8	1,600	1.11
6-14	Sh-Sh	10	1,695	0.90
10-14	TOC-Sh	12	1,960	1.11
10-40	TOC-TOC	25	2,565	1.91
14-40	Sh-TOC	27	2,745	2.15
14-40	Sh-Sh	27	2,185	29[d]

3B. Sites within oil field

50 E		500 E		6000 E	
Interval (cm)	Sed. rate (cm/100 yr)	Interval (cm)	Sed. rate (cm/100 yr)	Interval (cm)	Sed. rate (cm/100 yr)
0- 2.5	0.04	0- 2.5	0.10	0- 2.5	0.16
2.5- 7.5	0.41	2.5- 7.5	0.17	2.5- 7.5	0.52
7.5-12.5	10.	7.5-12.5	0.56	7.5-12.5	0.19
12.5-45	0.67	12.5-30	0.72	12.5-30	0.41
		30	-55	0.92	

[a]Interval between sample mid points.
[b]TOC = total disseminated organic carbon, Sh = shell.
[c]Mid point of interval to which sedimentation rate applies.
[d]This value appears to be based on a shell sample contaminated with some modern ^{14}C.

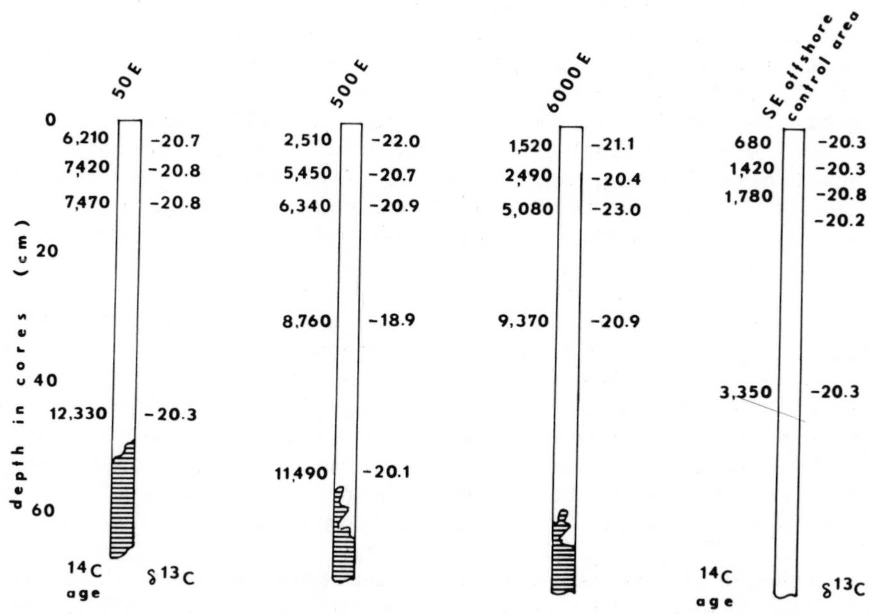

Figure 6. Observed carbon isotope distribution with depth. Lined pattern in lower part of some cores is Beaumont formation. Compare with predictive model of Figure 2. See Table 2 for more complete listing of ^{14}C dates and Table 3 for sedimentation rates calculated for intervals between ^{14}C data.

the core would appear to have been deposited in less time than it really was and thus would appear to have a high rate of deposition relative to unaffected, deeper parts of the core. Actually, the youngest date gives a slower rate of deposition than the rest of the dates in the core. Furthermore, each successivley deeper date gives a higher rate of sedimentation. This is generally true in the other cores dated and appears to be real and natural (Table 3).

A second basic component of the prediction model was that oil field activities would only add to the natural sedimentation taking place in the study area. The radiocarbon ages of surficial sediment indicate that erosion, not deposition, has taken place within the field. From the cores closest to platform 288-A (50E) to the cores from the offshore control area the uppermost 4 to 5 cm of the cores examined have successively younger ages (Figure 6). Using the depositional rates from the control area these ages indicate that from the core 15 m from platform 288-A one meter of sediment has been eroded, and from the core 1.8 km from platform 288-A 7 cm has been eroded.

Erosion of sediments affected by the field does not necessarily indicate that no polluting carbonaceous material has been added to the sediment, because quite temporary deposition of such material, or even suspension of it near the bottom, might enable detritus feeding organisms to scavenge it and incorporate it into the sediments through their natural life processes (e.g., defecation, burrowing, and interment after death). The occurrence of two $\delta^{13}C$ modes with one distinctively more negative than the other warrants further examination of the data for such a pollution effect.

The tails of the $\delta^{13}C$ frequency distribution curve are not particularly useful in this regard, because they can be related to other phenomena. The tail of isotopically heavier samples may reflect the occasional inclusion of fresh tissue of marine organisms in the sediment. The observed values would result from such inclusion, because live organisms are more positive than sedimented organic matter. An average value for six live shelf organisms is -18.1 (B. Fry, personal communication).

Although it makes up most of the more negative distribution tail, the Beaumont Formation is not considered a probable source for the carbon causing the more negative of the two large modes, because all but 2 of the 14 samples taken closest to the Beaumont contact in the cores have $\delta^{13}C$ values well within the more positive mode, and the average of all 14 samples (-20.1) falls near the more positive limit of that mode.

Both erosion and a contribution of ^{14}C-depleted carbon could cause the decrease in sedimentation rates from the control area towards platform 288-A (Tables 2-3, Figure 6). If contamination with dead carbon were the cause, the degree of negativity of the stable carbon isotope $\delta^{13}C$ should correlate with the increase in age of the surficial sediment toward the source. This correlation exists at sites 6000E and 500E, but the surficial $\delta^{13}C$ at 50E (-20.7) is far too positive to correspond to an age of 6210 B.P. produced by hydrocarbon contamination.

Finally, a statistical examination was made of $\delta^{13}C$ as a function of distance from a major production platform (288-A). This includes 75 samples taken at core depths of from zero to 20 cm. The summarized data (Table 4) show that only those stations within 15 m (15 ft) of the platform have an obvious predominance of negative mode values. The negative mode actually becomes strongest away from the platform at one nautical mile (1.8 km). The progression of values is only suggestive, because the standard deviations show that differences between successive distances from the platform have small statistical significance. However, confidence of significant difference does go over 90% for some adjacent pairs and is over 60% for all.

TABLE 4. $\delta^{13}C$ MODES AND VALUES WITH DISTANCE FROM PLATFORM 288-A.

Distance from platform (m)	Mean $\delta^{13}C$	% of samples in negative mode
15 (12)[a]	-20.90 (0.75)[b]	75
53 (12)	-20.27 (0.73)	25
76 (12)	-20.37 (0.64)	42
152 (13)	-20.64 (0.47)	54
460 (10)	-20.74 (0.51)	55
1800 (11)	-20.93 (0.81)	55
7000 SE (5)	-20.24 (0.43)	20
7000 SW (25)	-20.57 (0.40)	40
7000 NE (18)	-20.66 (0.37)	58

[a]Number of samples in data set.
[b]Standard deviation.

The suggestion that isotopically lighter carbon comes from the platform but bypasses all but the most immediate vicinity to be deposited only after several hundred or thousand meters, does not carry on to the control areas. The offshore area has only a trace of the more negative mode, while the alongshore (northeast and southwest) areas are remarkably similar to the overall area and, indeed, to the entire Texas - western Louisiana continental shelf.

The difference between $\delta^{13}C$ values for the continental shelves to either side (east and southwest) of the study area is significant at a high degree of confidence. There is no apparent reason why each should contribute a frequency mode to the distribution of values in the study area (rather than grade through the area); but if they did, it could account entirely for the stable carbon isotope values observed. It is interesting to note that $\delta^{13}C$ becomes more negative towards the shelf area that has one of the highest concentrations of production platforms in the world.

CONCLUSIONS

The most obvious effect of the oil field studied is the erosion of sediments within the field. This is indicated both by radiocarbon ages and sedimentation rates of surficial sediments. The trends in these parameters also fit a predictive model of carbonaceous contamination originating from a production platform. However, the magnitudes of stable isotope effects predicted from this model and the radiocarbon data are much more than those observed. Therefore, erosion of from one to two meters of sediment near the production platform has occurred, and the erosion may extend to as much as one mile from the platform.

A bimodal stable carbon isotope ratio distribution suggests one contaminated mode; and its statistical distribution relative to the production platform studied is compatible with the erosion effect. That is, the contaminated material does not diminish with distance from the source but bypasses a considerable distance before it is deposited in the sediment. However, in detail, the distribution of the possibly contaminated mode is quite irregular and does not correspond to any obvious two or three dimensional model of distribution from a production platform source.

The study area lies between two continental shelf areas with distinct stable carbon isotope differences of sufficient magnitude to account for the two frequency modes found in the Buccaneer Field area. However, it cannot be determined in a limited study such as this why a shelf gradient in stable carbon isotope values should produce a bimodal distribution in any one place nor if the source of the gradient is natural or related to a larger scale offshore oil production effect.

ACKNOWLEDGEMENTS

I sincerely thank Brian Fry and Patrick L. Parker for stable carbon isotope measurements and support in initiating this study. Short cores of surficial sediment were generously made available for subsampling by J. B. Anderson. Radiocarbon analyses were done at the University of Texas Radiocarbon Laboratory under the supervision of Salvatore Valastro, Jr. This work is a result of research sponsored by the Environmental Protection Agency and Department of Commerce, National Oceanic and Atmospheric Administration, National Marine Fisheries Service, Southeast Fisheries Center, Galveston Laboratory under Contract No. 03-6-042-35118. University of Texas Marine Science Institute Contribution No. 445, Galveston Geophysics Laboratory.

REFERENCES

Calder, J. A., 1969, "Carbon Isotope Effects in Biochemical Systems." Ph.D. Thesis, The University of Texas at Austin.

Gearing, P. J., 1975, "Organic Carbon Stable Isotope Ratios in Continental Margin Sediments." Ph.D. Thesis, The University of Texas at Austin.

Gearing, P. J., Plucker, F. E., and Parker, P. L., 1977, "Organic carbon stable isotope ratios of continental margin sediments," Mar. Chem., 5:251.

Land, L. S., Behrens, E. W., and Frishman, S. A., 1979, "The ooids of Baffin Bay, Texas," J. Sedimentary Petrol., 49:1269.

Nelson, H. F., and Bray, E. E., 1970, "Stratigraphy and history of the Holocene sediments in the Sabine - High Island area, Gulf of Mexico," in: Morgan, J. P. (ed.), "Deltaic Sedimentation." Soc. Economic Paleontologists and Mineralogists Special Publication No. 15, Tulsa, Oklahoma, p. 48.

Parker, P. L., Behrens, E. W., Calder, J. A., and Schultz, D., 1972, "Stable carbon isotope ratio variations in the organic carbon from Gulf of Mexico sediments," Contrib. Mar. Sci., 16:139.

Plucker, F. E., 1970, "Stable Carbon Isotope Ratios of Organic Matter of Gulf of Mexico Shelf Sediments." M.S. Thesis, The University of Texas at Austin.

DISTRIBUTION AND ABUNDANCE OF MACROBENTHIC AND MEIOBENTHIC ORGANISMS

Donald E. Harper, Jr., Deborah L. Potts,
Robert R. Salzer, Robert J. Case, Rebecca L. Jaschek,
and Constance M. Walker

Texas A&M Marine Laboratory
Building 311, Fort Crockett
Galveston, Texas 77550

INTRODUCTION

The increased need for energy self-sufficiency has caused U.S. petroleum companies to embark on an extensive search for oil and gas reserves. Much of the exploration will be conducted on the continental shelf of the Gulf of Mexico, and a possible consequence of energy production activities is an impact of unknown magnitude on the marine environment.

The Texas continental shelf has not been as extensively exploited as the Louisiana shelf and is still a relatively unimpacted area. But as the search for oil and gas increases, Texas shelf waters will be increasingly subjected to exploration, drilling, and production activities. One means of determining the possible impacts of future activities is by studying an existing offshore oil field. The Buccaneer Field was ideal for this type of study. In 1976 the field had been producing both oil and gas for 11 years. The field was geographically isolated from other fields, and far enough from Galveston to be outside much of the sphere of influence of Galveston pollution, yet near enough to begin sampling cruises on short notice when weather permitted. The field and surrounding area was in shallow enough water to permit sampling and direct observations of the bottom by divers with minimal risk of diving accidents.

Buccaneer Field was selected for study based on the results of a pilot study during which the areal distributions of benthic biota, sediments, heavy metals, and hydrocarbons in and around the field were investigated in November-December, 1975 (Harper et al., 1976). Oily sediments (306 ppm) were found adjacent to one of the production platforms, whereas mostly biogenic background levels of about 0.5 ppm were found at all other stations. The macrobenthic faunal densities in the oiled bottom were reduced and the meiobenthos had an inverse ratio of nematodes to foraminifera compared with other stations. There was also evidence of a depressed fauna to the northeast of the field which may have been caused by an increased percentage of clay in the bottom or by some aspect of the production activities.

Based on conclusions drawn by analysis of pilot study data, the National Marine Fisheries Service (NMFS), Galveston, began a monitoring study in May 1976 to obtain more detailed information on the biological and abiotic characteristics of the field and its surroundings. The objectives of the benthic portion of the study were:

(1) to determine the areal and seasonal distributions of the macrobenthic and meiobenthic assemblages in and around the Buccaneer Field, and

(2) to determine if the Buccaneer Field structures or associated production activities had an effect on the population size, faunal composition, or distribution of the benthic organisms.

We had expected to complete the survey during the first year of study, then conduct detailed studies in areas either shown or suspected to be influenced by production activities. The detailed studies were not done because the benthic portion of the research program was deleted by NMFS after the first year.

The slow moving and non-motile benthic fauna are ideal subjects for studying the effects of pollutants. Given a uniform habitat, unaffected benthic communities are relatively uniformly distributed. The introduction of toxic substances tends to reduce both numbers of individuals and numbers of species in severely affected areas. Oil field bleedwater brine caused such reductions in Trinity Bay, Texas that were attributed to dissolved hydrocarbons, specifically naphthalenes, contained in the brine (Mackin, 1971; Armstrong et al., 1979). The hydrocarbons, incorporated in the sediments around the outfall, decreased in concentration at increasing distances from the source. Beyond the depauperate zone, the populations increased, culminating in a "zone of enrichment" where the faunal densities were larger than in unaffected "control" areas more distant from the pollution source.

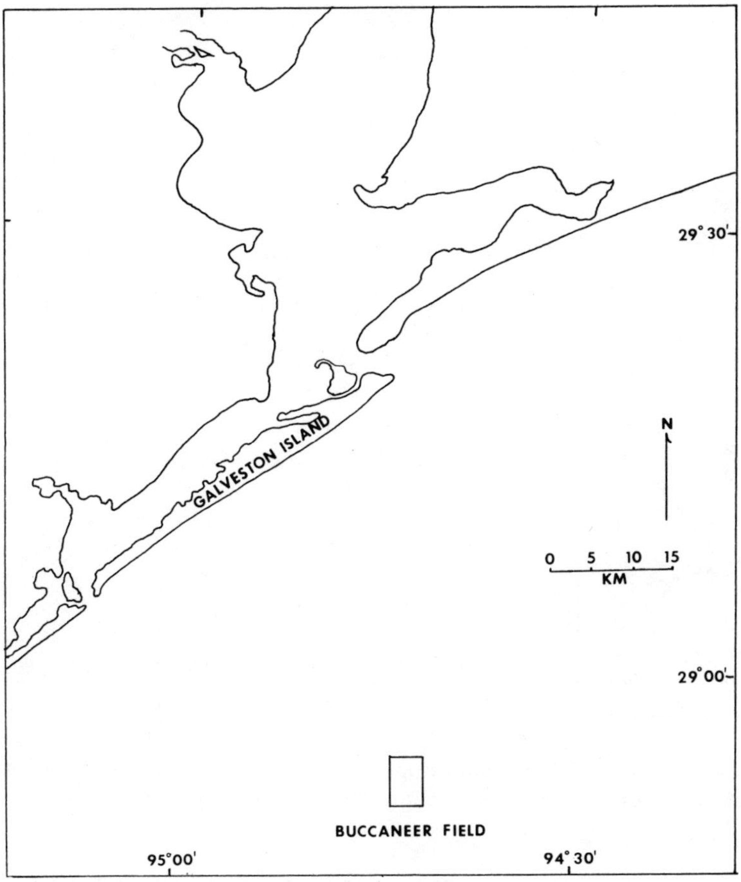

Figure 1. Map of the Gulf of Mexico offshore from Galveston, Texas showing the location of the Buccaneer Field.

In a study of oil production platforms located in deeper waters off Louisiana, Waller (1974) reported that his oil field stations were less productive than the control areas. However, Kritzler (1974), Farrell (1974), and Fish et al. (1974) found no differences in benthic fauna between production and control areas.

STUDY AREA

Buccaneer Field lay about 50 km almost due south of the south jetty lighthouse at Galveston, Texas (Figure 1). Structures in the field included two production platforms, 288-A and 296-B, each with an auxiliary crew quarters-helipad platform, and 13 well jackets, each of which supported one or more wells. The struc-

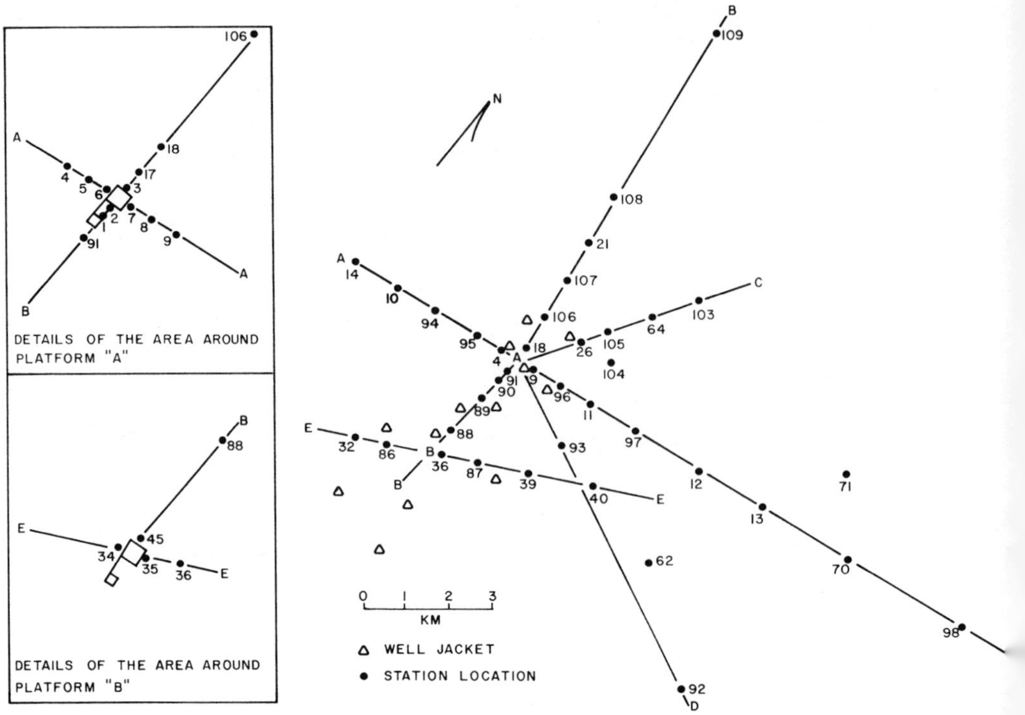

Figure 2. Map of the Buccaneer Field and vicinity showing the transects and macro- and meiobenthic sampling stations.

tures roughly outlined a north-south oriented rectangle. Three of the 34 active wells in the field produced oil.

The oil, mixed with brine when brought to the surface, was pumped into tanks on production platforms where the oil and brine density-separated. The oil was then pumped to shore, while the brine (bleedwater or produced water) was discharged into the ocean via a pipe that terminated about 1.5 m above the air-water interface. Shell Oil Company (owner of the field) data indicated that an average of 138 m^3 of brine was discharged per day between January, 1975 and February, 1976. The data also showed that oil-brine separation was incomplete; the average oil content of the brine was 25.1 ppm. We twice obtained samples of the bleedwater being discharged from platform 288-A. The samples had salinities (measured by refractometer) of 50 and 49°/oo on October 10 and November 3, 1976, respectively, and both had a strong oily odor. If the water was shallower, one would expect to find quantities of oil incorporated in the sediments as was found in Trinity Bay, Texas (Mackin, 1971; Armstrong et al., 1979) and Louisiana (Mackin and Hopkins, 1961). Because of the water depth in the Buccaneer

Field, the oil detected during the pilot study may have resulted from an accidental spill. The sea floor around platform 288-A was examined numerous times by divers after the pilot study samples were collected and no obvious trace of oil was found. However, Mr. Richard McGaffey of Galveston informed us orally that he had periodically observed oil on the sediments while diving around platform 288-A.

FIELD METHODS

Transects were established radiating from platforms 288-A and 296-B (Figure 2). Stations on the transects were located close together near the platforms and at increased inter-station distances away from the platforms. The rational for this design was that if there was an environmental stress associated with the platform, it would be manifest relatively close to the platforms and would decrease as the distance from the platforms increased, a method used successfully by Mackin (1971) to study the effects of bleedwater being discharged into Texas estuaries. Buccaneer stations were concentrated to the north and east because pilot study data indicated a possible depressed fauna to the northeast, and because the strongest bottom currents flowed to the northeast (see abiotic data).

Around platform 288-A, the stations also formed concentric rings of 15, 50, 100, 1000, and 2000 m distant from the platform. Beyond 2000 from the platform, there was no regular station spacing. The sampling effort was concentrated near platform 288-A and few stations were established in proximity to platform 296-B. LORAN A was not always reliable, so radar fixes on platforms 288-A and 296-B were used to locate stations. Station numbers and distances from platforms 288-A or 296-B are listed in Appendix A.

A quarterly sampling program was conducted; collections were made in July and October-November, 1976, and January-February and April, 1977. The second and third quarter collections were protracted because of bad weather and it is possible that some natural population changes occurred between the times of the first and last cruises.

The benthos were sampled in triplicate by divers who pushed Ekman grabs into the sediment, triggered them, and secured the jaws and vent flaps with elastic bands. The maximum penetration depth of the grabs was 15 cm but the actual depth depended on the consistency of the substrate. The diving method of sampling was quite efficient. The boat did not have to be anchored, and sampling could be completed in five to seven minutes at any station. While on the bottom, the divers also collected a water sample, observed the characteristics of the substrate, and noted the presence of macrofauna not likely to be sampled by the grab.

When the samples were returned to the vessel, a meiofaunal subsample was taken from each benthic sample with a 2.54-cm diameter plastic tube. The meiofaunal samples were preserved intact in 5% seawater formalin. The macrofaunal samples were washed on a 0.5-mm mesh sieve to remove sand and mud. The organisms and shell hash retained on the sieve were preserved in 5% seawater formalin.

Abiotic data recorded at each station included surface and bottom temperature and salinity, sediment temperature, and a visual analysis of the sediment composition. Temperature was measured with a Celcius thermometer, and salinity with a refractometer.

LABORATORY METHODS

In the laboratory the macrofaunal samples were washed with fresh water to remove formalin. The lighter organisms were separated from the shell hash by repeated decanting and were preserved in 70% ethanol. Rose Bengal stained 70% ethanol was added to the shell hash to aid in the detection of remaining organisms. After at least 24 hr had elapsed, the shell hash was examined microscopically and all organisms were removed and combined with the decanted material. The organisms were sorted, identified to the lowest possible taxonomic rank, and counted. Because initial sample analyses were performed by several persons, Harper examined all samples to correct misidentifications or at least to standardize taxonomic errors.

The meiofaunal samples were washed on 0.5 and 0.063-mm mesh sieves. Each 0.5-mm fraction was returned to the appropriate macrofaunal sample. Meiofaunal organisms retained on the 0.063-mm sieve were removed from the sediment by flotation with a saturated sucrose solution and were preserved in rose Bengal stained 70% ethanol. These organisms were identified to the lowest possible taxon and counted.

DATA ANALYSIS

The temperature and salinity values recorded during each one or two-day cruise were averaged. The averaged values were used to describe temperature and salinity trends and, unless otherwise indicated, the discussions herein will refer to averaged data.

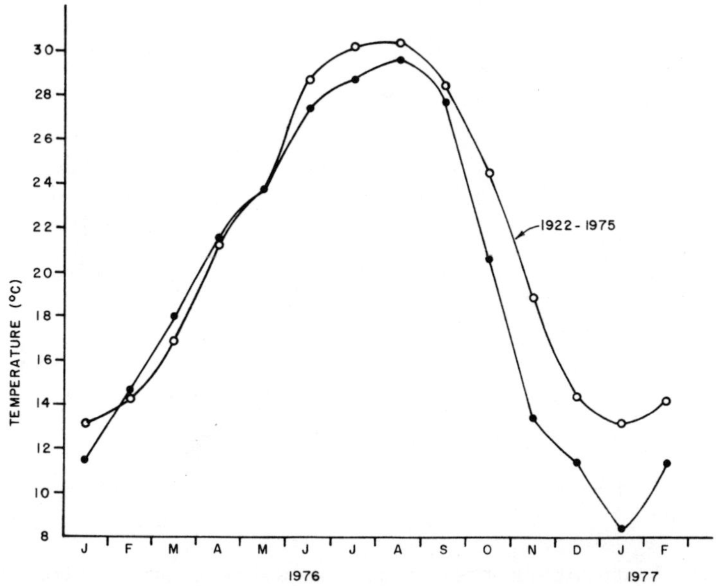

Figure 3. Comparison of the water temperature trends in the Galveston Channel, January 1976 through February 1977, with the average trend, 1922 through 1975.

Each Ekman grab sampled an area of 232 cm^2; multiplication by a factor of 43 extrapolated raw data to numbers of individuals per m^2. The meiofaunal core tube had a cross-sectional area of 5.07 cm^2; multiplication by 1970 will extrapolate raw data to numbers of individuals per m^2, resulting in probably unrealistic numbers. We preferred to evaluate the meiofaunal data in terms of numbers of individuals per cm^2.

The biological data were examined descriptively, comparing diversity and faunal densities between stations and over time. Ordination, Shannon-Weiner diversity index, Duncan's Muliple Range Test, and numerical analysis using the methods of Clifford and Stephenson (1975) and Smith (1976) were also used in the data analysis.

RESULTS

Abiotic Characteristics of the Study Area

Temperature. The water depth in the area of the field was sufficiently shallow (19-20 m) to allow complete overturn during

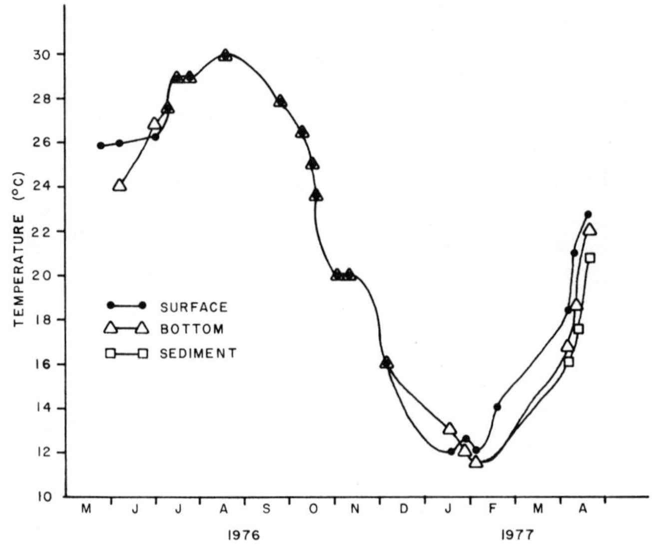

Figure 4. Temperature trends of the surface and bottom waters and the sediments in the Buccaneer Field study area, May 1976 through April 1977.

fall cooling, but the bottom water temperature probably lags behind the surface water temperature during most spring warming periods as it did in 1977 (see data in Temple et al., 1977).The winter of 1976-77 was unusually cold in the Galveston area as it was in much of the eastern and central United States. The mean monthly temperature trend in the Galveston Channel, January, 1976 through February, 1977 (Figure 3) is compared with the monthly mean trend for the period 1922-1975 (National Ocean Survey records). The 1976 water temperature was near average from January through May and was slightly cooler (about 1.5°C) than average from July through September. In October and November, however, the water temperature decreased much more rapidly than usual and was 4-6° cooler than average. Below average temperatures continued through February, 1977; the mean temperature in January (the coldest month) was about 4.6°C cooler than average.

The above data, while not directly comparable with offshore water temperatures, do indicate the water temperature trends of the area prior to and during the study. During the study, water temperatures were recorded by the biological field crew and by Martin (1977). In 1976, the water temperature increased between May, when the study began, and August (Figure 4). By late September the temperature had begun to decrease. National Ocean Survey air temperature data indicate that a series of cold fronts passed through Galveston in October and November coincident with

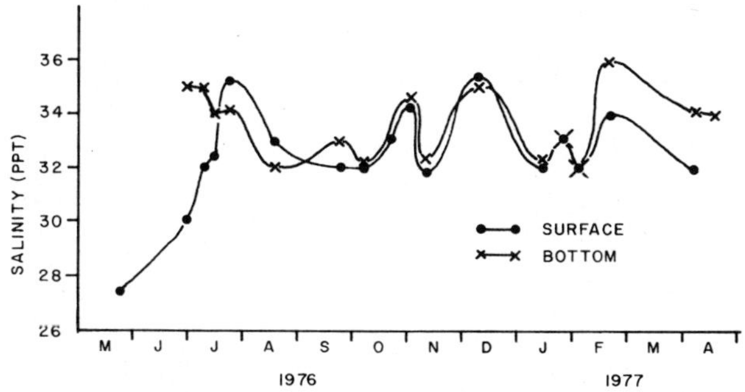

Figure 5. Salinity trends of the surface and bottom waters in the Buccaneer Field study area, May 1976 through April 1977.

the rapid cooling of Galveston Channel (Figure 3) and offshore waters (Figure 4). The lowest recorded Buccaneer Field water temperatures, about 12°C, occurred between mid-January and early February, 1977. The temperature began to increase in mid-February and continued to rise rapidly through April when sampling ceased. The sediment and bottom water temperature trends were virtually identical during most of the study. During the spring warming period (February-April), the surface water warmed more quickly and was 2-3°C warmer than either the bottom water or sediment.

Salinity. The bottom water salinity was always at or near oceanic values, fluctuating between 32 and 36°/oo (Figure 5). The surface salinity was depressed in May when sampling began, but increased rapidly and the water temperature was nearly isohaline during most of the study. In April, 1977 the surface salinity appeared to be decreasing, which conformed to the patterns of spring (May-June) salinity reductions due to increased freshwater discharge from both the Mississippi River and local rivers (Temple et al., 1977).

Bathymetry and Sediments. The water over most of the study area was 19-20 m deep. Proceeding north and northeast from platform 288-A to stations 109 and 64, respectively, the bottom shallowed to 16 m. About midway between stations 108 and 109 was a ridge, the crown of which was about 12 m deep.

Data collected during the pilot study (Harper et al., 1976) and this study (Anderson et al., 1977) indicated sediments in the study area graded from sand-silt-clay in the northeast through silty sand and clay to clayey sand in the southwest (Figure 6). These geological nuances were not readily apparent to the biologist/divers who classified most of the sediments as muddy

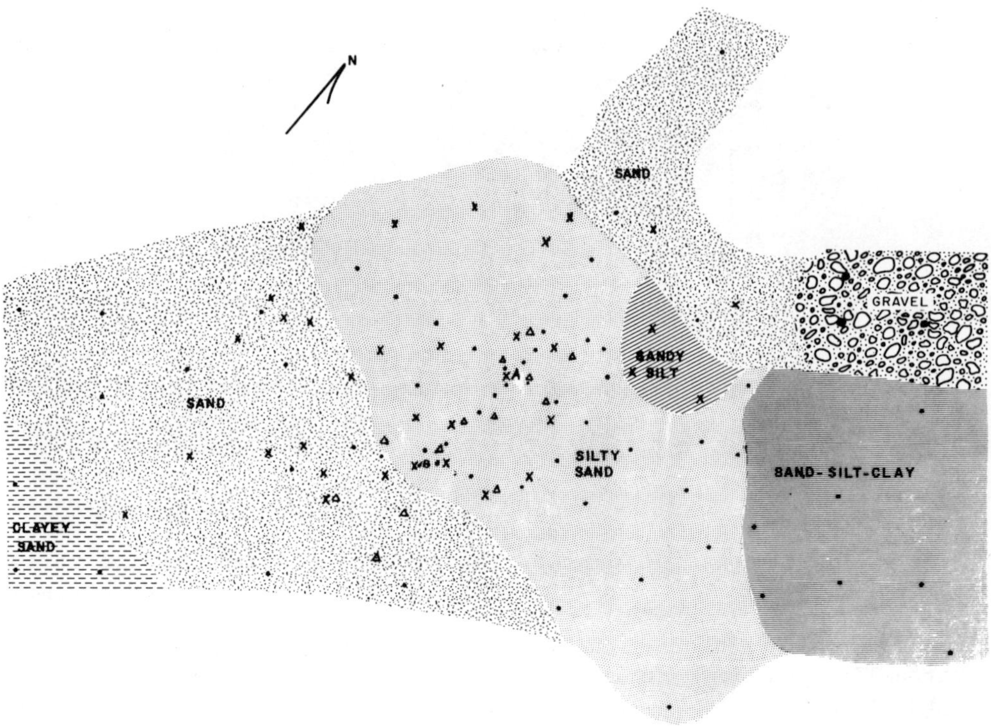

Figure 6. Distribution of sediments in the Buccaneer Field study area. x - pilot study station, • - Anderson et al. (1977) station, Δ - well jacket.

sand with shell hash. The sediments in the shallower northern part of the study area were sand. Coarse, gravel-sized, brown shell fragments occurred to the northeast; station 103 in this area was abandoned after the first collection because of its quite different substrate and fauna.

Around platform 288-A the bottom consisted of a thin veneer of silt covering a hard clay that was difficult to sample, and was littered with metal debris (pieces of catwalk, gears, welding rods, etc.). The bottom around platform 296-B was hard sandy mud and was also littered with metal debris. A short distance southwest of platform 296-B was a shelly bottom.

There is evidence that the loose sediments are periodically removed and subsequently redeposited or are redistributed by currents and/or waves. The sediments at station 12, to the east of the field, were soft muddy sand in July and October, but hard clay was exposed in January and April. This was the same area where exposed, hard clay was collected during the pilot study in

December, 1975 (Harper et al., 1976). Also, seismic records detected a structural offset that extended from the vicinity of platform 288-A through the northeast edge of the study area (Anderson et al., 1977). This offset was actually seen in January, apparently after softer sediments were removed; divers reported the presence of a 15-cm high scarp in the vicinity of station 8. It is probable that, during relatively calm weather, finer sediments accumulate and then are removed during the stormy winter months. On calm days, bottom visibility was often 2-3 m but, after a period of heavy seas, the divers reported zero visibility on the bottom. Danek and Tomlinson (this Volume) estimated that orbital velocities were sufficient to suspend sediments about 9% of the time.

Hydrocarbons and Organic Carbon in the Sediments. With the exception of one obviously oily sediment sample collected adjacent to platform 288-A during the pilot study (Harper et al., 1976), most sediments contained primarily biogenic hydrocarbons. A concentration gradient of relatively fresh Buccaneer crude oil was found in December, 1976 at platform 288-A; the highest concentration of alkanes was found below the platform, and the concentration decreased with increasing distance out to 23 m from the platform. Weathered Buccaneer crude oil was found in the sediments adjacent to platform 288-A in October, 1977 (Middleditch and Basile, 1978).

Three regions of high organic carbon were found, one adjacent to platform 288-A on the southwest side, one (a relatively small area) 1000-1500 m east-northeast of platform 288-A in the vicinity of a well jacket, and one east of platform 288-A in the Beaumont Clay area (Behrens, 1977).

Wind, Waves, and Currents. Wind-generated waves and currents are the agents most probably responsible for redistribution and/or transporation of sediments and hydrocarbons in the Buccaneer Field area. The seas are generally calm in the summer and fall while southeasterly winds blow (average speed of 5.7 m/sec), and moderate in winter and spring when short-lived northerly winds associated with cold fronts occur. (average speed of 6.7 m/sec, maximum speed of 22 m/sec). Wave periods of 5.1 sec will "feel" bottom in the Buccaneer Field area; wind speeds necessary to produce these periods, 9.1 m/sec, occur less than 25% of the time. Waves greater than 2 m in height occur only 20% of the time (Fisher et al., 1972; SEADOCK, 1975).

Current data from the Buccaneer Field were obtained from current meters installed near the surface and bottom (4 and 17-m depths, respectively) at a production platform. During the period October 25, 1971 through April 30, 1974, surface currents were predominantly westward in March through June and easterly to

TABLE 1. THE MACROBENTHIC TAXA OCCURRING IN THE BUCCANEER FIELD ARRANGED IN ORDER OF DECREASING ABUNDANCE. ALSO INDICATED ARE THE TOTAL NUMBER OF STATIONS (TOTAL OF 48) AT WHICH MEMBERS OF EACH TAXON WERE COLLECTED. P - <0.1%.

Taxon	% Occurrence	# Stations
Polychaeta	71.8	48
Amphipoda	14.4	48
Bivalvia	4.4	48
Nemertinea	2.0	48
Tanaidacea	1.3	39
Decapoda	1.1	48
Cumacea	1.1	48
Sipunculida	0.9	47
Gastropoda	0.6	48
Isopoda	0.6	41
Ophiuroidea	0.4	46
Ostracoda	0.3	44
Cephalochordata	0.3	17
Anthozoa	0.1	27
Nematoda	0.1	27
Hemichordata	0.1	15
Brachiopoda	P	13
Hydrozoa	P	11
Bryozoa	P	10
Platyhelminthes	P	9
Phoronida	P	8
Mysidacea	P	8
Nebaliacea	P	7
Osteichthys	P	7
Stomatopoda	P	4
Protozoa	P	3
Echinoidea	P	2
Scaphopoda	P	1
Echiuroidea	P	1
Pycnogonida	P	1

northeasterly from July through September. Bottom currents were predominantly westerly and northwesterly from February through May and predominantly easterly and northeasterly July, August, November, and December. The strongest bottom currents (greater than 50 cm/sec) occurred most frequently in July and August (northeasterly) (SEADOCK, 1975). Data obtained during the present study indicated that net currents were predominantly south to southeasterly in July and August, 1976, then shifted to south-

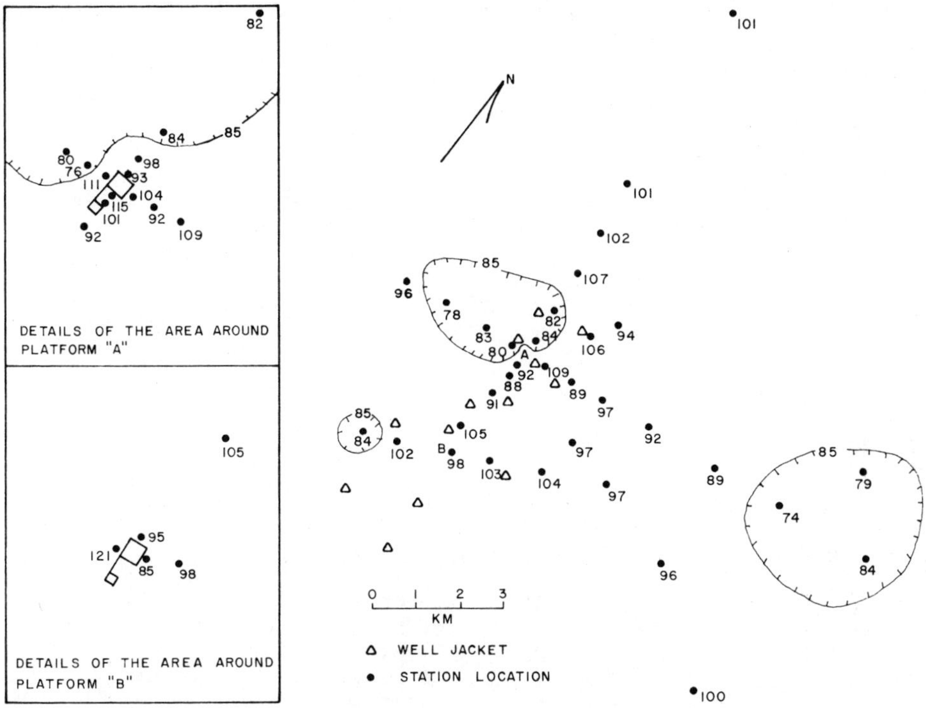

Figure 7. Macrobenthic species areal distribution. Each data point represents the total number of species collected at that station. Suggested isopleths are superimposed.

westerly from November, 1976 through February, 1977 (Martin, 1977). Several periodic current reversals were observed in the data.

Daily current records from the Buccaneer Field in August, September, and December, 1973 indicated that flow reversals occurred frequently at both the surface and bottom of the water column, and that the bottom current often flowed 180° relative to the surface current direction. The surface currents appeared to be responding to both wind stress and tides (SEADOCK, 1975).

Macrobenthic Population Densities and Diversity

About 82,700 individuals (raw numbers) of epi- and infaunal macroinvertebrates were collected during the study. We estimate that between 400 and 420 species were collected (Appendix B); the exact number is difficult to deduce because many young specimens could not be identified and were recorded as a separate species even though the adult might have been known.

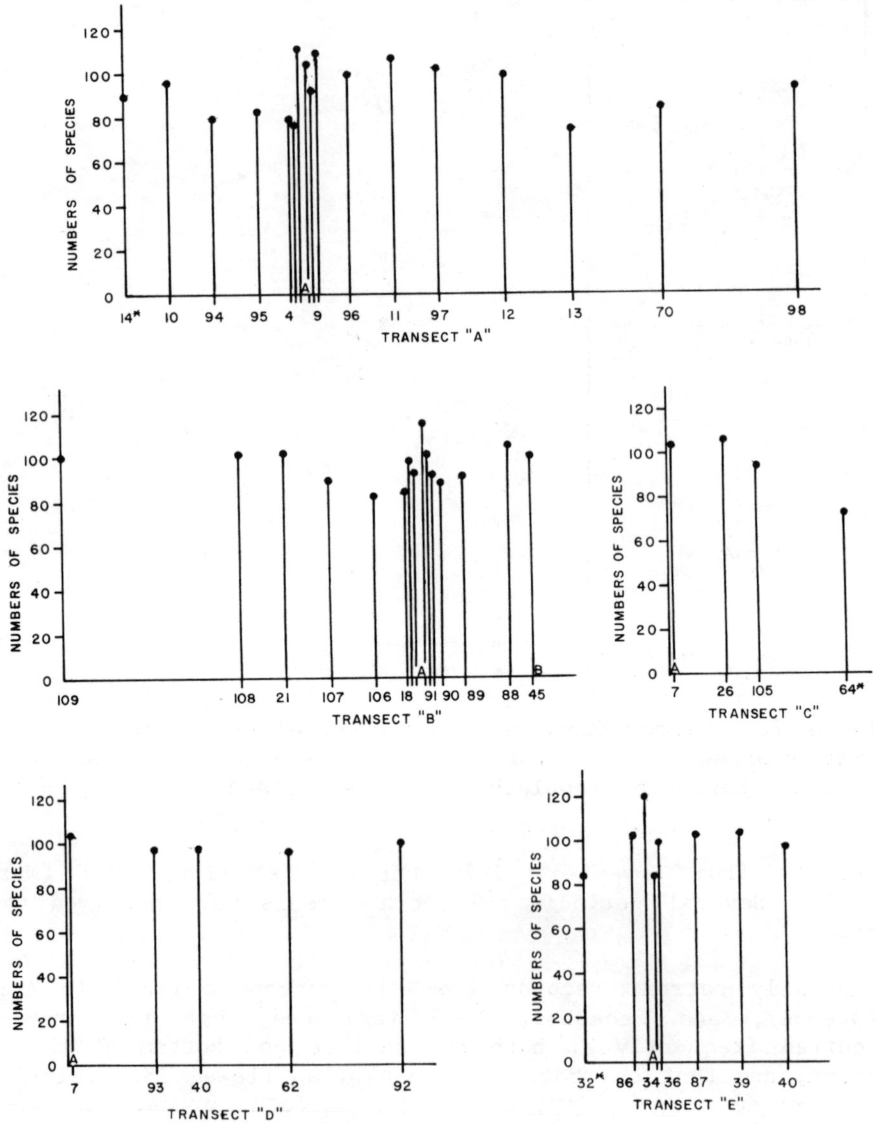

Figure 8. Distribution of species along each transect in the Buccaneer Field. Each data point represents the total species collected at that station. Stations not numbered (left to right) are: Transect A - 5, 6, 7, 8; Transect B - 17, 3, 2, 1; Transect E - 35. * - average of only three collections.

The Polychaeta was the dominant taxon in the Buccaneer Field benthic community, comprising 71.8% of all species collected during the study period (Table 1). Amphipods were the second dominant group (14.4%), and only two other taxa, Bivalvia (4.4%) and Nemertinea (2.0%), comprised 2% or more of the total population. The bivalves occurred predominantly at one station; a massive set of young bivalves (*Corbula operculata*?) occurred at station 64 in January. The majority of taxa that comprised between 1.0 and 0.1% of individuals occurred at most stations. The taxa comprising less than 0.1% of individuals occurred at relatively few stations.

In many benthic studies, one or two species will be overwhelmingly dominant, comprising 40-50% of all individuals in the population. This did not occur in the Buccaneer Field. The dominant species (a species of *Lumbrineris*, tentatively identified as *L. verrilli*) comprised only 14.2% of all individuals collected. The 23 most abundant species (i.e., 500 or more total individuals) comprised 76% of all specimens collected.

Areal Distributions. Most stations had between 90 and 110 total species during the study (Figure 7). Three areas had stations with less than 85 species, the largest consisting of 5 stations west and north of platform 288-A, and the smaller ones consisting of 3 stations in the eastern end of the study area, and one station (32) west of platform 296-B.

The distribution of total species along the transects indicates a high diversity around platform 288-A, followed by decreased diversity at distances of 1000-2000 m north, west, and south of the platform, beyond which were areas of increased diversity (Figure 8). The diversity was relatively constant on the eastern leg of tansect A only. There was no consistent trend around platform 296-B. There was no evidence that diversity was suppressed around any structure in any of the quarters.

The average quarterly faunal abundance at most stations was 5000-7000 individuals/m^2 (Figure 9). The fauna were depressed (fewer than 5000 individuals/m^2) in three areas; to the east of the field, corresponding with the area where clay was either exposed or near the surface, and in proximity to platforms 288-A and 296-B (inset, Figure 9). The densities within these latter two areas were between 2200 and 3800 individuals/m^2.

There were also three areas in which faunal densities were larger than 7000 individuals/m^2 ("enriched areas"), one on transect B, 2000-3000 m north of platform 288-A, one 1000 m east and north of platform 288-A, and one within a 1000 m radius of platform B (enclosing the depressed area). The second of these

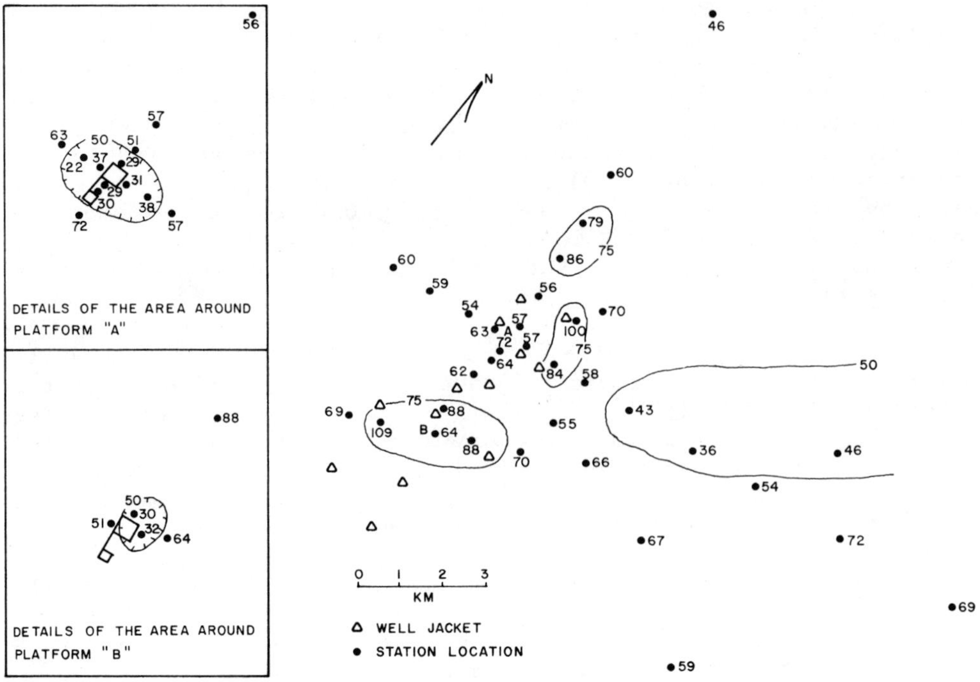

Figure 9. Areal distribution of macrobenthic abundances. Each data plot represents the average number of individuals/m^2 per quarter x 100. Suggested isodemes are superimposed.

enriched areas corresponded with an area of high total organic carbon in the sediments (Behrens, 1977).

The distribution of populations along each of the transects clearly depicts the depressed faunal zones close to platforms 288-A and 296-B as well as the enlarged populations away from the platforms (Figure 10). The depressed fauna at station 12 (Beaumont Clay) is also evident.

The two dominant taxa, Polychaeta and Amphipoda, were rather uniformly distributed throughout the study area. Polychaetes comprised 65-75% of the fauna at most stations, with only one area having less than 60%, a group of three stations at the east end of the field (Figure 11). The amphipods comprised 10-20% of the total assemblage at most stations (Figure 12). The percentage was less than 10% on the northern sandy bottoms and was in excess of 35% at the group of three stations where polychaetes were less abundant.

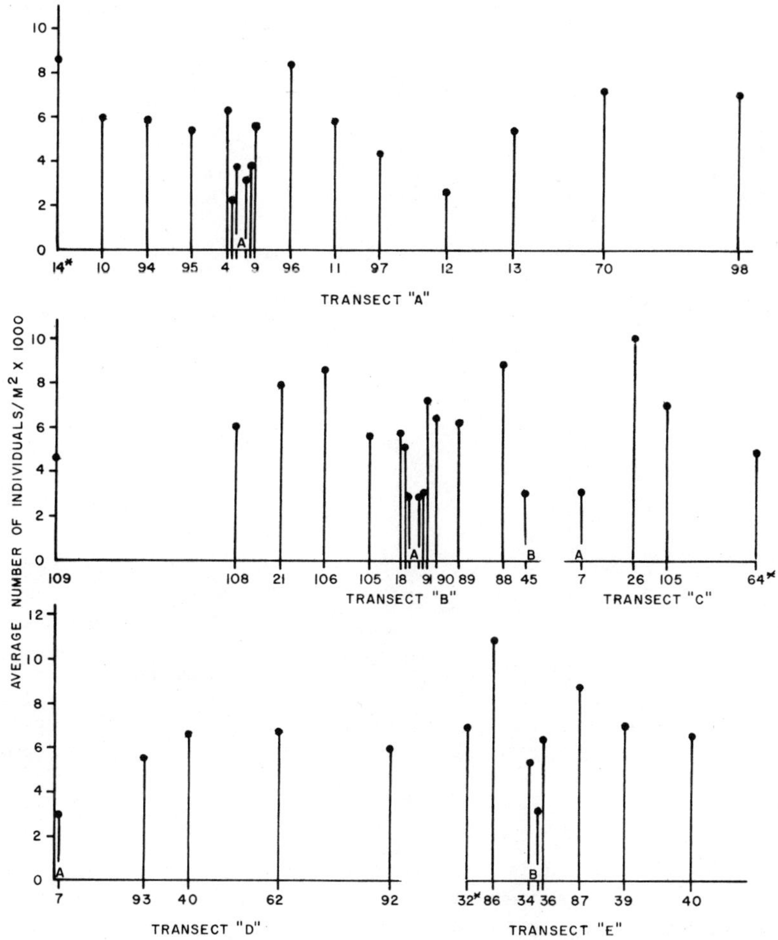

Figure 10. The distribution of macrobenthic abundances along each transect in the Buccaneer Field. Stations not numbered (left to right) are: Transect A - 5, 6, 7, 8; Transect B - 17, 3, 2, 1; Transect E - 35. * - average of only three collections.

Of particular interest among the amphipods are members of the genus *Ampelisca*, which have been used as indicators of pollution. The distributional pattern of the ampeliscids (*A. abdita*, *A. agassizi*, and *A. verrilli*) suggests a depressed area around platform 288-A and on the northeastern side of platform 296-B (Figure 13). Populations were also small at station 12 (Beaumont Clay) and in the northern sandy area.

Numerical analysis performed on combined data from the four quarters agreed closely with the distributional pattern of the

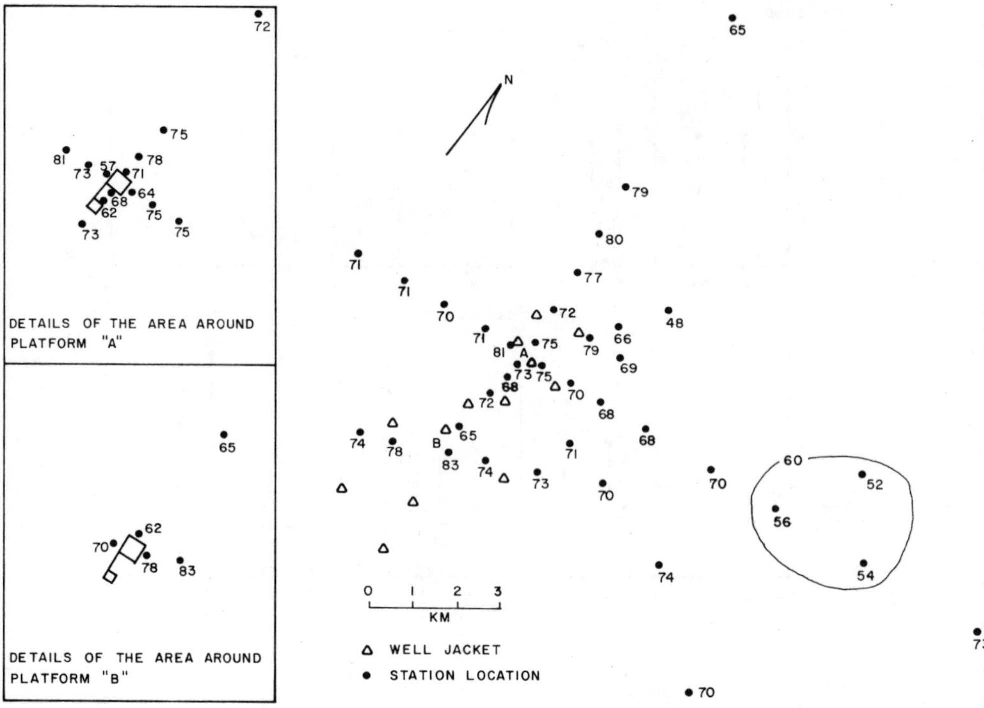

Figure 11. Areal distribution of the percentage of polychaetes in the total macrobenthic assemblage. Suggested isodemes are superimposed.

average abundances described above. Four station groups were produced in all analyses (Figure 14). Station group I consisted of the majority of stations in the study area; all had a high degree of species co-occurrence as indicated by the degree of ecological similarity. Station group II consisted of a cluster of two to five stations (depending on the number of species used in the analysis) with reduced populations, most of which were located in the vicinity of exposed clay toward the eastern end of the study area (Figure 15). Group III consisted of stations in proximity to platforms 288-A and 296-B. Group IV consisted of two stations, 64 and 109, which were unrelated to any of the other stations and relatively unrelated to each other.

Four to six species groups were produced by numerical analysis, again depending on the number of species used in the analysis. For this paper, 68 species comprising 95% of all individuals collected were used, producing four species groups. These groups are compared with the station groups in the two way table summary (Table 2). Animals of species group I were ubiquitous, occurring at almost every station. Station group II

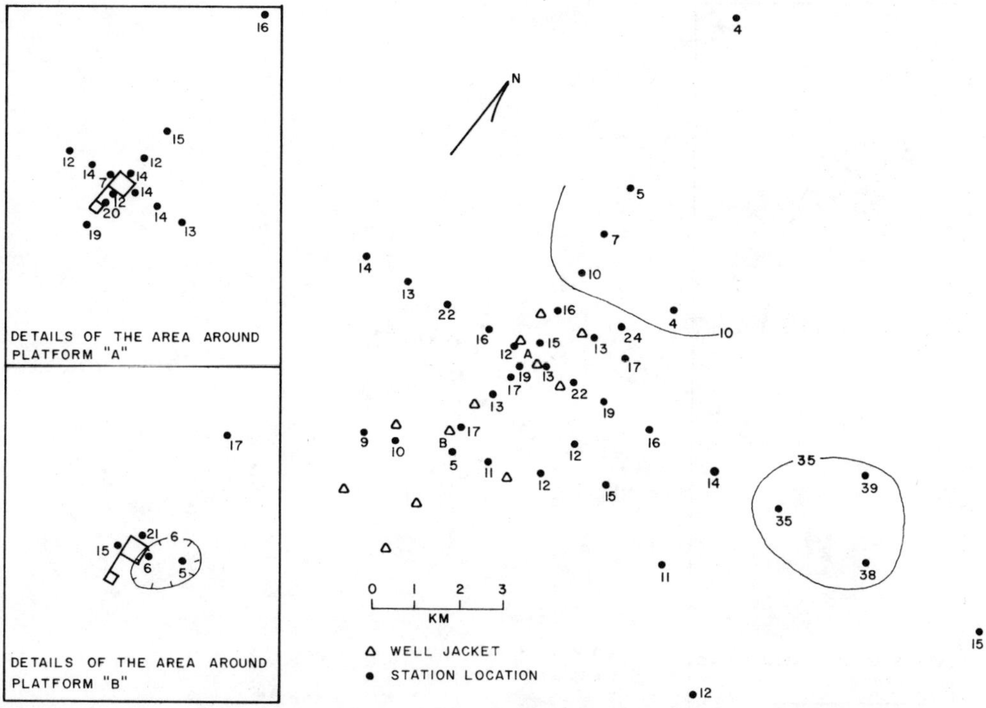

Figure 12. Areal distribution of the percentage of amphipods in the total macrobenthic assemblage. Suggested isodemes are superimposed.

(vicinity of clay bottoms) was characterized by infrequent occurrence and low numbers of the species in species groups III and IV. Station group II (around the platforms) was characterized by infrequent occurrence of species group II members and a high degree of constancy and fidelity of species in group IV (4 species) compared with other station groups. Station group IV was characterized by low representation of members of species groups II and IV.

Principal components analysis using the 25 most frequently occurring species produced a compact cluster containing most stations in the study area, corresponding closely with station group I above (Figure 16). The remaining stations were scattered and most were near-platform stations (station group III above). Stations 64 and 109 (station group IV above) again were widely separated from the other stations.

Assemblage Stability. The combination of high diversity and depressed populations around platform 288-A suggested an unstable assemblage with a high turnover rate. This possibility was

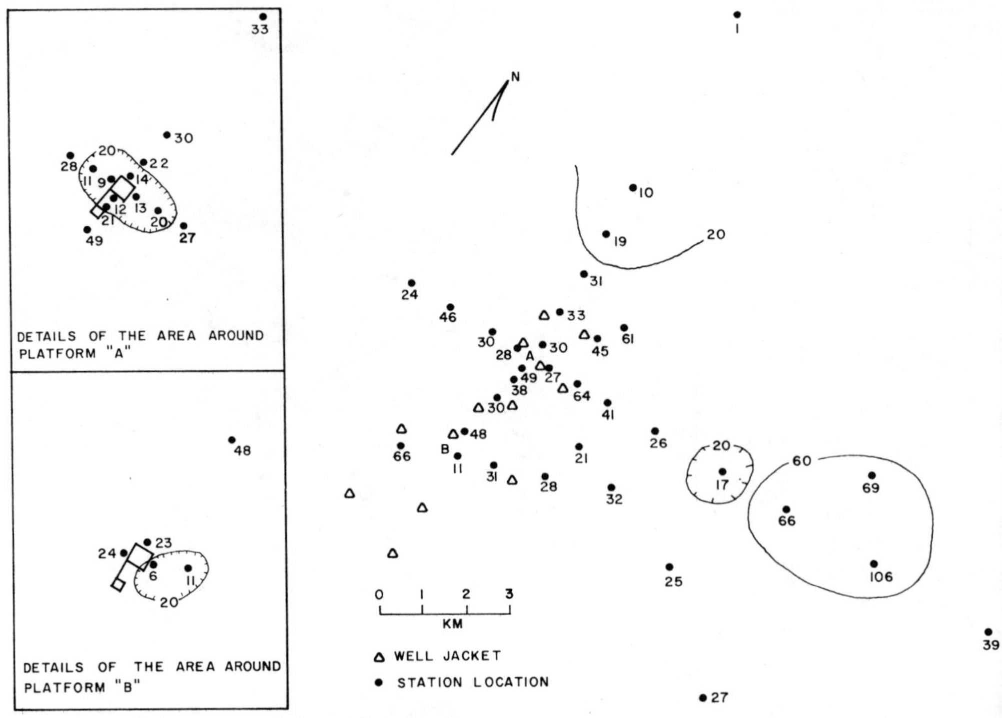

Figure 13. Areal distribution of the average number of ampeliscid amphipods/m^2 per quarter. Suggested isodemes are superimposed.

investigated in three ways: by calculating, for each station, the ratio of recurring species (three and four occurrences) to nonrecurring species (one and two occurrences); the ratio of species occurring four times to those occurring once; and the ratio of species occurring four times to the total number of species present. All three methods gave similar results (Figure 17). The ratios were lower in the vicinity of platforms 288-A and 296-B, and at stations 12 (exposed clay) and 109 (relatively shallow, sandy bottom), meaning that those stations had a higher turnover rate than the remainder of the study area. One might expect, then, that the Shannon-Weiner diversity index (H') would be different at those stations. This was not the case. An average H' was calculated for each station and tested using Duncan's Multiple Range test. The results of the test indicated that there was no significant difference between those stations

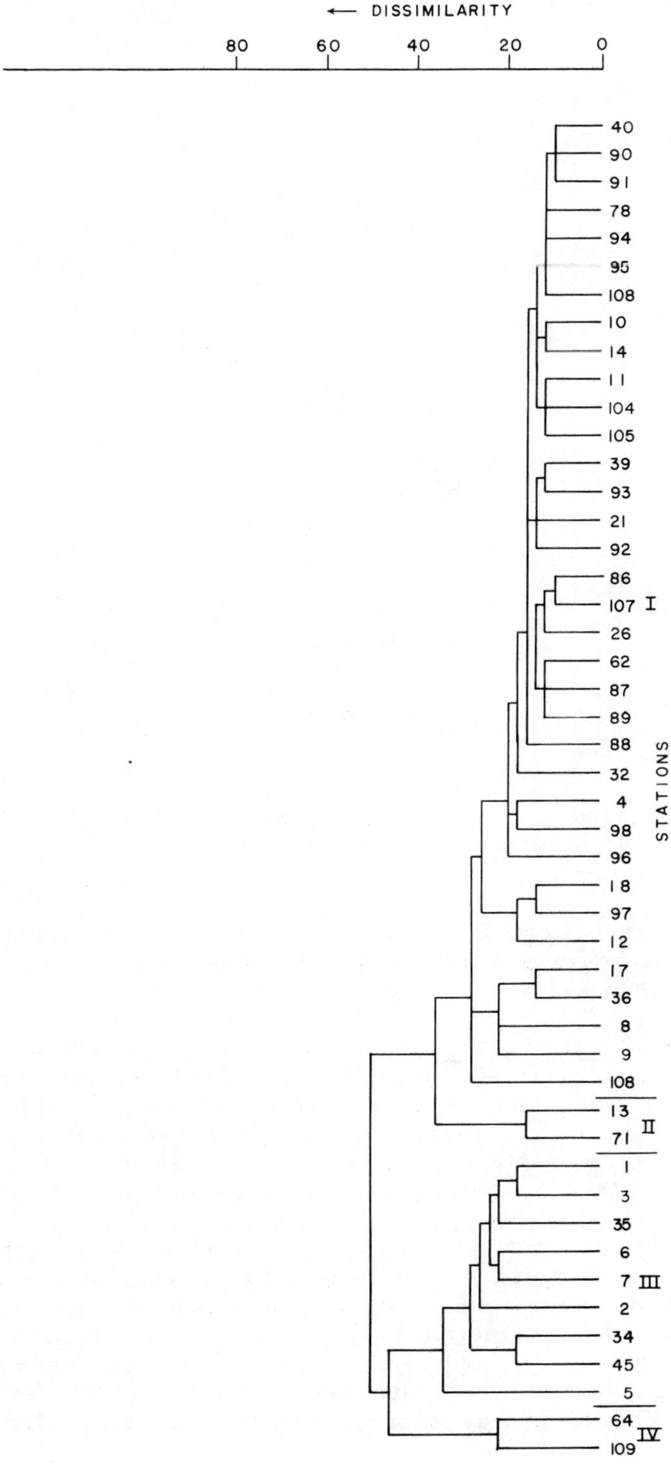

Figure 14. Station group dendrogram produced by cluster analysis of macrobenthic data.

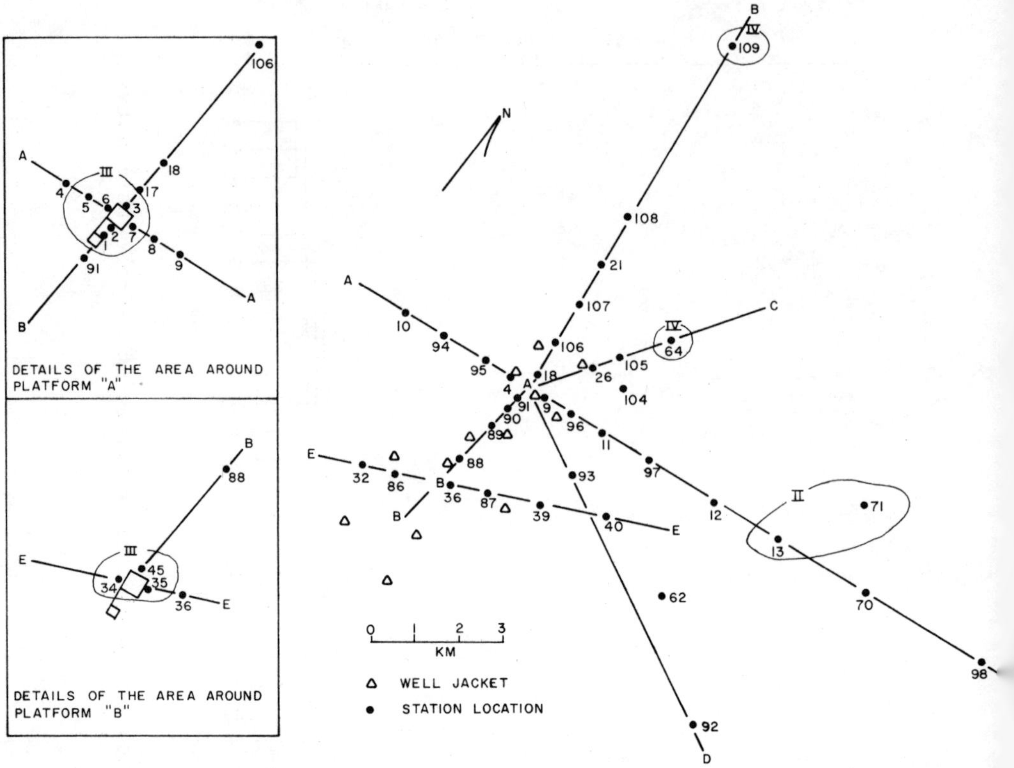

Figure 15. Station locations and the areal extent of the station groups derived by numerical analysis of the macrobenthic data.

with low ratios of species stability and other stations in the study area, although all low ratio stations were included within the group "C" bracket (Figure 18).

Pteropod Shells in the Sediments. When the benthic samples were being processed it was apparent that large numbers of empty, but intact, pteropod shells were present in the vicinity of the oil field structures. Pteropods, which are components of the plankton that swim by flapping a pair of lateral extensions of the foot (called "wings"), should not have been collected in bottom samples. The largest numbers were collected at stations 9 and 26, which were adjacent to well jackets. Stations 4 and 88, which were near but not close to well jackets, had higher numbers of shells in the sediments than more remote stations. Large numbers of shells were also collected in proximity to platforms 288-A and 296-B. At station 3, adjacent to the north side of platform 288-A, the July samples contained large numbers of *Creseis acicula* pteropods that were either alive or very recently killed at the

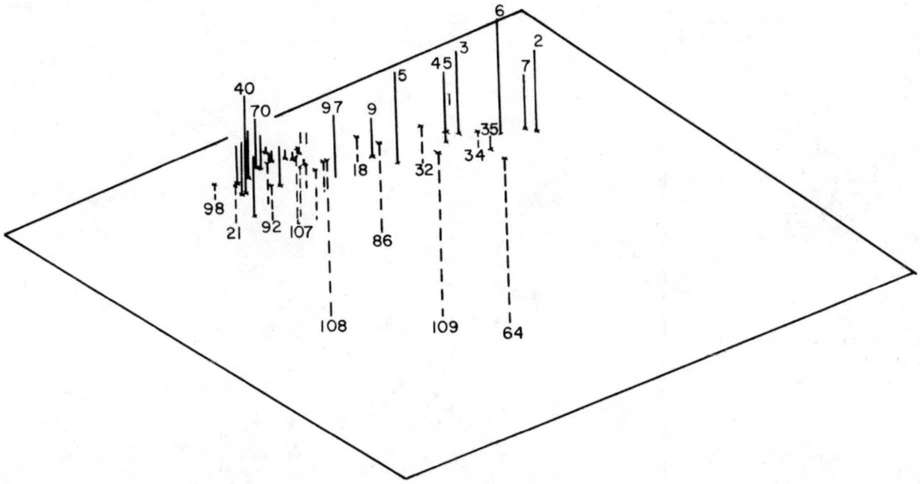

Figure 16. Principal components analysis of the macrobenthic data. Numbers represent stations. Only a few stations in the compact cluster are labeled. Dashed lines represent negative values on the Z axis.

TABLE 2. SUMMARY OF THE TWO-WAY TABLE PRODUCED BY CLUSTER ANALYSIS USING 68 SPECIES AND 48 STATIONS, INDICATING RELATIVE CONSTANCY AND FIDELITY OF MEMBERS OF SPECIES GROUPS IN EACH STATION GROUP.

		STATION GROUPS			
		I	II	III	IV
SPECIES GROUPS	I	+++	+++	+++	+++
	II	+++	++	+	+
	III	++	+	++	++
	IV	+	+	+++	+

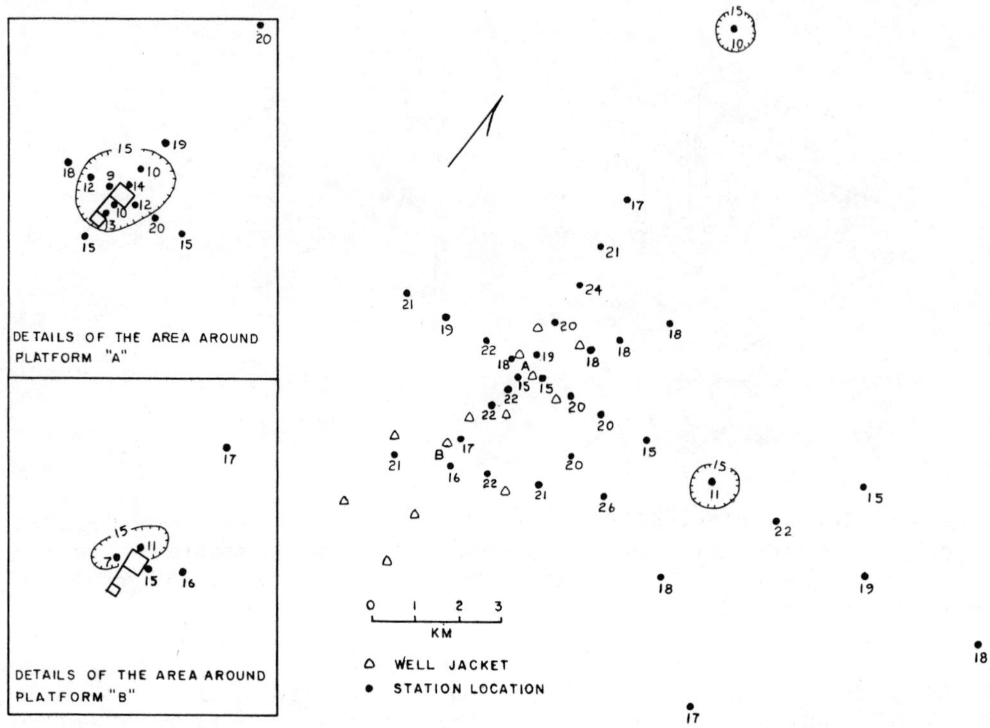

Figure 17. Areal distribution of the ratio of recurrent (each quarter) to total species. Suggested isopleths are superimposed.

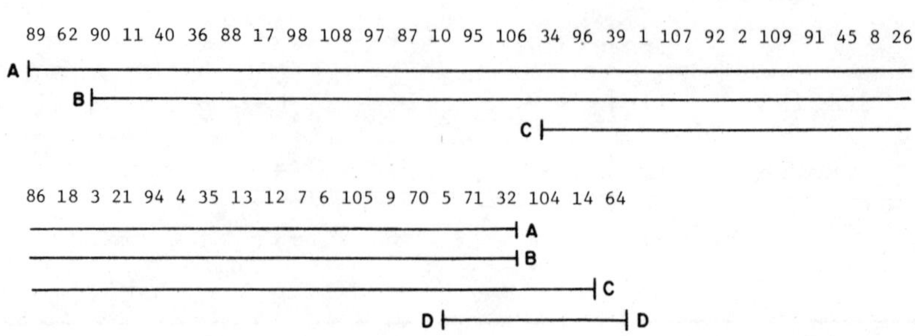

Figure 18. Results of Duncan's Multiple Range test on the mean Shannon-Weiner diversity index at each station. Stations within each bracket are not significantly different.

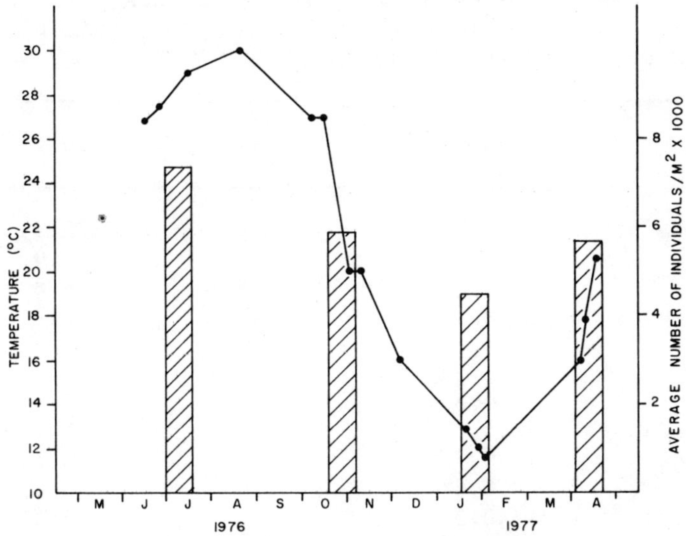

Figure 19. Comparison of the macrobenthic abundance and sediment temperature trends in the Buccaneer Field study area.

time of collection; the flesh was still in the shells and showed no signs of decay. A few intact specimens were collected at station 1, also.

Seasonal Trends. The average quarterly number of individuals/m^2 per station is compared with the sediment temperature trend in Figure 19 (some data temperature points were inferred from bottom water temperatures). The number of individuals decreased between July and January concomitant with the temperature decrease, then increased through April as the temperature increased. The trend did not agree with the seasonal patterns of macrobenthic assemblages inhabiting a sandy mud bottom offshore from Freeport, Texas, similar in composition to the Buccaneer Field sediments (Harper and McKinney, 1979). During the Freeport study, the largest populations occurred in the spring months (March-May) and the lowest populations usually occurred in late fall; it has, however, been found that many species' annual trends are somewhat variable, indicating that the observed population trends in the Buccaneer Field may not have been unusual. The Buccaneer Field abundance trends were also different from the nearshore (18-27 m deep) trends of benthic populations sampled during the South Texas Outer Continental Shelf (STOCS) study (Holland, 1976). The largest STOCS populations occurred in April-May and the lowest in August-September. While the Buccaneer Field and Freeport assemblages were relatively equal in size, the STOCS populations were about an order of magnitude smaller, possibly

TABLE 3. PERCENT COMPOSITION IN EACH QUARTER OF THE TAXA COMPRISING MORE THAN 0.1% OF THE TOTAL MACROBENTHIC INDIVIDUALS COLLECTED DURING THE BUCCANEER FIELD STUDY.

Taxon	July	Oct.	Jan.	April
Polychaeta	65.7	78.0	75.3	68.1
Amphipoda	19.1	7.5	9.9	21.0
Bivalvia	3.6	4.5	6.8	2.8
Nemertinea	2.4	2.4	1.7	1.6
Tanaidacea	2.6	0.7	0.9	1.1
Decapoda	1.2	1.6	0.9	0.5
Cumacea	0.8	1.7	0.3	1.5
Sipunculida	0.7	0.7	1.1	1.2
Gastropoda	0.7	0.6	0.7	0.5
Isopoda	0.6	0.5	0.8	0.5
Ophiuroidea	0.3	0.5	0.4	0.2
Ostracoda	0.3	0.5	0.2	0.2
Cephalochordata	0.9		0.2	
Anthozoa	0.1			
Nematoda	0.1	0.1	0.2	0.2
Hemichordata	0.3			
Brachiopoda		0.1		
Hydrozoa				0.1

reflecting latitudinal differences in the study areas (South Texas is in a more tropical climate).

The seasonal trend in number of species was similar to the abundance trend. The average numbers of species collected each quarter were: July, 53.7; October-November, 46.7; January-February, 39.2; April, 42.7.

The percentage of polychaetes increased by ten percentage points during the fall and winter, while the amphipods underwent a simultaneous decline (Table 3). The bivalves were most abundant in January, 1977 (because of the large number of young bivalves at station 64) and the nemertean population was rather stable.

Meiobenthic Communities

The meiofaunal data are not as extensive as the macrofaunal data; only the first of three replicate samples collected at each station was analyzed. About 81,000 meiofaunal specimens (raw numbers) were collected. Because of the considerable time

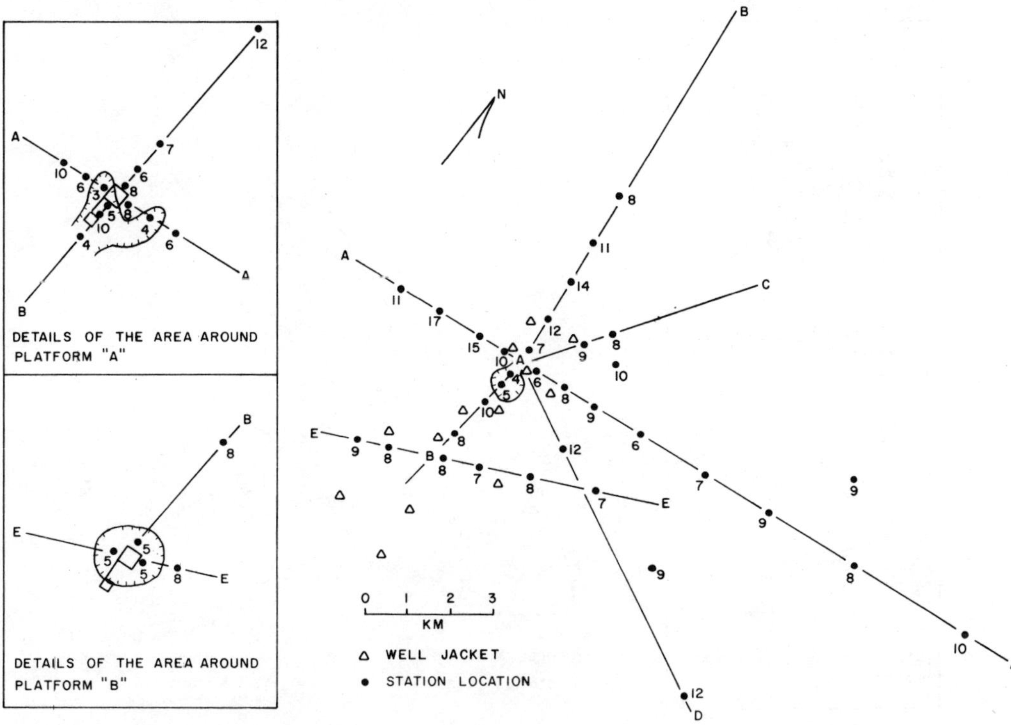

Figure 20. Areal distribution of meiobenthic abundances in the Buccaneer Field study area. Each data point represents the average number of individuals/cm^2 per quarter x 10. Suggested isodemes have been superimposed.

required to identify many meiofaunal organisms to species, many specimens were identified only to phylum level. A list of meiofaunal organisms is contained in Appendix C. Because many taxa were not identified to species, no diversity calculations were made.

Areal Distribution. The average quarterly meiobenthic density at each station was in excess of 110 individuals/cm^2 at most stations (Figure 20). Three areas contained lower numbers: the 15 m stations around platform 296-B, a crescent-shaped area west, south, and east of platform 288-A, and two stations south of platform 288-A on transect B. These data, while suggesting that the meiofaunal populations were depressed around the platforms, are erratic. On the eastern leg of transect A, the abundance at station 7 (adjacent to platform 288-A) was twice as large as at station 8 (50 m from the platform). On the southern leg of transect B, the densities at stations 2 and 91 (adjacent to and 100 m from platform 288-A, respectively) were low while the

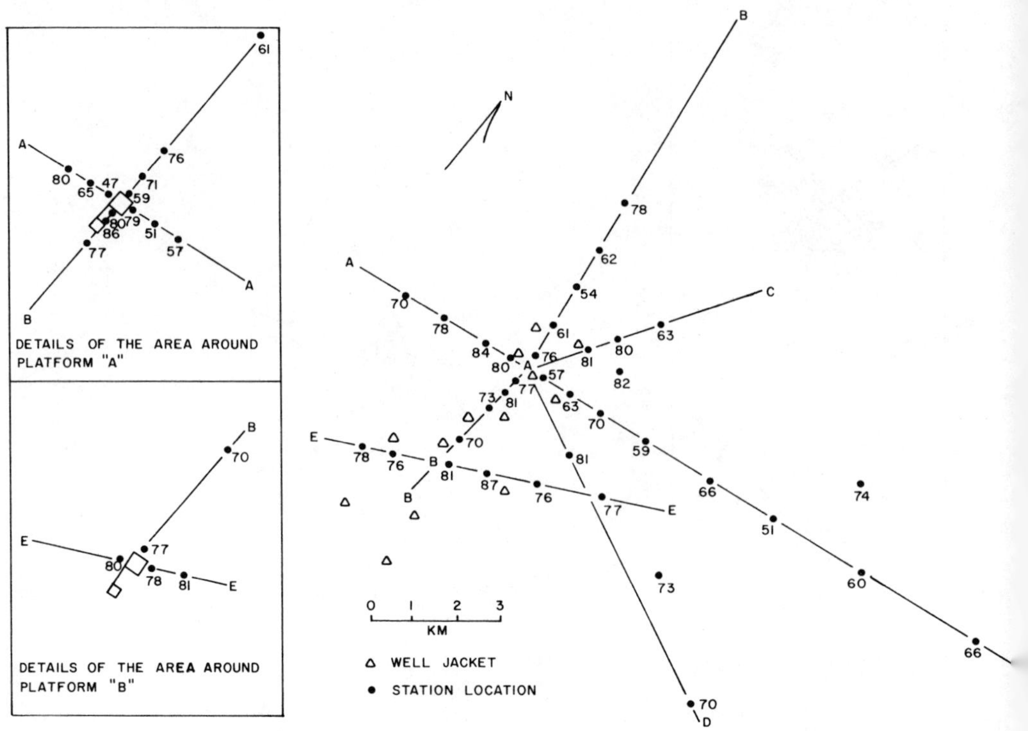

Figure 21. Percentage of nematodes in the meiobenthic assemblage, Buccaneer Field.

intermediate station (1) had large numbers. This variability makes it difficult to determine distributional patterns, and may be caused by the fact that only four meiofaunal data points were available per station, vs. 12 macrofaunal data points.

The dominant taxa among the meiofauna were the Nematoda, the Foraminifera, and the Harpacticoida. The majority of stations had 60-80% nematodes (Figure 21). The data suggest that slightly higher numbers occurred near the platforms on some transects, but the pattern is not consistent enough to draw a conclusion. Foraminiferans comprised 1-6% of the meiobenthos at all stations except at a group of six stations north and east of platform 288-A where the percentage was over 10% (Figure 22). Harpacticoid copepods were the only group that appeared to be depressed around a platform, and then only around platform 288-A (Figure 23). However, one of the largest percentages (27%) occurred at station 6 which was almost beneath the bleedwater outfall at platform 288-A.

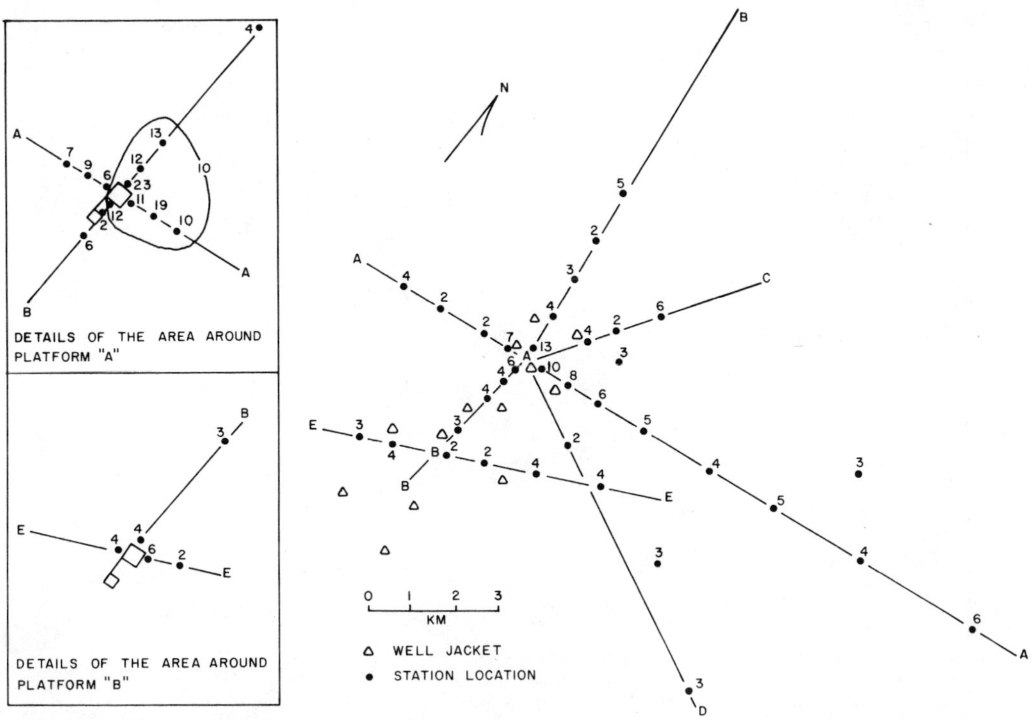

Figure 22. Percentage of foraminiferans in the meiobenthic assemblage, Buccaneer Field. Suggested isodemes have been superimposed.

Seasonal Trends. The average number of individuals/cm^2 per station collected in each quarter is compared with the sediment temperature trend in Figure 24. The smallest density occurred in October-November, followed by an increase through April. This trend was similar to the macrobenthic abundance trend (Figure 25), and does not indicate that predation by macrofauna controlled the size of meiobenthic populations. When the pilot study data were analyzed, the macro- and meiofaunal densities appeared to be inverse (Harper et al., 1976). However, during the present study, the coefficients of correlation between macro- and meiofaunal abundances were very low for each quarter and for all quarters combined (Table 4). Thus it appears that some factor other than predation by the macrofauna controlled the size of the meiofaunal populations during this study.

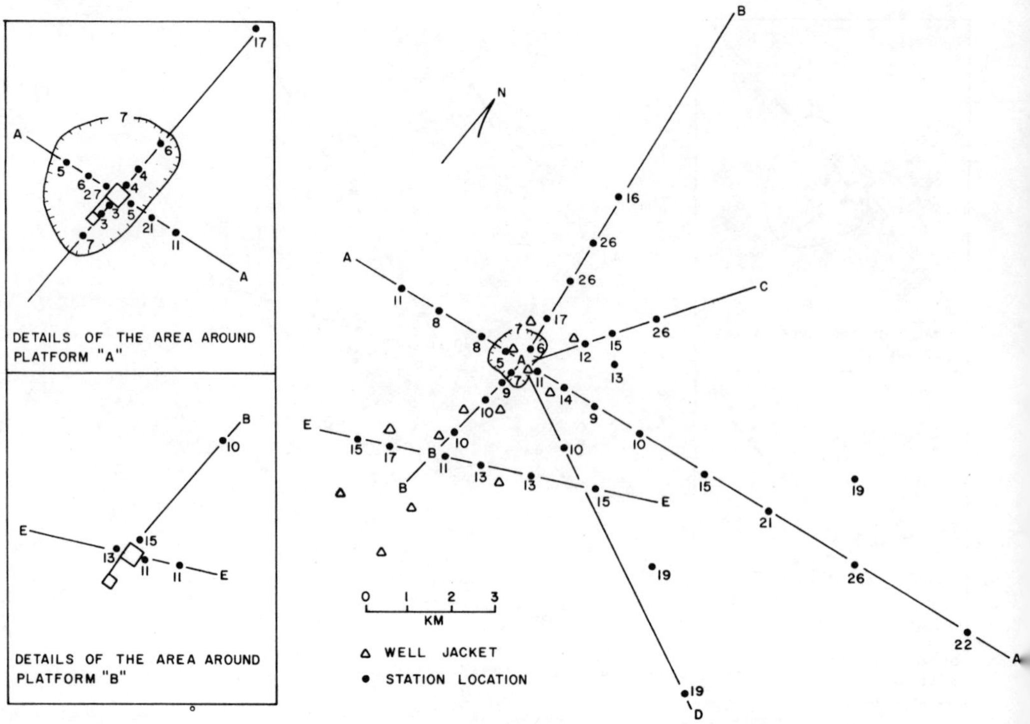

Figure 23. Percentage of harpacticoid copepods in the meiobenthic assemblage, Buccaneer Field. Suggested isodemes have been superimposed.

TABLE 4. COEFFICIENTS OF CORRELATION, MACROBENTHIC vs. MEIOBENTHIC DENSITIES FOR EACH QUARTER.

Quarter	Coefficient of correlation
July	-0.130
October-November	0.154
January-February	0.004
April	0.519

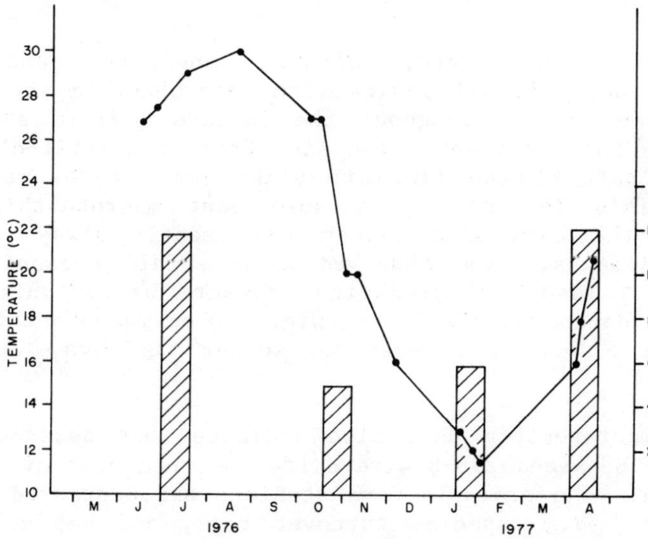

Figure 24. Comparison of the meiobenthic population and sediment temperature trends in the Buccaneer Field study area.

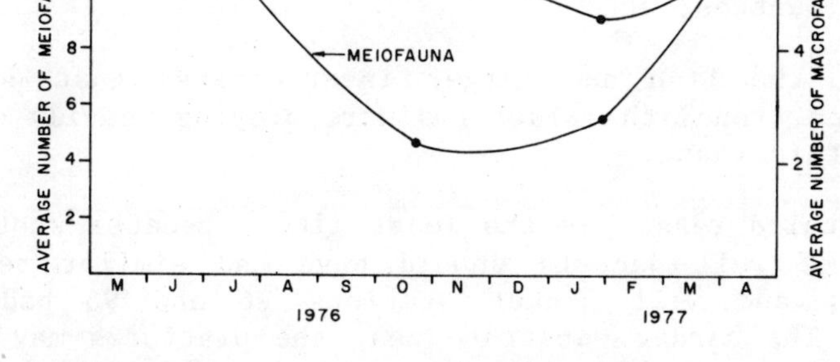

Figure 25. Comparison of the macrobenthic and meiobenthic population trends.

DISCUSSION

Distributional data, cluster analysis, and principal components analysis all indicate the macrobenthic abundances were relatively uniform throughout the Buccaneer Field study area in spite of the sediment gradient from sand-silt-clay in the northeast part of the field to silty sand in the center of the field. This is probably because most macrobenthic organisms inhabited the thin silt veneer that nearly always covered the firmer sediments. The observed macrobenthic seasonal trend may have been a result of predation pressure by nektonic organisms; nekton abundance was highest in winter when macrobenthic densities were lowest, and vice versa in summer (Gallaway et al., this Volume).

The macrobenthic data also indicate that assemblages around platforms 288-A and 296-B were different than most of those in the study area. The near-platform stations had depressed faunas, had a relatively high species turnover rate, and had a few species that were more frequently found there than in the remainder of the field. The meiobenthic assemblages may also have been depressed around the platforms, but the fewer data and large densities close to platform 288-A followed by smaller numbers away from the platform make the data difficult to interpret.

There are three main probable causes for the relatively sparse densities around the platforms:

(1) a toxic substance from the platforms was periodically killing organisms relatively quickly after they settled, causing a low abundance and a high turnover rate;

(2) the harder substrate in the vicinity of the platforms was unsuitable for habitation for many of the benthos; and

(3) the fish and larger invertebrates that congregated around the platforms were preying heavily on the benthic fauna.

The third reason is the least likely because stations in the vicinity of well jackets should have had similar reductions in abundance, and well jacket stations 26 and 96 had "enriched" numbers. The harder substrate near the platforms may account for part of the decrease in numbers, because the stations located in the vicinity of the exposed clay toward the eastern end of the study area also had reduced densities. Anderson et al. (1977) suggested that eddy turbulence created as currents flowed past structure legs scoured fine sediments which would remove the organisms with their habitat. An erosional environment was also indicated by Behrens' (1977) data. However, the zone of depression extended outward from the immediate vicinity of platform 288-A and onto softer bottoms where scour was less likely to occur. Also, scouring should have occurred to some degree near well jackets, yet large densities were found. Therefore, harder bottoms were probably not the definitive cause of the reduced numbers of individuals.

There is also a possibility that substances in the bleedwater being discharged from the platforms effected an abundance decrease. Concentration gradients of both relatively fresh and weathered crude oil were found in the sediments close to platform 288-A; aromatic hydrocarbons, including naphthalenes, were components of the oil (Middleditch and Basile, 1978; Middleditch, this Volume). It has been suggested that incorporation in copepod fecal pellets is a means of carrying oil from the water surface to the sediments (Nelson-Smith, 1972). There are, however, other possible causes of sediment contamination by oil, aside from accidental spills. Dilution and diffusion are probably rapid most of the time, but there may be hydrographic conditions, i.e., slack current and lack of stratification, in which the high salinity brine flows to the bottom as a density current. If this event occurred, a shallow brine layer would spread out on the bottom, exposing the sediments and organisms to dissolved hydrocarbons. Isothermal or nearly isothermal conditions are most likely to occur in late summer and fall as the water cools. Such an event would be periodic because there was no evidence for decreased numbers of species around the platforms in any quarter.

Also of interest are the areas of enriched macrofaunal abundances that occurred at stations 1000-2000 m north and east of platforms 288-A and 296-B. The occurrence of depressed faunas in the vicinity of the platforms and enriched areas at some distance from the platforms is similar to the results of Mackin's (1971) study of the effects of bleedwater discharge on the biota of Trinity Bay, Texas. In Mackin's study, the bottom close to the bleedwater outlet was saturated with oil and nearly azoic. At distances of about 15 to 100 m from the platform, species' populations were depressed compared with control stations, but no trace of oil could be detected in the sediments by sight or smell. The zone of inhibition was followed by a zone of enrichment, beyond which was an area of "control" populations intermediate in size between those of the enriched and depressed zones. The members of Mackin's research team (of which Harper was a member) speculated that, in the zone of enrichment, the level of toxic material in the bleedwater [identified as naphthalenes by Armstrong et al. (1979)] was reduced and that the remaining hydrocarbons were being degraded by bacterial populations, and that the bacterial populations were being fed upon by the benthic organisms. It is possible that the same events occurred at the Buccaneer Field also, with greater dispersion of the hydrocarbons due to stronger bottom currents. Relatively high concentrations of sulfur and sulfur-degrading bacteria have been detected in the sediments around the platforms (Middleditch, this Volume; Sizemore et al., this Volume). These bacteria may be an important food source to some benthic organisms, but in very large concentrations could produce toxic quantities of hydrogen sulfide.

It is also obvious that something associated with the structures caused an accumulation of pteropod shells. Predation by platform-associated fish is a highly probable cause. The pteropods *Clio* and *Cavolina* were found in the stomachs of several species of fish caught near structures, viz: jewfish (*Epinephelus itajara*), spadefish (*Chaetodipterus faber*), gray triggerfish (*Balistes capriscus*), blue runner (*Caranx crysos*), and red snapper (*Lutjanus campechanus*), according to Fotheringham (1977). Ingestion of the living pteropods and subsequent defecation of the shells by fish that are closely associated with the structures could account for many of the shells accumulated in the vicinity of the structures, both well jackets and platforms. This assumes that the shells are not crushed when eaten and pass intact through the fishes' digestive systems. However, many of the dead pteropod shells and all of the "live" pteropods collected at station 3 in July were *Creseis acicula*, which has a thin, fragile, acutely pointed conical shell. If eaten, the shells would tend to stick in the stomach or intestinal epithelium if not crushed or dissolved.

It is unlikely that the "live" pteropods were collected from the water column because the Ekman grabs had vent flaps to allow free passage through the water column. Shelled pteropods migrate vertically through the water column, rising to the surface at twilight and descending about midnight (Hyman, 1967). It is possible that swarms of pteropods being carried into the field by currents contacted the bottom during the descent phase of migration and became disoriented. However, it is also possible that, if the pteropods encountered a toxic substance near the surface that caused either mortality or inhibited nervous activity associated with swimming, the weight of the shells would cause the organisms to sink.

SUMMARY

Some aspect of the platforms effected a macrobenthic abundance decrease that was restricted to an area within 100 m of the platforms and possibly within 50 m. It is not clear whether the effect was due to periodic contact with toxic substances in the bleedwater, substrate disturbance due to currents eddying around the platform leg and removing the substrate, or some other cause. Toxic poisoning was certainly not a continual event because there was no evidence that diversity was depressed around the platforms. Whatever the cause, the area impacted is relatively small when compared with the total area of the field.

ACKNOWLEDGEMENTS

Numerous persons contributed to the completion of this study. Sampling cruises were made aboard the R/V TEXAS STAR, captained by Steven Mann, the M/V SCUBA DO, captained by Richard McGaffey, and the M/V MISS CHARLOTTE, captained by Richard Hering. Assisting in field and/or laboratory work were Mark Anderson, Fahmida Ansari, Russell Baier, Christine Becker, Shirley Benet, John Clary, Bruce Cunningham, Byron Giezentanner, Brandt Henningsen, Timothy Goings, Saleem Lalani, Gerald Livingston, John Machol, Dennis McLaughlin, David Overton, and Nick Vratis. John Clary prepared the figures. Taxonomic assistance was provided by Larry McKinney (Amphipoda), Linda Pequegnat (Caridea), Harold Harry (Mollusca), Richard Defenbaugh (Hydroida), and Geoffrey Matthews (Copepoda). This work is a result of research sponsored by the Environmental Protection Agency and Department of Commerce, National Oceanic and Atmospheric Administration, National Marine Fisheries Service, Southeast Fisheries Center, Galveston Laboratory under Contract Nos. 03-6-042-35110 and 03-6-042-35117.

REFERENCES

Anderson, J. B., Schwarzer, R. R., and Clark, H. C., 1977, "Sedimentology, geochemistry, and trace metal analysis," in: Jackson, W. B. (ed.), "Environmental Assessment of an Active Oil Field in the Northwestern Gulf of Mexico, 1976-1977." NOAA, pp. 89-174. Available from: NTIS, Springfield, Virginia, Accession No. PB283890.

Armstrong, H. W., Fucik, K., Anderson, J. W., and Neff, J. M., 1979, "Effects of oilfield brine effluent on sediments and benthic organisms in Trinity Bay, Texas," Mar. Environ. Res., 2:55.

Behrens, E. W., 1977, "Total organic carbon and carbon isotopes of sediments," in: Jackson, W. B. (ed.), "Environmental Assessment of an Active Oil Field in the Northwestern Gulf of Mexico, 1976-1977." Available from: NTIS, Springfield, Virginia, Accession No. PB283890.

Clifford, H. T., and Stephenson, W., 1975, "An Introduction to Numerical Classification." Academic Press, New York.

Farrell, D., 1974, "Benthic Ecology of Timbalier Bay, Louisiana, and Adjacent Offshore Areas in Relation to Oil Production." Ph.D. Thesis, Florida State University.

Fish, A. G., Massey, L. L., Inabinet, J. R., and Lewis, P. L., 1974, "A study of the effects of environmental factors upon the distribution of the foraminiferal, nematode and polychaetous annelid meiofauna of Timbalier Bay, Louisiana." GURC Offshore Ecology Investigation, unpublished manuscript.

Fisher, W. L., McGowen, J. H., Brown, L. F., Jr., and Groat, C. G., 1972, "Environmental Atlas of the Texas Coastal Zone - Galveston-Houston Area." Bureau of Economic Geology, University of Texas, Austin, Texas.

Fotheringham, N., 1977, "Effects of offshore oil field structures on their biotic environment: benthos and plankton," in: Jackson, W. B. (ed.), "Environmental Assessment of an Active Oil Field in the Northwestern Gulf of Mexico, 1976-1977." Available from: NTIS, Springfield, Virginia, Accession No. PB283890.

Harper, D. E., Jr., and McKinney, L. D., 1979, "Benthos," in: "Evaluation of Brine Disposal from the Bryan Mound Site of the Strategic Petroleum Reserve Program." Report to the Department of Energy on Texas A&M Research Foundation Project 4050.

Harper, D. E., Jr., Scrudato, R. J., and Giam, C. S., 1976, "Pilot Study of the Buccaneer Oil Field (Benthos and Sediments). A Preliminary Environmental Assessment of the Buccaneer Oil/Gas Field," Report to NMFS, Galveston, Texas, from Texas A&M University, College Station, Texas, 63 pp. plus appendices.

Holland, J. S., 1976, "Benthos project," in: "Environmental Studies, South Texas Outer Continental Shelf, 1975, Biology and Chemistry." Report to the Bureau of Land Management from the University of Texas, Texas A&M University, and Rice University on contract 08550-CT5-17.

Hyman, L., 1967, "The Invertebrates. Vol VI. Mollusca I." McGraw-Hill, New York.

Kritzler, H., 1974, "Oil production and polychaetous annelids in a Louisiana estuary." GURC Offshore Ecology Investigation, unpublished manuscript.

Mackin, J. G., 1971, "A Study of the Effects of Oil Field Brine Effluents on Biotic Communities in Texas Estuaries." Texas A&M Research Foundation Project 735 Report.

Mackin, J. G., and Hopkins, S. H., 1961, "Studies on oyster mortality in relation to natural environments and to oil fields in Louisiana," Pub. Inst. Mar. Sci. Univ. Texas, 7:11.

Martin, J. A., 1977, "Hydrography of the Buccaneer oil field and adjacent waters," in: Jackson, W. B. (ed.), "Environmental Assessment of an Active Oil Field in the Northwestern Gulf of Mexico, 1976-1977." NOAA, pp. 575-632. Available from: NTIS, Springfield, Virginia, Accession No. PB283890.

Middleditch, B. S., and Basile, B., 1978, "Alkanes in surficial sediments from the region of the Buccaneer oil field," J. Chromatog., 158:449.

Nelson-Smith, A. A., 1972, "Oil Pollution and Marine Ecology." Elek Science, London.

SEADOCK, 1975, "Environmental Report, Texas Offshore Crude Oil Unloading Facility." Texas A&M Research Foundation final report to SEADOCK, Inc.

Smith, R. W., 1976, "Numerical Analysis of Ecological Survey Data." Ph.D. Thesis, University of Southern California.

Temple, R. F., Harrington, D. L., and Martin, J. A., 1977, "Monthly Temperature and Salinity Measurements of Continental Shelf Waters of the Northwestern Gulf of Mexico, 1963-1965." NOAA Technical Report SSRF-207.

Waller, R. S., 1974, "Effects of Platforms on Biota (Invertebrates)." GURC Offshore Ecology Investigation, unpublished manuscript.

APPENDIX A. STATION LOCATIONS IN TERMS OF DIRECTION AND DISTANCE (IN METERS) FROM EITHER PLATFORM 288-A OR PLATFORM 296-B.

Transect A (platform 288-A)

Station	Distance	Direction
14	4000	263°
10	3000	263°
94	2000	263°
95	1000	263°
4	100	263°
5	50	263°
6	<15	263°
7	<15	83°
8	50	83°
9	100	83°
96	1000	83°
11	2000	83°
97	3000	83°
12	4500	83°
13	6000	83°
70	8000	83°
98	11000	83°
71	7250	71°

Transect B (platform 296-B)

Station	Distance	Direction
109	8000	353°
108	4000	353°
21	3000	353°
107	2000	353°
106	1000	353°
18	100	353°
17	50	353°
3	<15	353°
2	<15	187°
1	50	187°
91	100	187°
90	500	187°
89	1000	187°
88	2000	187°
45	2500	187°

Transect C (platform 288-A)

Station	Distance	Direction
7	15	34°
26	1250	34°
105	2000	34°
64	4000	34°

Transect D (platform 288-A)

Station	Distance	Direction
7	15	109°
93	2000	109°
40	3000	109°
62	5000	109°
92	7500	109°

Transect E (platform 296-B)

Station	Distance	Direction
32	1500	244°
86	900	244°
34	15	244°
35	15	64°
36	50	64°
87	1000	64°
39	2000	64°
40	3000	64°

APPENDIX B. MACROFAUNAL ORGANISMS COLLECTED DURING THE BUCCANEER FIELD STUDY.

Phylum Protozoa
  Foraminifera A

Phylum Cnidaria
  Class Hydrozoa
    Clytia coronata
    Clytia cylindrica
    Clytia elongata
    Lovenella grandis
    Obelia dichotoma

  Class Anthozoa
    Anemone (unidentified)
    Anemone, brown spotted rim
    Anemone, holothuroid-like
    Anemone, sand encrusted
    Anemone, tubular
    Astrangia astraeiformis
    Renilla mulleri

Phylum Platyhelminthes
  Class Turbellaria
    Polycladida*
    Turbellaria, brown stripes*
    Turbellaria, crossed eyes
    Turbellaria, black spots
    Turbellaria, dark margin
    Stylochus ellipticus

Phylum Nemertinea
    Cerebratulus lacteus
    Cerebratulus luridus
    Nemertean, broad brown bands*
    Nemertean, brown bands*
    Nemertean, gray bands*
    Nemertean, red banded
    Nemertean, red/yellow banded*
    Nemertean, rust colored*
    Nemertean, white
    Nemertean, yellow/brown bands
    Nemertean, yellow/purple bands
    Nemertean, yellow banded

Phylum Nematoda
    Nematoda (several spp.)

Phylum Bryozoa
    Bryozoan sp. A
    Bugula neritina
    Cupuladria canariensis
    Membranipora crustulenta

Phylum Phoronida
    Phoronis architecta

Phylum Brachiopoda
    Glottidea cf. *pyramidata*

Phylum Mollusca
    Acteocina canaliculata
    Anachis avera
    Anachis obesa
    Bubble shell*
    Bulla-like opisthobranch
    Caecum glabrum
    *Caecum imbricatum**
    Cantharus cancellarius
    Crepidula fornicata
    Crepidula sp. (young *C. fornicta*)?
    Cyclostremella humilis
    *Epitonium angulatum**
    *Epitonium novangeliae**
    Eulima bilineata
    *Kurtziella atrostyla**
    Mangelia sp.
    Melanella arcuata
    Melanellidae*
    *Nannodiella vespuciana**
    Nassarius acutus
    Natica pusilla
    Nudibranchia, black stripes*
    Nudibranchia A
    Nudibranchia B
    Olivella dealbata

(continued)

APPENDIX B (continued)

---

*Pyramidella crenulata**
Pyramidellidae*
Rictaxis punctostriatus
*Seila adamsi**
Sinum maculatum
Teinostoma parvicallum
Terebra protexta
Terebra sp.*
*Thais haemastoma**
Turbonilla cf. *portoricana*
Turbonilla interrupta
Turbonilla, brown spiral band
Vitrinella helicoidea
*Vitrinella texana**
Volvulella persimilis
Volvulella texasiana

Class Bivalvia
Abra aequalis
Anadara floridana
Anadara transversa
*Atrina seminuda**
Atrina sp.
Bivalvia, dirty ventrum
Bivalvia, flat, ridged
Bivalvia, larger than shell
*Calliocardia texasiana**
Cardiomya costatella?
Cardiomya ornatissima?*
Chama congregata
Chione chenchi
Chione grus
Corbula barrattiana
*Corbula contract**
Corbula krebsiana
Corbula operculata
Corbula swiftiana
Crassinella lunulata
Cyclinella tenuis
Diplodonta sp.
Diplodonta semiaspera
Dosinia elegans
Gemma purpurea
Jouannetia quillingi?
Laevicardium mortoni
Lima pellucida
Linga amiantus
Lioberis castaneus
Lyonsia hyalina floridana
Macoma sp.*
Macoma tageliformis
Macoma tenta
*Mactra fragilis**
*Mercenaria campechensis**
*Mulinia lateralis**
Mysella planulata
Noetia ponderosa
Nucula proxima
Nuculana acuta
Nuculana concentrica
Parvilucina multilineata
Pholadidae
Semele bellastriata
Semele proficua
Semele purpuracens
Strigilla mirabilis
Tellina aequistriata
Tellina versicolor
Trachycardium muricatum
Veneridae

Class Scaphopoda
*Dentalium texasianum**

Phylum Annelida

Class Polychaeta
Aglaophamus verrilli
Amaeana trilobata
Ampharete acutifrons
Ampharete americana
Amphictis scaphobranchiata
Anaitides erythrophyllus
*Anaitides mucosa**
Ancistrosyllis hartmanae
Ancistrosyllis jonesi
Ancistrosyllis papillosa
Aricidea fragilis
Aricidea cf. *lopezi*
Armandia maculata
Asychis elongata
Autolytus sp.
Capitella capitata
Ceratocephale sp.

(continued)

APPENDIX B (continued)

Ceratonereis irritabilis
Chaetopterus variopedatus
*Chaetozone setosa**
Chloeia viridis
*Chone duneri**
Cirratulid, ventral groove
Cirratulus hedgpethi
Clymenella torquata calida
Clymenella zonalis
Cossura delta
Diopatra cuprea
Drilonereis longa
Drilonereis magna
Eteone heteropoda
Eteone lactea
Eumida sanguinea
Eunice antennata
Eunoe sp.
Glycera americana
Goniada teres
Grubeulepis sp.
Grubeulepis fimbriata
Gyptis vittata
*Harmothoe aculeata**
Hydroides dianthus
Hydroides protulicula
Jasmineria bilobata
*Laonice cirrata**
Lepidasthenia sp.
Lepidonotus sublevis
Lumbrineris ernesti?
Lumbrineris cf. *verrilli*
Litocorsa stemma
Magelona cincta
Magelona filiformis
Magelona pettiboneae
Magelona phyllisae
Magelona riojai
Malacoceros vanderhorsti
Marphysa, cmpd. falc., spin.*
Marphysa regalis
Marphysa sanguinea
Mediomastus californiensis
Megalomma bioculatum
Melinna maculata
Mesochaetopterus taylori
Microspio pigmentata
Myriowenia californiensis
Nephtys bucera
Nephtys incisa
Nephtys, subdermal eyes
Nephtys picta
Nereiphylla fragilis
Nereis micromma
Nereis succinea
Ninoe nigripes
Notomastus hemipodus
Notomastus latericeus
Onuphis eremita oculata
Onuphis nebulosa
*Ophelina cylindrica**
Ophiodromus obscura
Owenia fusiformis
Paleonotus heteroseta
*Parahesione luteola**
Paranaitis speciosa
Paraprionospio pinnata
Pectinaria gouldii
Pherusia inflata
*Phyllodoce castenea**
Pilargis pacifica
Pista palmata
Poecilochaetus johnsoni
Polydora ligni
Polydora socialis
Polydora sp.
Polygordius appendiculatus
Polynoidae, bristly parapodia
Polyodontes lupina
Prionospio cirrifera
Prionospio cirrobranchiata
Prionospio dayi
Prionospio heterobranchia
Pseudeurythoe ambigua
*Sabellaria heteroseta**
Sabellaria vulgaris vulgaris
Schistomeringos rudolphi
Scoloplos foliosus
Scoloplos fragilis
Scoloplos rubra

(continued)

APPENDIX B (continued)

Sigambra tentaculata
Sigambra wassi
Spiochaetopterus oculatus
Spionidae, 4 branchiae*
Spiophanes bombyx
*Spio setosa**
Sthenelais limicola
Streblospio benedicti
Syllidae
*Syllis gracilis**
Syllis hyalina
Terebellides stroemi
Tharyx marioni
Tharyx setigera
Thelepis setosus

Phylum Sipunculida
Aspidosiphon cf. *speciosus*
Phasocolion strombi

Phylum Echiuroidea
Thalassema sp.?*

Phylum Crustacea
Class Ostracoda
Ostracod B
Ostracod C
Ostracod D
Ostracod E
Ostracod F
Ostracod G
Ostracod H
Ostracod I
Ostracod J
Ostracod K
Ostracod L
Ostracod M*
Ostracod N
Ostracod O
Class Copepoda
Canuella sp.*
Class Malacostraca
Subclass Phyllocarida
Order Nebaliacea
Nebalia sp.
Subclass Eumalacostraca
Order Stomatopoda
Squilla empusa
Order Amphipoda
Ampelisca abdita
Ampelisca agassizi
Ampelisca verrilli
Amphilocus neopolitanus
Amphipoda, black coxa*
*Argissa hamatipes**
Batea cartharinensis
Caprella equilibra
Corophium acherusicum
*Jassa falcata**
Lembos brunneomaculatus mackinneyi
Leucothoidae*
Listriella barnardi
Listriella sp.
Lysianassidae*
Lysina sp. A*
Monoculodes edwardsi
Monoculodes, brown spots*
Paracaprella tenuis
Parametopella texensis
Paraphoxis sp.*
Photis macromanus
Photis melanicus
Platyischnopus sp.
*Podocerus brasiliensis**
Synchhelidium americanum
*Tiron tropakus**
Unciola irrorata
Order Isopoda
Anthuridae A
Anthuridae, eyes + long antennae
Edotea sp.?*
Erichthonius brasiliensis
Order Tanaidacea
Apseudes sp.
Tanaid, rounded rostrum
Order Cumacea
Cumacea A
Cumacea B
Cumacea C
Cumacea D
Cumacea E

(continued)

APPENDIX B (continued)

Cumacea F
Cumacea, frontal spines
Oxyurostylis salinoi
Order Mysidacea
Bowmaniella brasiliensis
Mysidopsis bigelowi
Promysis atlantica
Order Decapoda
Albunea paretii
Alpheus floridanus
Automate evermanni
Callianassa sp.
Callinectes sapidus
Callinectes similis
Caridea A
*Chasmocarcinus mississippiensis**
Euceramus praelongus
Euryplax nitida
Goneplacidae*
Goneplax sp.
Hepatus pudibundus
Hexapanopeus paulensis
Hippidea*
Isocheles wurdemanni
Latreutes parvulus
Lepidopa benedicti
Leptochela serratorbita
Majidae*
Neopanope texana
Ogyrides limicola
*Ovalipes guadulpensis**
Paguristes hummi
Pagurus turgidus
Parthinopidae, spiny
*Pelia mutica**
*Penaeus setiferus**
Persephona aquilonaris
Persephona crinata
Pinnixa sayana
*Pinnixa lunzi**
*Pinnotheres ostreum**
Polyonyx gibbesi
*Porcellana sayana**
Processa hemphilli
Raninoides louisianensis
Sicyonia dorsalis
Solenocera sp.
Spiocarcinus lobatus
Trachypenaeus similis
Upogebia affinis

Phylum Cheliciformes
Class Pycnogonida
Pycnogonids

Phylum Echinodermata
Class Ophiuroidea
Hemipholus elongata
Micropholis atra
Ophiolepsis paucispina
Ophiophragmus moorei
Ophiuroid B
Class Echinoidea
Echinoid young
Echinoid, lanceolate spines

Phylum Hemichordata
Balanoglossus sp.

Phylum Chordata
Bascanichthys teres
*Bregmaceros atlanticus**
*Microdesmus lanceolatus**

*Single occurrence.

APPENDIX C. LIST OF MEIOFAUNAL ORGANISMS COLLECTED DURING THE BUCCANEER FIELD STUDY.

Phylum Protozoa
- Class Sarcodina
  - Order Foraminiferida
    - *Ammoscalaria* sp.
    - *Angulogerina bella*
    - *Ammonia beccarii*
    - *Bolivina barbata*
    - *Bolivina fragilis*
    - *Caneris* spp.
    - *Cibicides* sp.
    - *Hanzawaia* spp.
    - *Planulina* sp.
    - *Proteonina difflugiformis*
    - *Elphidium* spp.
    - *Bigeneria irregularis*
    - *Buliminella* sp.
    - *Virgulina* sp.
    - *Reusella atlantica*
    - *Reophax* sp.
    - *Quinqueloculina* spp.
    - *Nonionella* sp.
    - *Proteonina comprina*
    - *Spiroloculina* sp.
    - Unidentified foram.

Phylum Cnidaria
- Class Hydrozoa
  - Hydroid

Phylum Nemertinea
- Nemertean spp.

Phylum Nematoda
- Nematode A
- Nematode B
- Nematode C
- Nematode D
- Nematode E
- *Desmoscolex* sp.

Phylum Kinorhyncha
- *Echinoderes remanei*
- *Semioderes* sp.
- *Pycnophyes* sp.
- *Trachydemus langi*

Phylum Gastrotricha
- Gastrotrich spp.

Phylum Phononida
- *Phoronis architecta*

Phylum Mollusca
- Class Gastropoda
  - *Caecum glabrum*
  - *Vitrinella* sp.
- Class Bivalvia
  - *Abra aequalis*
  - *Tellina versicolor*
  - Disc-shaped bivalve
  - Bivalve, unid.

Phylum Annelida
- Class Polychaeta
  - Cirratulidae
  - *Jasminiera bilobata*
  - *Lumbrineris* cf. *tenuis*
  - *Magelona phylissae*
  - *Mediomastus californiensis*
  - *Pisione remota*
  - Polychaete D
  - Polychaete E
  - *Prionospio* spp.
  - *Sigambra* sp.
  - Spionidae
  - Syllidae A
  - Syllidae B
  - Syllidae C
  - *Nerillidae* sp.
  - *Polygordius appendiculatus*

Phylum Sipunculida
- *Aspidosiphon speciosus*
- *Golfingia* sp.
- Sipunculida C

Phylum Crustacea
- Class Copepoda
  - *Scotolana canadensis*
  - *Ectinosoma* sp.
  - Ectinosomidae sp.

(continued)

Appendix C (continued)

Harpacticoid sp.
Copepodite stage
Harpacticoid nauplius
Cyclopoid copepod
Class Ostracoda
Ostracod A
Ostracod B
Ostracod C
Ostracod D
Class Malacostraca
Subclass Eumalacostraca
Order Tanaidacea
Tanaid, unid.
Order Amphipoda
Ampelisca abdita
Ampelisca agassizi
Ampelisca verrilli
Amphipod B
Order Isopoda
Anthuridae A
Isopod B
Order Cumacea
Cumacea B
Oxyurostylis salinoi

Phylum Cheliceriformes
Class Arachnida
Order Acarina
Acarina A
Acarina B
Acarina C
Acarina D
Acarina E
Acarina F
Class Pycnogonida
Pycnogonid

Phylum Hemichordata
Hemichordate sp.

OBSERVATIONS ON THE EFFECTS OF OIL FIELD STRUCTURES ON THEIR BIOTIC ENVIRONMENT: PLATFORM FOULING COMMUNITY

Nick Fotheringham[a]

Department of Biology
University of Houston
Houston, Texas 77004

INTRODUCTION

Fouling organisms include nearly 2000 species of sessile algae and invertebrates (WHOI, 1952) that have great economic importance to owners and operators of ships and fixed offshore structures. They increase the fuel consumption of ships and affect the design of offshore oil platforms. The fouling community has been described in many parts of the world and on many substrates (McDougall, 1943; Scheer, 1945; Fuller, 1946; WHOI, 1952; Daniel, 1954; Raja, 1959, 1963; Skerman, 1959; Kawahara, 1962; Nair, 1962; Stubbings and Houghton, 1964; Haderlie, 1969; Pearse and Chess, 1971; Lee and Trott, 1973), including Texas coast oil field structures (Gunter and Geyer, 1955). However, the variation in this community is large, even over relatively short distances (Weiss, 1948), and alternative stable communities may be established at the same site on structures added at different times of the year (Sutherland, 1974).

The fouling community, its planktonic larvae, and its predators were investigated at the Buccaneer Field from May, 1976 to April, 1977 in an effort to determine the structure and the factors regulating the structure of the fouling community. Offshore structures increase the availability of a limiting

[a]Present address: Dames & Moore, 4321 Directors Row, Houston, Texas 77092.

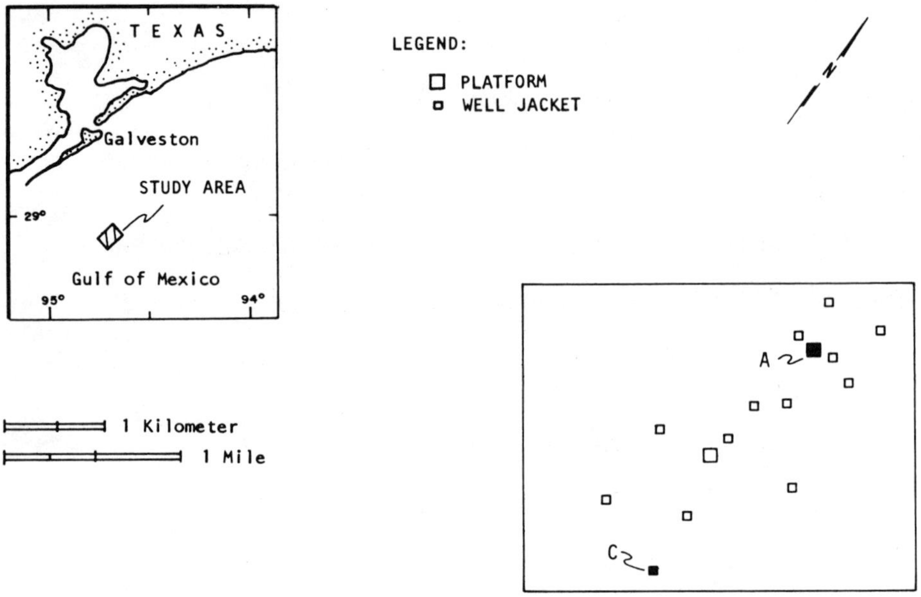

Figure 1. Location of Buccaneer Field sampling stations. A = production platform 288-A, C = well jacket 296-12.

resource by providing a substrate for numerous sessile species and associated motile species. This substrate may represent a refuge for undesirable fouling organisms and so may enhance the fouling of shipping in the region. The structures may also provide an invasion route ("stepping stones," in the sense of MacArthur and Wilson, 1967) for undesirable species, such as may enter the Gulf of Mexico if a sea-level canal were constructed across Central America.

MATERIALS AND METHODS

The Buccaneer Field is located approximately 50 km (30 miles) south of the lighthouse on Galveston's south jetty. Shell Oil Company has constructed 16 structures in the field: 2 production platforms and 14 well jackets. Two structures, platform 288-A and well jacket 296-12 (Figure 1), were selected for study. Platform 288-A was selected for its proximity to several of the well jackets, which are potential sources of colonists and predators. For contrast, well jacket 296-12 was selected for its relative isolation. Both structures are located in approximately 21.5 m (70 ft) of water overlying a silty sand substrate.

Most of the field work was conducted with SCUBA, using 2- and 3-man dive teams to collect samples and carry out field experiments on the structures. The field experiments were designed to examine the dynamic properties of the fouling community. The macroscopic component of the community was scraped away from selected portions of the structures. This component was removed with a rock hammer or chisel and sledge from zones 0.5 m (1.6 ft) wide around each of three legs on each structure at the 3 m (10 ft), 8 m (25 ft), and 15 m (45 ft) depths.

The legs selected were from the eastern sides of the structures. A portion of each zone approximately 20 x 30 cm was cleaned thoroughly with a wire brush, and a stainless steel mesh cage measuring 16 x 12 x 2 cm (similar to that of Connell, 1970) was placed over the upper half of this area and fastened with stainless steel or plastic bands around the leg. A wire cloth with 3.15 meshes to the centimeter was used for the cages.

Each cage enclosed two 8 x 12 cm quadrats and protected them from fish grazing. Two matching quadrats below the cage were thus exposed to fish predation. Each quadrat was sampled photogrammetrically using a Nikonos II underwater camera and strobe. Photographs of good quality were obtained even when the water was very turbid (visibility of less than 2 m). Photographs of identical format were taken of undisturbed quadrats for comparison.

Large qualitative (10-20 l) samples were collected from the littoral zone, 8 m, and 15 m depths on each structure to provide a taxonomic base for the photogrammetric surveys. These samples were supplemented by qualitative samples from 5, 11, 18, and 21 m depths and by miscellaneous collections of species not previously encountered.

Color transparencies of the 8 x 12 cm quadrats were projected onto a wall from a distance calibrated to produce an image 16 x 24 cm in size. Identifiable organisms and colonies or clusters (e.g., hydroids) were traced onto a piece of white paper, and the area of each object was measured with a compensating polar planimeter to the nearest 0.4 cm^2 (i.e., to the nearest 0.1 cm^2 on the original scale). Areas occupied by each species were summed and converted to a proportion of the total area.

Samples of fouling organisms were sorted and the specimens identified to the lowest practical taxon. Presence of ovigerous females or juveniles was noted. Spicules were isolated, mounted, and measured as an aid in identifying sponges.

TABLE 1. LIST OF ALGAE AND BENTHIC INVERTEBRATES COLLECTED AT THE BUCCANEER FIELD.

Chlorophyceae
  Cladophoraceae
    Cladophora delicatula
    Cladophora vagabunda
    Cladophora catenifera
    Chaetomorpha media
  Ulvaceae
    Enteromorpha lingulata
    Enteromorpha flexuosa
  Derbesiaceae
    Derbesia vaucheriaeformis
  Bryopsidaceae
    Bryopsis plumosa

Phaeophyceae
  Ectocarpaceae
    Giffordia mitchelliae

Rhodophyceae
  Gelidiaceae
    Gelidium pusillum
  Rhodomelaceae
    Polysiphonia denudata
    Polysiphonia subtilissima
    Centrocerus clavulatum
    Bangia fuscopurpurea
  Ceramiaceae
    Callithamnion byssoides
    Ceramium byssoideum

Porifera
  Calcarea
    Homocoelidae
      Clathrina sp.
      Leucosolenia botryoides
    Heterocoelidae
      Scypha sp.
      Leuconia sp.
  Demospongiaria
    Spongiidae
    Haliclonidae
      Haliclona permollis
      Haliclona viridis
      Haliclona loosanoffi
      Haliclona sp.
    Renieridae
    Halichondriidae
      Halichondria bowerbanki
    Hymeniacidonidae
      Hymeniacidon heliophila
    Suberitidae
      Prosuberites microsclerus
    Clionidae
      Cliona celata
      Cliona vastifica

Hydrozoa
  Clavidae
    Turritopsis nutricula
  Campanularidae
    Obelia dichtoma
    Obelia hyalina
    Clytia cylindrica
  Hydractinidae
    Stylactis sp.
  Corynidae
    Syncoryne eximia
  Eudendridae
    Eudendrium eximium
  Plumularidiae
    Plumularia diaphana
  Tubularidae
    Tubularia crocea

Anthozoa
  Ceriantharia
  Actiniaria
    Actiniidae
    Aiptasiidae
    Aiptasiomorphidae
  Gorgonacea
    Leptogorgia virgulata
  Madreporaria
    Astrangia sp.

Rhynchocoela

Turbellaria

(continued)

TABLE 1 (continued)

Polychaeta
- Nereidae
  - *Neanthes succinea*
- Polynoidae
  - *Lepidonotus sublevis*
  - *Lepidonotus variabilis*
- Syllidae
  - *Trypanosyllis gemmipara*
  - *Haplosyllis spongicola*
- Eunicidae
  - *Eunice* sp.
- Lumbrineridae
  - *Lumbrineris* sp.
- Serpulidae
  - *Eupomatus dianthus*
  - *Eupomatus protulicola*
  - *Spirobranchus giganteus*
- Terebellidae
  - *Thelepus setosus*
- Sabellidae
  - *Sabella microphthalma*

Gastropoda
- Dotoidae
  - *Doto* sp.
- Calyptraeidae
  - *Crepidula plana*
  - *Crepidula fornicata*
- Columbellidae
  - *Mitrella lunata*
- Olividae
  - *Oliva sayana*
- Fasciolariidae
  - *Pleuroploca gigantea*
- Buccinidae
  - *Cantharus cancellarius*
- Muricidae
  - *Thais haemastoma*
  - *Murex fulvescens*
- Cerithiopsidae
  - *Seila adamsi*
- Cypraeidae
  - *Cypraea spurca acicularis*
- Ovulidae
  - *Neosimnia acicularis*
- Epitoniidae
  - *Epitonium albidum*
- Turridae
  - *Mangelia* cf. *M. oxytata*

Pelecypoda
- Mytilidae
  - *Lithophaga bisulcata*
  - *Lithophaga aristata*
  - *Brachiodontes exustus*
  - *Musculus lateralis*
- Arcidae
  - *Anadara transversa*
  - *Arca imbricata*
- Chamidae
  - *Chama macerophylla*
- Pteridae
  - *Pteria colymbus*
- Ostreidae
  - *Ostrea equestris*
  - *Crassostrea virginica*
- Pectinidae
  - *Aequipecten gibbus*
- Isognomonidae
  - *Isognomon bicolor*
- Veneridae
  - *Chione grus*
- Limidae
  - *Lima pellucida*
- Pinnidae
  - *Atrina seminuda*
- Pholadidae
  - *Diplothyra smythi*

Cephalopoda
- Octopodidae
  - *Octopus* sp.

Pycnogonida
- Tanystylidae
  - *Tanystylum orbiculare*

Amphipoda
- Ischyroceridae
  - *Jassa falcata*
- Leucothoidae
  - *Leucothoe spinicarpa*

(continued)

TABLE 1 (continued)

Corophiidae	Portunidae
Erichthonius brasiliensis	*Portunus gibbesii*
Corophium sp.	*Callinectes similis*
Caprellidae	*Cronius ruber*
Caprella equilibra	
Gammaridae	Xanthidae
Elasmopus sp.	*Menippe mercenaria*
Stenothoidae	*Hexapanopeus paulensis*
Stenothoe gallensis	*Pilumnus dasypodus*
	Pilumnus pannosus
Isopoda	*Pseudomedaeus agassizi*
Tanaidae	*Panopeus turgidus*
Tanais sp.	
Sphaeromidae	Bryozoa
Dynamene perforata	Bugulidae
	Bugula neritina
Cirripedia	*Bugula rylandi*
Chthamalidae	Savignyellidae
Chthamalus fragilis	*Savignyella lafonti*
Balanidae	Schizoporellidae
Balanus tintinnabulum	*Schizoporella errata*
Balanus improvisus	Scrupariidae
Balanus (*amphitrite* complex)	*Eucratea loricata*
Chelonibia patula	Smittinidae
	Parasmittina trispinosa
Decapoda	Membraniporidae
Alpheidae	*Membranipora tenuis*
Synalpheus fritzmuelleri	*Membranipora savartii*
Diogenidae	Aeteidae
Petrochirus diogenes	*Aetea anguina*
Dardanus fucosus	
Paguristes puncticeps	Ophiuroidea
Paguridae	Ophiothricidae
Pagurus impressus	*Ophiothrix angulata*
Pylopagurus holthuisi	Ophiactidae
Majidae	*Ophiactis savignyi*
Stenorhynchus seticornis	
Porcellanidae	Echinoidea
Porcellana sayana	Arbaciidae
Dromiidae	*Arbacia punctulata*
Dromidia antillensis	
Grapsidae	Ascidiacea
Pachygrapsus transversus	Enterogona

RESULTS AND INTERPRETATION

A list of the algae, sessile invertebrates, and associated motile invertebrates identified from the Buccaneer Field samples is presented in Table 1. Sixteen algal samples and 101 invertebrates have been identified from the structures. Among taxa common to both habitats, the species composition of the biota on the structures differed markedly from that associated with adjacent soft bottom habitats, as described by Harper (1977).

The depths and structures at which each of the identified species were found are presented with their relative abundances in Table 2. Since the largest samples were collected in the littoral zone and at 8 m and 15 m, these depths are disproportionately represented in the vertical profiles. These large samples were taken in June-July, 1976 and January, 1977; hence Table 2 is subdivided into summer and winter seasons.

A more quantitative view of the natural community was derived from the interpretation of photographs of undisturbed quadrats. Four photographic transects were run, each consisting of six quadrats at each of six depths (3, 6, 9, 12, 15, and 18 m). A starting point was selected haphazardly and the quadrats at each depth were adjacent to one another to avoid unnecessary bias on the part of the photographer. Transects were run at platform 288-A on November 1, 1976 and January 21, 1977; transects were run at well jacket 296-12 on January 22, 1977. Results are shown in Tables 3-6. This technique has the limitation that only sessile species recognizable in photographs were included; however, these collectively represented most of the biomass on the structures.

Perhaps the most conspicuous structural feature of the Buccaneer Field fouling community is the abundance of the large barnacle *Balanus tintinnabulum*. This barnacle is often 3-4 cm in diameter and, when crowded, may be twice that high. Moreover, individuals were often found growing on top of one another, especially near the sea surface, yielding a three-dimensional microhabitat which may be 10-15 cm thick. This undoubtedly contributed to the abundance of snapping shrimp, xanthid crabs, ophiuroids, etc. on the upper portions of the structures and may affect colonization rates on new structures (Schoener, 1974). Twenty years ago *B. tintinnabulum* was an incidental species on Texas offshore oil field structures (Gunter and Geyer, 1955), and it has remained so on similar structures in Louisiana which are older than those at Buccaneer Field (Thomas, 1975). However, it occupied as much as 77% of the original substrate on Buccaneer Field structures (Table 6).

Unoccupied habitat was extremely rare in natural quadrats and consisted almost entirely of scars left by dislodged barnacles or

TABLE 2. SEASONAL AND VERTICAL DISTRIBUTION OF ALGAE AND INVERTEBRATES.

Depth interval[a]	lit.				3-6 m				6-9 m				9-12 m				12-15 m				15-18 m				18-21 m				btm.			
Season/structure[b]	1	2	3	4	1	2	3	4	1	2	3	4	1	2	3	4	1	2	3	4	1	2	3	4	1	2	3	4	1	2	3	4
Cladophora delicatula	P	P	P	P							P																					
Cladophora vagabunda		P	P																													
Cladophora catenifera			P																													
Chaetomorpha media	P																															
Enteromorpha lingulata	P		P																													
Enteromorpha flexuosa	P																															
Derbesia vaucheriaeformis	P		P								P						P															
Bryopsis plumosa	P																															
Giffordia mitchelliae	P																															
Gelidium pusillum	P	P																														
Polysiphonia denudata	P		P								P	P																				
Polysiphonia subtilissima	P								P		P	P											P									
Centrocerus clavulatum			P	P																												
Bangia fuscopurpurea		P		P																												
Callithamnion byssoides		P	P						P		P	P					P				P											
Ceramium byssiodeum	P		P																													
Clathrina sp.	P																															
Leucosolenia botryoides	P		P						P																							
Scypha sp.	P		P						P												P											
Leuconia sp.	P		P																													
Haliclona sp.											P																					
Haliclona permollis									P												P											
Haliclona viridis					P				P																							
Haliclona loosanoffi	P				P				P								P				P		P		P							

(continued)

TABLE 2 (continued)

Depth interval[a]	lit.				3-6 m				6-9 m				9-12 m				12-15 m				15-18 m				18-21 m				btm.			
Season/structure[b]	1	2	3	4	1	2	3	4	1	2	3	4	1	2	3	4	1	2	3	4	1	2	3	4	1	2	3	4	1	2	3	4
Halichondria bowerbanki	P								P								P				P											
Hymeniacidon helophila											P																					
Prosuberites microsclerus											P																					
Cliona celata									P																							
Cliona vastifica									P																							
Turritopsis nutricula			C		C				C	U	A	U									C	U	C		R							
Obelia dichtoma				C								U					U				R											
Obelia hyalina			C																													
Clytia cylindrica																	A															
Stylactis sp.	R		U																													
Syncoryne eximia		A		A																												
Eudendrium eximium																					U					R						
Plumularia diaphana																					R				R		R					
Tubularia crocea		C								A							C															
Leptogorgia virgulata									R						R												R					
Astrangia sp.																									R	R	U	U				
Neanthes succinea	C	C	C	U	C				A	C	A	A					U				A	A	C	U	R							
Lepidonotus sublevis																															R	
Lepidonotus variabilis									R												R		R									
Trypanosyllis gemmipara	C	R	U	U	A				A	U	A	U					U				C	R	C	U	U							
Haplosyllis spongicola	A		A		A				A	R	A	C					A					R	C	U	U							
Eunice sp.	R										R												C	R								

(continued)

TABLE 2 (continued)

Depth interval[a]	lit.				3-6 m				6-9 m				9-12 m				12-15 m				15-18 m				18-21 m				btm.			
Season/structure[b]	1	2	3	4	1	2	3	4	1	2	3	4	1	2	3	4	1	2	3	4	1	2	3	4	1	2	3	4	1	2	3	4
Lumbrineris sp.											R												R									
Eupomatus dianthus									R														U								C	
Eupomatus protulicola	U								R		R												R									
Spirobranchus giganteus																											R					
Thelepus setosus																					R											
Sabella microphthalma																																
Doto sp.			C								C												R									
Crepidula plana									R																						C	
Crepidula fornicata																													C	R	C	
Mitrella lunata																							R						R	R		
Oliva sayana																															U	
Pleuroploca gigantea																							R							R		
Cantharus cancellarius																															U	
Thais haemastoma																						R							R	U		
Murex fulvescens																													R		R	
Seila adamsi																															U	
Cypraea spurca acicularis																														R		
Neosimnia acicularis																											R					
Epitonium albidum																														R		
Mangelia cf. *M. oxytata*																														R		
Lithophaga bisulcata		R	R	R							R		C								U	U	U								U	
Lithophaga aristata			R	R						R			C								R		R									
Brachiodontes exustus	R																															
Musculus lateralis	R		R	C					R	A	R	C										C	U	C								

(continued)

TABLE 2 (continued)

Depth interval[a]	lit.				3–6 m				6–9 m				9–12 m				12–15 m				15–18 m				18–21 m				btm.			
Season/structure[b]	1	2	3	4	1	2	3	4	1	2	3	4	1	2	3	4	1	2	3	4	1	2	3	4	1	2	3	4	1	2	3	4
Anadara transversa	U		R		C				U	R	R		C										U								U	
Arca imbricata									U		R																					
Chama macerophylla	R		U	R					U	R	R	R	R									U	U	R							C	
Pteria colymbus									R	R	R						R						R									
Ostrea equestris	R		R			A			R	U		R	C								R	U	U	R	U						U	
Crassostrea virginica											R																					
Aequipecten gibbus									R																							
Isognomon bicolor	A	C	A	A	C				A	C	U	U	C									U		R								
Chione grus			R						U		R												R						U			
Lima pellucida									R																							
Atrina seminuda																								R								
Diplothyra smythi																															C	
Octopus sp.																														R		
Tanystylum orbiculare	A		A								C																					
Jassa falcata		A	A	A	C				R	U	R										R											
Leucothoe spinicarpa					C				A	C			R								A	C	A	C								
Erichthonius brasiliensis				U						C		C	C								C	C	C	U								
Corophium sp.																					C											
Caprella equilibra	A	A	C	A	A				A	A	A	R	C								C	U			U							
Elasmopus sp.		A	A	C						R	C	R																				
Stenothoe gallensis		C	C	R						U	C	R														R		R				

(continued)

TABLE 2 (continued)

Depth interval[a]	lit.				3-6 m				6-9 m				9-12 m				12-15 m				15-18 m				18-21 m				btm.			
Season/structure[b]	1	2	3	4	1	2	3	4	1	2	3	4	1	2	3	4	1	2	3	4	1	2	3	4	1	2	3	4	1	2	3	4
Tanais sp.	A	A	A	U	C						A																					
Dynamene perforata	A	C	A	C																												
Chthamalus fragilis			R																													
Balanus tintinnabulum	A	A	A	A	A	A	A	A	A	A	A	A	A	A	A	A	A	A	A	A	A	A	A	A	U	U	U	U				
Balanus improvisus											R																					
Balanus (*amphitrite* complex)			C						U				R												U						R	
Chelonibia patula																															U	
Synalpheus fritzmuelleri			U	C	A				A	A	A		U				R				A	A	A	C	U							
Petrochirus diogenes																															C	
Dardanus fucosus																															C	
Paguristes puncticeps																															U	
Pagurus impressus																															C	
Pylopagurus holthuisi																															U	
Stenorhynchus seticornis																							R	R								
Porcellana sayana																															C	
Dromidia antillensis											C										C		U									
Pachygrapsus transversus	U	U	U	C																												
Portunus gibbesii			U																													
Callinectes similis			U																													
Cronius ruber											R	R																				

(continued)

TABLE 2 (continued)

Depth interval[a]	lit.				3-6 m				6-9 m				9-12 m				12-15 m				15-18 m				18-21 m				btm.			
Season/structure[b]	1	2	3	4	1	2	3	4	1	2	3	4	1	2	3	4	1	2	3	4	1	2	3	4	1	2	3	4	1	2	3	4
Menippe mercenaria	R	U	C	R	A				C	R	C	R	U				C				R	R			C							
Hexapanopeus paulensis	R								C		C										C		A							R	C	
Pilumnus dasypodus	R								R				R								R	R	R								U	
Pilumnus pannosus													R																			
Pseudomedaeus agassizi													A								C	A	A	C	C					R	C	
Panopeus turgidus											R																			R		
Bugula neritina		R		U	R				A	C			A								A	C			R							
Bugula rylandi									U	U			C								U	U										
Savignyella lafonti		R											C								A	R	A				R					
Schizoporella errata									C		C	U	U												U							
Eucratea loricata																											U					
Parasmittina trispinosa										R												U					R					
Membranipora tenuis	U																															
Membranipora savartii									R			R																				
Aetea anguina	A	C			U				C				C								C	U			C							
Ophiothrix angulata									R		R	R																				
Ophiactis savignyi	A	U		C					A	A	R		C								A	C		R								
Arbacia punctulata												U												U								

[a]lit. = littoral, btm. = bottom; [b]1 = platform 288-A in summer, 2 = platform 288-A in winter, 3 = well jacket 296-12 in summer, 4 = well jacket 296-12 in winter.
P = present, A = abundant, C = common, U = uncommon, R = rare.

TABLE 3. PROPORTION OF SPACE OCCUPIED BY VARIOUS FOULING SPECIES IN UNDISTURBED PHOTOGRAMMETRIC QUADRATS ON NOVEMBER 1, 1976 ON PLATFORM 288-A.

	Depth									
	6 m		9 m		12 m		15 m		18 m	
Species	$\bar{x}$[a]	s.e.[b]	$\bar{x}$	s.e.	$\bar{x}$	s.e.	$\bar{x}$	s.e.	$\bar{x}$	s.e.
Balanus tintinnabulum	0.079	0.050	0.233	0.083	0.592	0.073	0.293	0.098	0.004	0.004
Bugula neritina	0.023	0.020	0.015	0.010	0.040	0.028	0.014	0.009	0	
Savignyella lafonti	0		0		0.974	0.017	0.982	0.008	0	
Ostrea equestris	0.318	0.162	0		0		0		0	
Astrangia sp.	0		0		0		0		0.002	0.002
Haliclona permollis	0.015	0.015	0.015	0.015	0.014	0.014	0		0	
Haliclona loosanoffi	0.160	0.160	0.245	0.245	0.473	0.213	0.991	0.006	0	
Haliclona sp.	0.797	0.161	0.595	0.595	0.155	0.155	0		0.824	
Hydroids	0.932	0.031	0.984	0.984	0		0		0	
Green algae	0.932	0.031	0.958	0.958	0.974	0.017	0.982	0.008	0	
Serpulids	0		0.017	0.017	0		0		0	
Sponge (red)	0		0		0		0		0.172	0.069
Bare substrate	0.006	0.006	0		0		0		0.002	0.002

[a]Mean of six replicates; [b]Standard error.

TABLE 4. PROPORTION OF SPACE OCCUPIED BY VARIOUS FOULING SPECIES IN UNDISTURBED PHOTOGRAMMETRIC QUADRATS ON JANUARY 1, 1977 ON PLATFORM 288-A.

	Depth											
	3 m		6 m		9 m		12 m		15 m		18 m	
Species	$\overline{x}$[a]	s.e.[b]	$\overline{x}$	s.e.	$\overline{x}$	s.e.	$\overline{x}$	s.e.	$\overline{x}$	s.e.	$\overline{x}$	s.e.
B. tintinnabulum	0.389	0.115	0.324	0.072	0.145	0.052	0.208	0.040	0.121	0.047	0.033	0.023
B. neritina	0.028	0.021	0.119	0.038	0.147	0.059	0.094	0.050	0.071	0.028	0.037	0.037
B. rylandi	0		0.001	0.001	0.076	0.055	0.185	0.130	0.020	0.015	0.015	0.015
S. lafonti	0		0		0.003	0.003	0		0		0	
P. trispinosa	0		0.001	0.001	0.004	0.004	0		0		0	
O. equestris	0		0		0		0.001	0.001	0		0	
Astrangia sp.	0		0		0		0		0		0.002	0.002
T. crocea	0.090	0.035	0.187	0.048	0.025	0.021	0		0		0	
H. permollis	0.662	0.151	0		0		0		0		0	
H. loosanoffi	0		0		0		0		0.731	0.151	0.800	0.163
Haliclona sp.	0.128	0.128	0.637	0.039	0.721	0.063	0.428	0.186	0.167	0.167	0.167	0.167
Hydroids	0.759	0.072	0.663	0.061	0.696	0.074	0.645	0.126	0.731	0.151	0.856	0.068
Red algae	0.014	0.009	0		0		0		0.012	0.012	0	
Green algae	0.022	0.022	0		0		0		0		0	
Ascidian[c]	0		0.013	0.013	0		0		0		0	
Ascidian[d]	0		0.003	0.003	0		0.407	0.189	0		0	
Sponge (red)	0		0		0		0		0		0.031	0.027

[a]Mean of six replicates; [b]Standard error; [c]Compound ascidian (transparent); [d]Compound ascidian (orange).

TABLE 5. PROPORTION OF SPACE OCCUPIED BY VARIOUS FOULING SPECIES IN UNDISTURBED PHOTOGRAMMETRIC QUADRATS ON JANUARY 22, 1977 ON LEG B1 OF WELL JACKET 296-12.

	Depth											
	3 m		6 m		9 m		12 m		15 m		18 m	
Species	$\overline{x}$[a]	s.e.[b]	$\overline{x}$	s.e.	$\overline{x}$	s.e.	$\overline{x}$	s.e.	$\overline{x}$	s.e.	$\overline{x}$	s.e.
B. tintinnabulum	0.485	0.130	0.583	0.111	0.190	0.103	0.433	0.142	0.175	0.066	0.105	0.048
P. trispinosa	0.003	0.003	0.011	0.011	0.027	0.027	0		0		0.006	0.006
O. equestris	0.030	0.019	0		0.015	0.015	0		0.006	0.006	0	
H. permollis	0		0		0.027	0.027	0		0.125	0.125	0	
H. loosanoffi	0		0.582	0.160	0.392	0.164	0.498	0.167	0.709	0.171	0.762	0.194
Haliclona sp.	0.839	0.095	0.381	0.156	0.404	0.163	0.385	0.198	0.100	0.100	0.181	0.181
Hydroids	0.949	0.051	0.908	0.042	0.741	0.151	0.691	0.114	0.932	0.027	0.908	0.092
Red algae	0.063	0.039	0.006	0.006	0		0		0		0	
Green algae	0.059	0.034	0		0		0		0		0	
Ascidian[c]	0		0		0		0		0.025	0.025	0.032	0.032
Sponge (orange)	0		0.007	0.006	0		0.117	0.063	0		0	

[a]Mean of six replicates; [b]Standard error; [c]Compound ascidian (transparent).

TABLE 6. PROPORTION OF SPACE OCCUPIED BY VARIOUS FOULING SPECIES IN UNDISTURBED PHOTOGRAMMETRIC QUADRATS ON JANUARY 22, 1977 ON LEG B2 OF WELL JACKET 296-12.

	Depth											
	3 m		6 m		9 m		12 m		15 m		18 m	
Species	$\bar{x}$[a]	s.e.[b]	$\bar{x}$	s.e.	$\bar{x}$	s.e.	$\bar{x}$	s.e.	$\bar{x}$	s.e.	$\bar{x}$	s.e.
B. tintinnabulum	0.392	0.151	0.682	0.112	0.529	0.083	0.771	0.098	0.386	0.072	0.255	0.161
B. neritina	0		0		0		0		0.052	0.052	0.144	0.119
P. trispinosa	0		0.007	0.004	0.006	0.006	0		0		0	
O. equestris	0		0		0		0.006	0.006	0.013	0.013	0	
H. permollis	0.318	0.318	0.694	0.188	0.332	0.210	0		0		0	
H. loosanoffi	0.156	0.156	0		0.055	0.052	0.029	0.018	0.651	0.186	0.405	0.167
Haliclona sp.	0.456	0.259	0		0.491	0.220	0.904	0.030	0.124	0.124	0.257	0.170
Hydroids	0.930	0.017	0.948	0.033	0.941	0.047	0.942	0.030	0.646	0.191	0.765	0.119
Red algae	0.076	0.013	0.010	0.006	0.051	0.048	0.005	0.004	0.002	0.002	0	
Ascidian[c]	0		0		0		0		0.170	0.098	0.193	0.113
Sponge (orange)	0.007	0.007	0		0		0.057	0.031	0.002	0.002	0	

[a]Mean of six replicates; [b]Standard error; [c]Compound ascidian (transparent).

oysters. Occupied habitat was typically exploited by several layers of organisms. Barnacles appear to be the most perennial inhabitants. A variety of sponges, hydroids, and algae appear to be among the earliest colonizers. The species richness on Buccaneer Field structures was apparently intermediate between Louisiana platforms of similar age (Thomas, 1975) and the fouled floats studied off the Florida Gulf coast by Pequegnat and Pequegnat (1968). However, sampling as extensive as that of the latter study may yield a longer species list for Buccaneer Field.

Recolonization patterns were studied through the use of denuded quadrats and denuded and caged quadrats. Results from nine quadrats representing summer colonization at 15 m on well jacket 296-12 (Table 7) indicate that sponges, hydroids, and the erect bryozoan *Savignyella lafonti* were the major colonists at this time. The hydroids and bryozoans were typically found growing over the sponges.

Two series of experiments were completed during the winter. Sets of experimental quadrats were established on both structures on September 30 - October 1, 1976. Those on well jacket 296-12 were monitored on January 21 and March 8, 1977. These quadrats were established only at the 8 m and 15 m depths. A second set of quadrats was established at the 3, 8, and 15 m depths on both structures on January 21-22, 1977. Results are presented in Tables 8-13.

Initial colonists on uncaged quadrats on well jacket 296-12 were green algae, hydroids, and sponges, all of which were less abundant in caged quadrats, presumably due to reduced light and planktonic food levels within the cages. Only *Savignyella lafonti* was noticeably more abundant in the cages, where it may have been encouraged by reduced fish predation. *Savignyella lafonti* was found in the gut contents of *Chaetodipterus faber*, one of the most abundant resident fishes.

The major barnacle colonists observed during the study were *Balanus improvisus* and *Balanus* (*amphitrite* complex). Juveniles of *Balanus tintinnabulum* were rarely observed. In general, barnacle colonists were rare, as were barnacle larvae in the plankton samples. Since these observations encompass nearly a year, they suggest that successful sets by the dominant barnacle may not occur annually. Studies of fouling communities (Osman, 1977) and coral reefs (Jackson, 1977) suggest that colonial animals may be better competitors for space than solitary species. In light of this observation, it is surprising that solitary species were not more prominant as opportunistic early exploiters of denuded space.

Subsequent (March 8) observations of the first set of well jacket 296-12 quadrats indicated an increased abundance of red

TABLE 7. PROPORTION OF SPACE OCCUPIED BY VARIOUS FOULING SPECIES IN EXPERIMENTAL PHOTOGRAMMETRIC QUADRATS ON WELL JACKET 296-12 AT THE 15 m DEPTH ON SEPTEMBER 13, 1976 (56 DAYS AFTER START OF THE TREATMENT).

Species	$\bar{x}^a$	s.e.[b]
Bugula neritina	0.005	0.005
Savignyella lafonti	0.349	0.091
Astrangia sp.	0.002	0.002
Haliclona permollis	0.085	0.032
Haliclona viridis	0.003	0.003
Haliclona sp.	0.456	0.114
Hydroids	0.110	0.076
Red algae	0.003	0.003
Green algae	0.039	0.026
Bare substrate	0.189	0.076

[a]mean of nine replicates; [b]standard error.

algae on the uncaged quadrats and of the encrusting bryozoan *Parasmittina trispinosa* at the 15 m depth in both caged and uncaged quadrats. Relatively little colonization occurred on the second set of quadrats, that which occurred being primarily hydroids and green algae.

The first set of experimental quadrats on platform 288-A were also colonized by hydroids, sponges, and green algae, with *Parasmittina trispinosa* and the erect bryozoan *Bugula neritina* becoming more abundant with time. The second set of quadrats indicated a similar pattern at lower intensity, except without *Bugula rylandi* and with a greater abundance of *Tubularia crocea*.

Several interesting differences between the two structures were observed. *Savignyella lafonti* was seasonally common on well jacket 296-12 but rare on platform 288-A. Similarly, *Bugula neritina* and *Bugula rylandi* were seasonally abundant on platform 288-A but very rare on well jacket 296-12. During the winter a large bloom of the large hydroid *Tubularia crocea* occurred on platform 288-A but not on well jacket 296-12. Similarly, the urchin *Arbacia punctulata* appeared and became moderately common on jacket 296-12 during the winter but was not observed on platform 288-A. Such spatial variation is often observed and may be attributed to stochastic processes (Fager, 1971).

TABLE 8. PROPORTION OF SPACE OCCUPIED BY VARIOUS FOULING SPECIES IN EXPERIMENTAL PHOTOGRAMMETRIC QUADRATS ON DECEMBER 29, 1976 ON WELL JACKET 296-12 (90 DAYS AFTER START OF THE TREATMENT).

	Scraped				Scraped and caged			
	8 m		15 m		8 m		15 m	
Species	$\overline{x}$[a]	s.e.[b]	$\overline{x}$	s.e.	$\overline{x}$	s.e.	$\overline{x}$	s.e.
Balanus (*amphitrite* complex)	0.049	0.002	0.007	0.004	0.065	0.023	0.002	0.002
Balanus tintinnabulum	0		0		0.002	0.002	0	
Savignyella lafonti	0.003	0.003	0.002	0.002	0.159	0.054	0.185	0.089
Parasmittina trispinosa	0.005	0.004	0.088	0.048	0.100	0.045	0.095	0.049
Ostrea equestris	0.004	0.004	0		0		0	
Astrangia sp.	0		0.001	0.001	0		0	
Haliclona permollis	0.052	0.052	0		0		0.004	0.004
Haliclona loosanoffi	0		0.036	0.022	0		0.025	0.020
Haliclona sp.	0.225	0.109	0.063	0.052	0.016	0.016	0	
Leuconia sp.	0		0		0.003	0.002	0	
Serpulids	0		0.001	0.001	0.001	0.001	0.004	0.002
Hydroids	0.661	0.149	0.303	0.189	0.018	0.018	0.186	0.124
Red algae	0.001	0.001	0		0		0	
Green algae	0.694	0.145	0.303	0.189	0.012	0.012	0.186	0.124
Sponge (orange)	0		0		0		0.002	0.002
Ascidian[c]	0		0		0.401	0.190	0.008	0.005
Bare substrate	0.041	0.041	0.466	0.148	0.384	0.062	0.331	0.136

[a]Mean of six replicates; [b]standard error; [c]compound ascidian (transparent).

TABLE 9. PROPORTION OF SPACE OCCUPIED BY VARIOUS FOULING SPECIES IN EXPERIMENTAL PHOTOGRAMMETRIC QUADRATS ON MARCH 8, 1977 ON WELL JACKET 296-12 (159 DAYS AFTER START OF THE TREATMENT).

	Scraped				Scraped and caged			
	8 m		15 m		8 m		15 m	
Species	$\overline{x}$[a]	s.e.[b]	$\overline{x}$	s.e.	$\overline{x}$	s.e.	$\overline{x}$	s.e.
Balanus (*amphitrite* complex)	0		0.002	0.001	0.026	0.015	0.001	0.001
Parasmittina trispinosa	0.001	0.001	0.272	0.101	0.058	0.053	0.240	0.090
Aetea anguina	0		0		0.189	0.147	0	
Haliclona permollis	0		0		0		0.006	0.006
Haliclona viridis	0		0		0.010	0.006	0.056	0.036
Haliclona loosanoffi	0		0.013	0.013	0		0.128	0.070
Haliclona sp.	0.336	0.208	0.103	0.062	0		0.017	0.017
Hydroids	0.509	0.196	0.296	0.144	0.067	0.067	0.281	0.130
Red algae	0.341	0.221	0.134	0.058	0		0.020	0.013
Green algae	0.696	0.118	0.049	0.045	0.009	0.009	0	
Serpulids	0		0.002	0.002	0		0.002	0.002
Corophiid tubes	0		0		0.067	0.067	0.153	0.059
Ascidian[c]	0.036	0.022	0.227	0.102	0.481	0.273	0	
Bare substrate	0.003	0.003	0.132	0.084	0.161	0.066	0.115	0.048

[a]Mean of six replicates; [b]standard error; [c]compound ascidian (transparent).

TABLE 10. PROPORTION OF SPACE OCCUPIED BY VARIOUS FOULING SPECIES IN EXPERIMENTAL PHOTOGRAMMETRIC QUADRATS ON MARCH 8, 1977 ON WELL JACKET 296-12 (46 DAYS AFTER START OF THE TREATMENT).

	Scraped						Scraped and caged					
	3 m		8 m		15 m		3 m		8 m		15 m	
Species	$\bar{x}$[a]	s.e.[b]	$\bar{x}$	s.e.	$\bar{x}$	s.e.	$\bar{x}$	s.e.	$\bar{x}$	s.e.	$\bar{x}$	s.e.
B. improvisus	0		0		0		0		0		0.001	0.001
Balanus sp.	0		0		0		0		0.001	0.001	0	
P. trispinosa	0		0.023	0.023	0.024	0.020	0		0		0.050	0.039
H. permollis	0.022	0.022	0		0		0		0		0	
H. loosanoffi	0		0		0.185	0.128	0		0		0	
Hydroids	0.525	0.201	0		0.493	0.150	0.063	0.050	0.189	0.144	0.132	0.084
Red algae	0.001	0.001	0.020	0.020	0.151	0.066	0		0.063	0.043	0.002	0.002
Green algae	0.080	0.044	0.619	0.068	0		0.008	0.007	0.236	0.133	0	
Bare substrate	0.444	0.205	0.338	0.110	0.423	0.136	0.926	0.047	0.675	0.111	0.815	0.083

[a]Mean of six replicates; [b]Standard error.

TABLE 11. PROPORTION OF SPACE OCCUPIED BY VARIOUS FOULING SPECIES IN EXPERIMENTAL PHOTOGRAMMETRIC QUADRATS ON JANUARY 21, 1977 ON PLATFORM 288-A (113 DAYS AFTER START OF THE TREATMENT).

	Scraped				Scraped and caged			
	8 m		15 m		8 m		15 m	
Species	$\overline{x}$[a]	s.e.[b]	$\overline{x}$	s.e.	$\overline{x}$	s.e.	$\overline{x}$	s.e.
Balanus (*amphitrite* complex)	0		0.002	0.001	0.026	0.015	0.001	0.001
Parasmittina trispinosa	0.001	0.001	0.272	0.101	0.058	0.053	0.240	0.090
Aeta anguina	0		0		0.189	0.147	0	
Haliclona permollis	0		0		0		0.006	0.006
Haliclona viridis	0		0		0.010	0.006	0.056	0.036
Haliclona loosanoffi	0		0.013	0.013	0		0.128	0.070
Haliclona sp.	0.336	0.208	0.103	0.062	0		0.017	0.017
Hydroids	0.509	0.196	0.296	0.144	0.067	0.067	0.281	0.130
Red algae	0.341	0.221	0.134	0.058	0		0.020	0.013
Green algae	0.696	0.118	0.049	0.045	0.009	0.009	0	
Serpulids	0		0.002	0.002	0		0.002	0.002
Corophiid tubes	0		0		0.067	0.067	0.153	0.059
Ascidian[c]	0.036	0.022	0.227	0.102	0.481	0.273	0	
Bare substrate	0.003	0.003	0.132	0.084	0.161	0.066	0.115	0.048

[a]Mean of six replicates; [b]standard error; [c]compound ascidian (transparent).

TABLE 12. PROPORTION OF SPACE OCCUPIED BY VARIOUS FOULING SPECIES IN EXPERIMENTAL PHOTOGRAMMETRIC QUADRATS ON MARCH 8, 1977 ON PLATFORM 288-A (159 DAYS AFTER START OF THE TREATMENT).

Species	Scraped				Scraped and caged			
	8 m		15 m		8 m		15 m	
	$\bar{x}$[a]	s.e.[b]	$\bar{x}$	s.e.	$\bar{x}$	s.e.	$\bar{x}$	s.e.
Balanus (*amphitrite* complex)	0.005	0.002	0		0.006	0.006	0	
Balanus improvisus	0		0		0.001	0.001	0	
Bugula neritina	0.296	0.065	0.483	0.125	0.252	0.094	0.434	0.056
Bugula rylandi	0.001	0.001	0.031	0.025	0.041	0.013	0.122	0.059
Parasmittina trispinosa	0.246	0.078	0		0.074	0.060	0.010	0.010
Ostrea equestris	0.005	0.002	0		0.002	0.002	0	
Chama macerophylla	0		0		0.001	0.001	0	
Tubularia crocea	0.019	0.006	0		0.001	0.001	0.003	0.002
Haliclona loosanoffi	0.019	0.019	0.412	0.127	0		0	
Haliclona viridis	0		0		0.002	0.002	0	
Haliclona sp.	0		0.043	0.030	0.061	0.050	0	
Hydroids	0.411	0.070	0.444	0.013	0.541	0.066	0.435	0.032
Red algae	0.032	0.016	0		0		0	
Green algae	0.174	0.102	0		0		0	
Corophiid tubes	0		0		0.108	0.065	0	
Bare substrate	0.006	0.005	0		0		0	

[a]Mean of six replicates; [b]standard error.

TABLE 13. PROPORTION OF SPACE OCCUPIED BY VARIOUS FOULING SPECIES IN EXPERIMENTAL PHOTOGRAMMETRIC QUADRATS ON MARCH 8, 1977 ON PLATFORM 288-A (46 DAYS AFTER START OF THE TREATMENT).

	Scraped						Scraped and caged					
	3 m		8 m		15 m		3 m		8 m		15 m	
Species	$\overline{x}$[a]	s.e.[b]	$\overline{x}$	s.e.	$\overline{x}$	s.e.	$\overline{x}$	s.e.	$\overline{x}$	s.e.	$\overline{x}$	s.e.
B. improvisus	0		0		0		0		0.004	0.002	0	
Balanus sp.	0.001	0.001	0		0		0.002	0.001	0		0	
B. neritina	0		0.031	0.026	0.030	0.008	0.002	0.002	0.011	0.007	0.038	0.013
P. trispinosa	0		0.016	0.011	0		0		0		0.004	0.004
T. crocea	0.002	0.002	0.003	0.003	0		0.079	0.033	0.038	0.020	0.009	0.009
Haliclona sp.	0		0		0.013	0.008	0		0		0	
Hydroids	0		0.087	0.058	0		0.137	0.050	0.066	0.047	0	
Red algae	0.021	0.021	0.006	0.006	0		0.054	0.033	0.010	0.010	0	
Green algae	0.172	0.056	0.011	0.006	0		0.174	0.046	0.002	0.002	0	
Corophiid tubes	0.005	0.005	0		0		0		0		0	
Bare substrate	0.801	0.059	0.847	0.084	0.957	0.007	0.742	0.044	0.870	0.038	0.949	0.018

[a]Mean of six replicates; [b]Standard error.

Among the motile invertebrates, *Pachygrapsus transversus* and the isopod *Dynamene perforata* were distinctly littoral, and the tanaids and pycnogonids were predominantly so. *Neanthes succinea*, *Trypanosyllis gemmipara*, *Turritopsis nutricula*, *Anadara transversa*, *Ostrea equestris*, *Leucothoe spinicarpa*, *Caprella equilibra*, *Balanus tintinnabulum*, *Synalpheus fritzmuelleri*, *Menippe mercenaria*, *Bugula neritina*, and *Aetea anquina* had rather broad vertical distributions. Hermit crabs, associated shell-inhabiting species, most gastropods, and *Octopus* sp. were largely limited to the bottom. *Erichthonius brasiliensis* was the dominant tube-building amphipod observed, and it appeared to fare better under the protection of the cages than in the open (Tables 9 and 11).

Another relevant feature of this bottom fauna is that most of the gastropod shells, many of which contain hermit crabs, were heavily encrusted by fouling organisms. One *Thais haemastoma* shell occupied by *Dardanus fucosus* contained over 20 commensal or fouling species. These included some which have not been found on the structures, such as *Chelonibia patula*, *Lepidonotus sublevis*, and *Porcellana sayana*. These shells are a potential source of immigrants to the structures, but some species apparently have not yet made the transition.

Very few organisms (e.g., barnacles) whose growth could be followed in the photographs settled on the experimental quadrats. The mean area covered by 11 *Balanus* (*amphitrite* complex) individuals measured 159 days after the quadrat was scraped was 0.134 ± 0.014 cm^2, suggesting a minimum growth rate of 0.026 cm^2 per month.

The cages on well jacket 296-12 were changed on December 29, 1976, and those on platform 288-A were changed on January 21, 1977. Several species had fouled the cages, and thus their minimum growth rates could be estimated from measurements of individuals returned to the laboratory. The sizes and estimated minimum growth rates of several mollusks and one bryozoan are presented in Table 14. These data are compared in the table to the size of the largest individual collected at Buccaneer Field and to the adult size reported in the literature (Andrews, 1977) to determine what proportion of the adult size had been attained.

The presence of ovigerous amphipods, isopods, and decapods in the samples was noted in an effort to determine breeding seasons. Ovigerous *Jassa falcata* were observed in October and January, *Leucothoe spinicarpa* in June, *Erichthonius brasiliensis* in July, December, and January, *Caprella equilibria* in June and January, and *Stenothoe gallensis* in January. Ovigerous *Tanais* sp. were observed in July, October, and January. Ovigerous *Synalpheus fritzmuelleri* were observed in May, July, September, and October, and numerous juveniles were observed in January. Ovigerous

TABLE 14. MEASUREMENTS AND ESTIMATED MINIMUM GROWTH RATES OF SELECTED INVERTEBRATES FOULING CAGES AT BUCCANEER FIELD.

Species	Structure	Date	n	$\bar{x}$[a]	s.e.	Maximum size	Maximum growth rate[b]	Maximum size observed	Reported size (Andrews, 1977)
P. colymbus	288-A	1/21	23	1.69	0.09	2.56	0.69	8.92	3.8 - 7.6
P. colymbus	296-12	12/29	12	1.59	0.15	2.48	0.83	8.92	3.8 - 7.6
M. lateralis	288-A	1/21	22	0.65	0.02	0.82	0.22	0.76	1.00
M. lateralis	296-12	12/29	10	0.59	0.03	0.69	0.23	0.76	1.00
A. transversa	288-A	1/21	1	0.52	-	0.52	0.14	0.97	1.2 - 3.6
A. gibbus	288-A	1/21	2	1.60	0.09	1.69	0.46	-	2.5 - 5.2
O. equestris	288-A	1/21	1	1.60	-	1.60	0.43	7.12	2.5 - 5.2
O. equestris	296-12	12/29	1	0.78	-	0.78	0.26	7.12	2.5 - 5.2
P. trispinosa	296-12	12/29	7	7.11	1.76	13.45	4.50	-	-

[a]Mean length in cm for mollusks, mean area in cm^2 for *P. trispinosa*; [b]Units are cm $month^{-1}$ for mollusks, cm^2 $month^{-1}$ for *P. trispinosa*.

Hexapanopeus paulensis were observed in September, *Pilumnus dasypodus* in September, *Porcellana sayana* in September, *Pachygrapsus transversus* in June and July, *Menippe mercenaria* included numerous juveniles throughout the year but was never observed with eggs, *Pseudomedaeus agassizi* in September, November, and January, *Stenorhynchus seticornis* in October, and *Panopeus turgidus* in January. Thus a large portion of these taxa breed during the winter months.

ACKNOWLEDGEMENTS

This work is a result of research sponsored by the Environmental Protection Agency and Department of Commerce, National Oceanic and Atmospheric Administration, National Marine Fisheries Service, Southeast Fisheries Center, Galveston Laboratory under Contract No. 03-6-042-35119. Drs. R. A. Bagnall, M. F. Johnson, and G. J. Penn, Mr. W. D. Quast, and several others assisted in the collection of samples and the establishment and monitoring of field experiments. Dr. M. F. Johnson identified the sponges, and Dr. G. J. Penn identified the algae.

REFERENCES

Andrews, J., 1977, "Shells and Shores of Texas." University of Texas Press, Austin, Texas.

Connell, J. H., 1970, "A predator-prey system in the marine intertidal region. I. *Balanus glandula* and several predatory species of *Thais*," Ecol. Monogr., 40:49.

Daniel, A., 1954, "The seasonal variation and the succession of the fouling communities in the Madras harbour waters," J. Madras Univ., 24B:189.

Fager, E. W., 1971, "Pattern in the development of a marine community," Limnol. Oceanogr., 16:241.

Fuller, J. L., 1946, "Season of attachment and growth of sedentary marine organisms at Lamoine, Maine," Ecology, 27:150.

Gunter, G., and Geyer, R. A., 1955, "Studies on fouling organisms of the northwest Gulf of Mexico," Publ. Inst. Mar. Sci., 4:37.

Haderlie, E. C., 1969, "Marine fouling and boring organisms in Monterey Harbor. II. Second year of investigation," Veliger, 12:182.

Harper, D. E., Jr., 1977, "Distribution and abundance of macrobenthic and meiobenthic organisms," in Jackson, W. B. (ed.), "Environmental Assessment of an Active Oil Field in the Northwestern Gulf of Mexico, 1976-1977." NOAA, pp. 175-273. Available from: NTIS, Springfield, Virginia, Accession No. PB283890.

Jackson, J. B. C., 1977, "Competition on marine hard substrata: the adaptive significance of solitary and colonial strategies," Amer. Natur., 111:743.

Kawahara, T., 1962, "Studies on the marine fouling communities. I. Development of a fouling community," Rep. Fac. Fish. Prefect. Univ. Mie, 4:27.

Lee, S. W., and Trott, L. B., 1973, "Marine succession of fouling organisms in Hong Kong, with a comparison of woody substrates and common, locally-available, antifouling paints," Mar. Biol., 20:101.

MacArthur, R. H., and Wilson, E. O., 1967, "The Theory of Island Biogeography." Princeton University Press, Princeton, New Jersey.

McDougall, K. D., 1943, "Sessile marine invertebrates of Beaufort, North Carolina," Ecol. Monogr., 13:321.

Nair, N. B., 1962, "Ecology of marine fouling and wood-boring organisms of western Norway," Sarsia, 8:1.

Osman, R. W., 1977, "The establishment and development of a marine epifaunal community," Ecol. Monogr., 47:37.

Pearse, J. B., and Chess, J. R., 1971, "Comparative investigations of the development of epibenthic communities from Gloucester, Massachusetts to St. Thomas, Virgin Islands," in: Crisp, D. J. (ed.), "Fourth International Marine Biology Symposium." Cambridge University Press, New York, p. 55.

Pequegnat, W. E., and Pequegnat, L. H., 1968, "Ecological Aspects of Marine Fouling in the Northeastern Gulf of Mexico." Ref. No. 68-22T, Department of Oceanography, Texas A&M University, College Station, Texas, 80 pp.

Raja, B. T. A., 1959, "Studies on the distribution and succession of sedentary organisms of the Madras Harbour," J. Mar. Biol. Assoc. India, 1:180.

Raja, B. T. A., 1963, "Observations on the rate of growth, sexual maturity, and breeding of four sedentary organisms from the Madras Harbour," J. Mar. Biol. Assoc. India, 5:113.

Scheer, B. T., 1945, "The development of marine fouling communities," Biol. Bull., 89:103.

Schoener, A., 1974, "Experimental zoogeography: colonization of marine mini-islands," Amer. Natur., 108:715.

Skerman, T. M., 1959, "Marine fouling at the Port of Auckland," New Zealand J. Sci., 2:57.

Stubbings, H. G., and Houghton, D. R., 1964, "The ecology of Chichester Harbour, S. England, with special reference to some fouling species," Int. Rev. ges. Hydrobiol., 49:233.

Sutherland, J. P., 1974, "Multiple stable points in natural communities," Amer. Natur., 108:859.

Thomas, P. J., 1975, "The Fouling Community on Selected Oil Platforms off Louisiana, with Special Emphasis in the Cirripedia Fauna." M.Sc. Thesis, Florida State University.

Weiss, C. M., 1948, "The seasonal occurrence of sedentary marine organisms in Biscayne Bay, Florida," Ecologia, 29:153.

WHOI, 1952, "Marine Fouling and its Prevention." prepared for the Bureau of Ships, Navy Department, Contribution No. 580, Woods Hole Oceanographic Institute, Woods Hole, Massachusetts, 388 pp.

THE EFFECT OF STRUCTURES ON MIGRATORY AND LOCAL MARINE BIRDS

Glenn D. Aumann

Department of Biology
University of Houston
Houston, Texas 77004

INTRODUCTION

The Buccaneer Field structures represent one of many man-made changes in the waters of the northwestern Gulf of Mexico. This area comprises a well-known and long-used flyway for migratory birds which pass from southern Mexico and the Yucatan Peninsula northward to the shores of the Gulf states in the spring and pass southward along the same route in the fall. Prior to the establishment of man-made structures in the Gulf, the migratory route would necessarily have been made across a featureless expanse of water 480 nautical miles wide at the narrowest point, with the complete trip being made non-stop (Pettingill, 1970). Birds migrating across water fly at heights of approximately 150-300 m (Bellrose, 1971). Flying at this altitude, birds can readily see structures the size of the Buccaneer platforms. It has been noted that birds will land on ships and structures during the migratory season, particularly under adverse weather conditions.

The Buccaneer Field is approximately 25 nautical miles from nearest landfall. On clear days, the structures can be seen quite readily over that distance from an elevation of 100 m or more. The presence of the field has resulted in increased boat traffic between shore and the field. The combination of visibility and boat traffic presumably can affect flight patterns of migratory birds and feeding habits of marine birds along the near coastline. This study was undertaken to determine the effects, if any, on migratory birds passing through in the spring and fall, and to determine the effects of the structures on feeding habits of resident marine birds.

MATERIALS AND METHODS

An annual schedule of observation periods was established to coincide with spring and fall migration periods and with summer and winter resident birds. Observations were conducted from Platform 288-A within the Buccaneer Field on four days of each quarter and from boat deck in a control area five to seven nautical miles west of the Field on two days of each quarter. Observations were made with the aid of binoculars within approximately 500 m of the platform and the boat used in the control area. Guides compiled by Peterson (1960), Robbins (1966), and Oberholser (1974) were used as references for identification and information pertaining to habits of each species.

Data collected consists of identification to species, numbers of individuals, flight directions of those birds observed flying by, feeding activities of those marine birds foraging over the surface of the water, and the behavior and condition of birds on or near the platform. The data were tabulated in four four-hour time segments, beginning at 0600 hr and ending at 2200 hr.

RESULTS

Observations were made during the four quarters extending from May, 1976 through April, 1977. The data have been analyzed to the following extent.

Spring Quarter

Seventy-four man-hours of observation were conducted from Platform 288-A on May 15, 16, 22, and 23. Table 1 lists the 59 species of birds noted either on or near the platform area. A total of 2177 birds were identified and counted during the 74 hr observation period. One species, *Chlidonias niger* (black tern), accounted for 1769 of that total. Sixty-one dead birds were found on Platform 288-A during the observation period, representing members of 27 species. One hundred and sixty-seven birds were observed on the platform for varying periods of time.

Twelve man-hours of observation were conducted in the control area during the spring quarter. No birds were sighted.

Summer Quarter

Seventy-five man-hours of observation were conducted from Platform 288-A on July 10, 11, 24, and 25. Table 2 lists the eight species of birds noted either on or near the platform area.

TABLE 1. PLATFORM 288-A SPECIES LIST: SPRING QUARTER

Butorides virescens	Green heron
Bubulcus ibis	Cattle egret
Porzana carolina	Sora
Porphyrula martinica	Purple gallinule
Larus delawarensis	Ring-billed gull
Larus atricilla	Laughing gull
Sterna hirundo	Common tern
Chlidonias niger	Black tern
Coccyzus americanus	Yellow-billed cuckoo
Tyrannus tyrannus	Eastern kingbird
Muscivora forficata	Scissor-tailed flycatcher
Myiarchus crinitus	Great crested flycatcher
Empidonax sp.	Flycatcher
Contopus virens	Eastern wood pewee
Riparia riparia	Bank swallow
Hirundo rustica	Barn swallow
Petrochelidon pyrrhonota	Cliff swallow
Progne subis	Purple martin
Mimus polyglottos	Mockingbird
Dumetella carolinensis	Gray catbird
Catharus ustulatus	Swainson's thrush
Bombycilla cedrorum	Cedar waxwing
Vireo griseus	White-eyed vireo
Vireo flavifrons	Yellow-throated vireo
Vireo olivaceus	Red-eyed vireo
Vermivora peregrina	Tennessee warbler
Vermivora celata	Orange-crowned warbler
Vermivora ruficapilla	Nashville warbler
Dendroica petechia	Yellow warbler
Dendroica magnolia	Magnolia warbler
Dendroica tigrina	Cape May warbler
Dendroica virens	Black-throated green warbler
Dendroica fusca	Blackburnian warbler
Dendroica dominica	Yellow-throated warbler
Dendroica pennsylvanica	Chestnut-sided warbler
Dendroica striata	Bay-breasted warbler
Seiurus aurocapillus	Ovenbird
Seiurus noveboracensis	Northern waterthrush
Seiurus motacilla	Louisiana waterthrush
Oporornis formosus	Kentucky warbler
Oporornis philadelphia	Mourning warbler
Wilsonia citrina	Hooded warbler
Wilsonia pusilla	Wilson's warbler
Setophaga ruticilla	American redstart
Xanthocephalus xanthocephalus	Yellow-headed blackbird
Agelaius phoeniceus	Red-winged blackbird

(continued)

TABLE 1 (continued)

Icterus spurius	Orchard oriole
Euphagus cyanocephalus	Brewer's blackbird
Molothrus ater	Brown-headed cowbird
Pheucticus ludovicianus	Rose-breasted grosbeak
Passerina cyanea	Indigo bunting
Passerculus sandwichensis	Savannah sparrow
Chondestes grammacus	Lark sparrow
Spizella passerina	Chipping sparrow
Melospiza lincolnii	Lincoln's sparrow
Melospiza melodia	Song sparrow

A total 261 birds were identified and counted. Of the total, 178 were *Larus atricilla* (laughing gull) and 69 were *Thalasseus maximus* (royal tern). Six birds were observed on the platform during the observation period, representing members of five species. No dead birds were found on the platform.

Thirty-seven man-hours of observation were conducted in the control area on July 22 and 24 during the summer quarter. Table 3 lists the five species of birds identified. Two hundred and forty-seven individual birds were counted during the observation time. Of the total, 148 were feeding over the surface of the water. The other 99 were observed flying in various directions.

Fall Quarter

Seventy-nine man-hours of observation were conducted on Platform 288-A on September 4, 5, 11, and 12. Table 4 lists the 30 species of birds noted either on or near the platform area. A total of 175 birds were identified and counted. Of the total, 110 were observed on the platform for varying periods of time, and 65 were observed flying by the platform. No dead birds were observed on the platform during the fall quarter observation period.

Thirty-three man-hours of observation were conducted in the control area on September 8 and 17 during the fall quarter. Table 5 lists the ten species of birds identified. One hundred and two individual birds were counted during the observation time. Of that total, 30 were feeding over the surface of the water, and 72 were observed flying in various directions.

TABLE 2. PLATFORM 288-A SPECIES LIST: SUMMER QUARTER

Porphyrula martinica	Purple gallinule
Larus atricilla	Laughing gull
Sterna forsteri	Forster's tern
Thalasseus maximus	Royal tern
Zenaida macroura	Mourning dove
Contopus virens	Eastern wood pewee
Icterus spurius	Orchard oriole
Molothrus ater	Brown-headed cowbird

Winter Quarter

One hundred man-hours of observation were conducted from Platform 288-A on December 18 and 19 and January 8 and 9. Table 6 lists the 14 species of birds noted either on or near the platform area. A total of 798 birds were identified and counted. Two species, *Larus atricilla* (laughing gull) and *Thalasseus maximus* (royal tern), accounted respectively for 540 and 186 of these birds. Ten dead birds were found on Platform 288-A during the observation period, representing members of five species. Seventeen birds were observed on the platform for varying periods of time.

Fourteen man-hours of observation were conducted in the control area on December 16, 1976. Table 7 lists the seven species of birds identified. One hundred and fifty-one individual birds were counted. Three species, *Larus atricilla* (laughing gull), *Thalasseus maximus* (royal tern), and *Chen caerulescens* (snow goose), accounted respectively for 83, 12, and 44 of these birds.

INTERPRETATION

Conclusions based on the data collected and analyzed to date must be viewed as tentative at best. The Buccaneer Field structures appear to have effects on birds both in relationship to the species involved and the season of the year. Thus, the following tentative interpretations are made on a seasonal basis.

Spring Quarter

Over 100 species of birds are known to migrate over the Gulf of Mexico along a route which includes the Buccaneer Field. Approximately another 100 species migrate over routes which lie to

TABLE 3. CONTROL AREA SPECIES LIST: SUMMER QUARTER

Fregata magnificens	Magnificent frigatebird
Larus atricilla	Laughing gull
Sterna forsteri	Forster's tern
Thalasseus maximus	Royal tern
Chlidonias niger	Black tern

the east or west of the Yucatan Peninsula - Galveston line, with occasional overlap of routes due to weather patterns or disorientation of the birds. The migratory period extends from early April for early migrants to early June for late migrants (Pettingill, 1970).

Northward migration over the Gulf in spring is initiated by change in length of day. However, the exact time of departure from the Mexican coast is more closely regulated by frontal weather patterns. The migratory departures from Mexico are heaviest during warm fronts moving in a northerly direction and lightest or non-existent during periods of cold fronts moving in a southerly direction (Gauthreaux, 1971).

Under favorable weather conditions at the point of departure and continuing favorable across the Gulf and along the coastline, small migratory birds will make the flight non-stop until reaching some point inland as far as 25 to 75 nautical miles (Lowery, 1945). If adverse weather conditions, that is, a frontal system with northerly winds, are encountered during the last portion of the flight, birds arriving on the U.S. Gulf Coast are exhausted and will make land-fall at the earliest opportunity (Gauthreaux, 1971).

Small passerine birds have an average flight speed, time of departure from the Yucatan peninsula, and distance to travel that they arrive on the northern coast during daylight hours (Gauthreaux, 1971). Flight elevation during daylight hours has been recorded from aircraft and radar observations to be approximately 350 to 500 m (Bellrose, 1971). Thus, on clear days, migratory birds can readily sight structures as large as those in the Buccaneer Field.

Data collected on Platform 288-A during the spring quarter indicate that the first observation period was conducted during a rather heavy northward flight of migrants, but the presence of birds on the platform also indicates that the several days prior to May 15 was most likely heavier. Migration near the Gulf of Mexico was considered unusual in the spring of 1976 by several

TABLE 4. PLATFORM 288-A SPECIES LIST: FALL QUARTER

Ardea herodias	Great blue heron
Bubulcus ibis	Cattle egret
Larus atricilla	Laughing gull
Sterna forsteri	Forster's tern
Thalasseus maximus	Royal tern
Chlidonias niger	Black tern
Zenaida macroura	Mourning dove
Caprimulgus carolinensis	Chuck-will's-widow
Archilochus carolinensis	Ruby-throated hummingbird
Colaptes auratus	Common flicker
Tyrannus tyrannus	Eastern kingbird
Tyrannus verticalis	Western kingbird
Empidonax sp.	Flycatcher
Hirundo rustica	Barn swallow
Lanius ludovicianus	Loggerhead shrike
Mniotilta varia	Black-and-white warbler
Vermivora ruficapilla	Nashville warbler
Dendroica petechia	Yellow warbler
Seiurus noveboracensis	Northern waterthrush
Geothlypis trichas	Common yellowthroat
Icteria virens	Yellow-breasted chat
Wilsonia pusilla	Wilson's warbler
Wilsonia canadensis	Canada warbler
Setophaga ruticilla	American redstart
Dolichonyx oryzivorus	Bobolink
Agelaius phoeniceus	Red-winged blackbird
Icterus spurius	Orchard oriole
Icterus galbula	Northern oriole
Molothrus ater	Brown-headed cowbird
Spiza americana	Dickcissel

onshore observers (Smith, 1976; Webster, 1976). Northerly winds from May 13 through 15 caused a great fallout of migrants along the coast. In the absence of such winds, migrants tend to continue inland for distances of 25 to 100 miles before making landfall.

Sixty-one dead birds were recovered from the platform during the spring observation period. Of these, 21 died on the platform between 2200 hr on May 15 and 1700 hr on May 16. Death appeared to result from exhaustion or lack of food and water. No birds were observed colliding with the structure or coming into contact with any substance. The mortality rate cannot be considered excessive as there are documented cases of large mortality over natural habitats during migratory periods. Smith (1976) notes

TABLE 5. CONTROL AREA SPECIES LIST: FALL QUARTER

Morus bassanus	Gannet
Fregata magnificens	Magnificent frigatebird
Larus argentatus	Herring gull
Sterna forsteri	Forster's tern
Thalasseus maximus	Royal tern
Chlidonias niger	Black tern
Caprimulgus carolinensis	Chuck-will's-widow
Archilochus colubris	Ruby-throated hummingbird
Hirundo rustica	Barn swallow
Molothrus ater	Brown-headed cowbird

that 200,000 migrants were estimated to have died over Lake Huron on a single day in late May of 1976. Most birds which landed on the platform stayed for varying periods of time ranging from several minutes to one or two hours and then flew north toward the coast. In several instances it was possible to keep track of individual birds for long periods of time. On the basis of minimal data, it appeared that any bird staying on the platform for more than six hours after arrival would eventually die there. It is assumed that exhaustion from the trans-Gulf flight and energy consumed from basal metabolism after reaching the platform caused death from starvation.

The large numbers of black terns recorded during the spring quarter were observed feeding over the water within visual range of the platform. The black tern migrates across the Gulf during May and June (Peterson, 1960) and the observation periods on May 15 and 16 were apparently during a peak migratory wave of this species. Neither the Buccaneer Field structures nor associated boat activity seemed to have any particular attraction for this species, as they were also observed over a wide area of the Gulf away from the platforms. Laughing gulls, royal terns, and common terns appeared to be attracted to the structures and the associated human activity. Flight directions and numbers of birds of these three species appear to be directly related to feeding activity in the vicinity of the Buccaneer Field.

Observation in the control area during the spring quarter yielded no data. The first control area selected was approximately five nautical miles northeast of Platform 288-A. Weather conditions were clear, with visibility over 15 miles and wind speed of approximately 2 knots. Large birds, most likely gulls and terns, were observed at a distance of over three nautical miles, well out of the area considered for a control. Their general flight directions were between the coastline and the

TABLE 6. PLATFORM 288-A SPECIES LIST: WINTER QUARTER

Sula dactylatra	Blue-faced booby
Bubulcus ibis	Cattle egret
Stercorarius pomarinus	Pomarine jaeger
Larus argentatus	Herring gull
Larus atricilla	Laughing gull
Thalasseus maximus	Royal tern
Zenaida macroura	Mourning dove
Turdus migratorius	American robin
Bombycilla cedrorum	Cedar waxwing
Molothrus ater	Brown-headed cowbird
Dendroica coronata	Yellow-rumped warbler
Junco hyemalis	Dark-eyed junco
Zonotrichia leucophrys	White-crowned sparrow
Melospiza lincolnii	Lincoln's sparrow

Buccaneer Field. Adverse weather conditions and problems with scheduling ship-time prevented completion of the second observation period in the control area during the spring quarter.

Summer Quarter

The number of species observed on the platform during the summer quarter was 8, with laughing gulls and royal terns comprising 247 of the 261 birds seen. The presence of the Buccaneer Field structures and associated boat traffic can quite safely be considered to be beneficial for these two species, as most individual birds were observed either following, going toward, or returning from man-made objects. The other species observed were few in number and remained on the platform for only short periods of time. They all represent summer terrestrial residents along the upper Gulf coast of Texas and possible represent individuals which sighted the Buccaneer Field structures from land and flew out.

Birds observed in the control area during the summer quarter were all local summer resident marine birds which feed over open water along the coastline. Magnificent frigatebirds were seen in the control area, but not over the Buccaneer Field area. However, not a large enough sample was made to warrant conclusions that this species would avoid the man-made structures. Many of the gulls and terns observed appeared to be flying from one vessel to another as the ships passed through the general vicinity of the control area and the relatively high count of individual birds

TABLE 7. CONTROL AREA SPECIES LIST: WINTER QUARTER

Chen caerulescens	Snow goose
Larus argentatus	Herring gull
Larus atricilla	Laughing gull
Larus philadelphia	Bonaparte's gull
Thalasseus maximus	Royal tern
Bombycilla cedrorum	Cedar waxwing
Molothrus ater	Brown-headed cowbird

(247) can perhaps be accounted for by reason of their foraging flights.

Fall Quarter

Fall migration, like spring migration, is initiated by change in day-length. Migratory birds move southward over the central portion of the U.S. The direct flight over the Gulf of Mexico is more closely regulated by frontal patterns on the U.S. mainland. Able (1972) states that the conditions most conducive to autumnal trans-Gulf flight occur shortly after the passage of a massive cold front which penetrates far into the Gulf. A less favorable, but still conducive, condition is the presence of high pressure cells over the central and eastern U.S. with airflow ocross the Gulf generally from the northeast or east. Weather conditions other than the above two result in bird populations building up along coastal land areas until favorable conditions for trans-Gulf migration occur. Fall migration extends from late September through mid-November, depending upon the species involved.

Observations on the platform during the fall quarter resulted in recognition of 175 birds of 30 different species. No dead birds were found and all birds on the platform appeared alert and healthy. Insectivorous birds were observed feeding on moths, butterflies, small beetles, and other flying insects which were in abundance on the platform. Nearly one-third of the birds were observed flying south or southwest past the platform and did not land. Approximately one-third of those which landed on the platform left within a few minutes. On the basis of the data collected, it appears reasonable to speculate that the platform structures have no adverse effect on migratory birds during the fall flight. The relative proximity of the Buccaneer Field to the mainland would require only an hour's flight for most small birds and if they continued on their southerly flight within a short period of time it should not make a significant difference in arrival time on the northern Mexican coastline.

Winter Quarter

The number of species observed on or near Platform 288-A during the winter quarter was 14, and all of them consisted of winter residents along the adjacent coastline. Laughing gulls and royal terns accounted for over 90% of the total birds observed. These two species are closely associated with the structures and related human activity. Flights between shore and the structures were observed, but sizeable numbers of both species remained on the jackets overnight on both observation periods. Eighty-seven laughing gulls and aver 40 royal terns were on the jackets at nightfall on January 8, and the same number were counted at daylight on January 9. Both species were observed feeding on wet garbage from the platforms and the tending vessels. The other species observed were few in number and remained on the platform for only short periods of time. Flight directions when leaving the platform were usually northward toward the coast.

Birds observed in the control area during the winter quarter were all winter residents along the coast. Three small flocks of snow geese passed over the contol area. Laughing gulls and royal terns constituted the other two species with large numbers of individuals.

CONCLUSIONS

The collection and analysis of data for the four quarters of observation during 1976-77 does not warrant definite conclusions concerning the overall effects of the Buccaneer Field structures on avian species. Migratory periods during spring and fall cover time segments of over a month each. Within those time periods, weather factors, particularly storm front winds, have an effect on when birds start their migratory flights, how long it takes to make the trans-Gulf crossing, where landfall is made, and the physical condition of the individual birds at the time they reach landfall. Observation periods during the four quarters of this study were scheduled around the availability of transportation to the platform and control areas. Consequently, data collections on any given day during the migratory seasons may or may not be representative of the season in total and could vary considerably from year to year.

Analysis of the data from the summer and winter quarters (non-migratory seasons) leads to a reasonable conclusion that the effect of the structures on birds ranges from insignificant for some species to beneficial for others. Species such as the black tern and snow goose pass through the area and apparently are not influenced negatively or positively by the structures. Terrestrial passerine birds observed on the platform were all representative of winter residents along the coast. The small number of dead birds recovered during these periods was insignificant compared to the number which spent short periods on the structures and left again on a northerly course. The presence of the field can reasonably be considered beneficial for species such as the laughing gull, ring-billed gull, herring gull, and royal tern. Laughing gulls and royal terns in particular make extensive use of the structures and surrounding water for resting and feeding areas.

It appears reasonable to speculate from the data that fall migrants suffer no adverse effects from the presence of the structures, and possibly may benefit from them. Insect-feeding birds in particular foraged to a considerable extent on the diverse and ample supply of insects present on the platform during the observation periods. Physical condition of the birds appeared to be excellent and departure time from either the platform or coastal area should have little or no effect on the duration of the flight across the Gulf.

The presence of large numbers of dead birds on the platform during the spring quarter could be interpreted as a detrimental effect of the Buccaneer Field structures. Mortality was never observed to be a direct result of either collisions with the structures or contamination by petrochemical substances. Dead and dying birds on the platform appeared to be a result of exhaustion and lack of food and water. If the platform structures were not present, an additional hour of flight time would put birds on land where resources would be available. However, it is well known that mortality from predators and other factors on land also contribute quite heavily to mortality of spring migrants. It should also be noted that dead birds on the platform are readily apparent, whereas dead birds on land quite often go undetected.

ACKNOWLEDGEMENTS

This work is a result of research sponsored by the Environmental Protection Agency and Department of Commerce, National Oceanic and Atmospheric Administration, National Marine Fisheries Service, Southeast Fisheries Center, Galveston Laboratory under Contract No. 03-6-042-35119

REFERENCES

Able, K. P., 1972, "Fall migration in coastal Louisiana and the evolution of migration patterns in the Gulf region," Wilson Bull., 84:231.

Bellrose, F. C., 1971, "The distribution of nocturnal migrants in the air space," Auk, 88:397.

Gauthreaux, S. A., 1971, "A radar and direct visual study of passerine spring migration in southern Louisiana," Auk, 88:343.

Lowery, G. H., Jr., 1945, "Trans-Gulf spring migration of birds and the coastal hiatus," Wilson Bull., 57:92.

Oberholser, H. C., and Kincaid, E. B., Jr., 1974, "The Bird Life of Texas." Univ. Texas Press, Austin, Texas.

Peterson, R. T., 1960, "A Field Guide to the Birds of Texas." Houghton-Mifflin Co., Boston, Massachusetts.

Pettingill, O. S., 1970, "Ornithology in Laboratory and Field." Burgess Publishing Co., Minneapolis, Minnesota.

Robbins, C. S., Bruun, B., and Zim, H. S., 1966, "Birds of North America." Golden Press, Racine, Wisconsin.

Smith, P. W., 1976, "The changing seasons," American Birds, 30:805.

Webster, F., 1976, "The spring migration: South Texas region," American Birds, 30:863.

BACTERIAL COMMUNITY COMPOSITION AND ACTIVITY

Ronald K. Sizemore, Chi-Hsin Hsu, and Kurt D. Olsen

Department of Biology
University of Houston
Houston, Texas 77004

INTRODUCTION

Microorganisms, particularly bacteria, constitute the largest numerical population in most aquatic environments and are often the most metabolically active aquatic organisms. The short generation time and great physiological capacity of bacteria as a group enable them to survive and flourish in environments that are inhospitable to other organisms. Indeed, bacterial activity can alter an aquatic environment thereby affecting the ability of the environment to maintain nonmicrobial populations. Often bacterial activity renders recalcitrant materials more palatable or less noxious, thus improving the environment for other organisms. Petroleum in general represents a form of organic pollution that may have adverse effects on aquatic habitats. In the marine environment, microorganisms are the major biotic factors responsible for recycling petroleum (LaRock and Severance, 1973; Walker and Colwell, 1974). Therefore, any comprehensive study of the effects of petroleum on the aquatic biosphere should include a study of bacterial adaptation to petroleum pollution.

Petroleum and associated pollutants have a number of effects on the microbial population. Some components of oil are recalcitrant and may be inhibitory to some portions of the bacterial population (Atlas, 1975; Walker and Colwell, 1974). One of the objectives of this study was to determine the effects of oil pollutants on the normal bacterial population of uncontaminated sites and to estimate the success of the microbial population around the production platforms to cope with the pollutants.

Microorganisms are not passive individuals which merely respond to habitat changes but are actively altering the aquatic environment. Estimations of the rate of attack on petroleum products by bacteria combined with an appreciation of the by-products resulting from microbial degradation are vital to determine the effect of oil pollution in the marine environment. This information can be used to estimate the fate of petroleum products in the ocean and therefore is vital in assessing the impact of oil field contaminants. To date, bacteriologists have concentrated a large portion of their studies on unexpected large oil spills occurring near coastlines (Kerr, 1977). These acutely oil-polluted sites have been found to support relatively large bacterial populations containing an abnormally high proportion of bacteria capable of degrading oil. Production platforms in the Gulf of Mexico afford a unique opportunity to study the microbial ecology of oil degradation because they represent an area of long-term exposure to low levels of oil pollutants. These sites of chronic oil contamination may be the most severely damaged environments (Loughry, 1977) but the autochthonous bacterial population of these areas has had a maximal opportunity to adapt to the presence of the platforms. This study was designed to determine the extent of microbial adaptation around the oil drilling site.

MATERIALS AND METHODS

Most of the techniques in this study have been described in detail elsewhere (Sizemore et al., 1979; Sizemore and Olsen, 1980) and are summarized below.

Samples for bacterial analysis were collected using sampling devices which are designed specifically to collect bacteria from the aquatic environment. Water samples were collected with a Niskin sterile bag sampler at a mid-water depth and from a depth approximately one to two meters above the bottom. Surface water samples were collected with sterile glass bottles. Surface sediment samples were routinely collected with a Petite Ponar Grab. Subsamples of sediments were taken with a sterile spatula from the center of the least disturbed portion of the grab sample. All samples during the entire study were plated or fixed immediately upon recovery. This insured that any changes in bacterial population such as would occur after even brief storage were avoided.

Sampling sites included a variety of stations around platforms 288-A and 296-B. During the first portion of the study, sites were sampled at intervals up to several kilometers from the rigs. However, as the study advanced it became obvious that any platform effects on bacterial communities would only be apparent in close proximity to the platform. Subsequent sampling sites were then chosen at intervals as close as feasible to the platform.

For the purpose of comparison, a control site five nautical miles due north of the platform site was chosen. This site is of a similar depth and bottom stratum as the platform area and is free from any petroleum production structures.

Bacterial enumeration utilized two techniques. Traditional plate counts were made using a broad spectrum medium, Modified Sea Water Yeast Extract (MSWYE) medium. This medium consists of 1 g of proteose peptone and 1 g of yeast extract dissolved in 1 l of 3 salts solution at pH 7.2-7.4. If MSWYE agar was required, 20 g of Bacto agar was added to each liter of MSWYE broth before autoclaving. Three salts solution, which is composed of 0.4 M NaCl, 0.028 M $MgSO_4$ and 0.01 M KCl, was used as artificial seawater throughout this study. To minimize sampling variance in this study the subsampling scheme and plating technique of Kaper et al. (1978) was utilized. This technique required quadruplicate subsampling of the original sample and duplicate plating of each subsample. This technique was used primarily to obtain cultures and to provide total viable counts to compare with the selective media used. The other enumeration technique involved epifluorescent microscopic examination of acridine orange stained bacterial cells. This technique generally avoids problems inherent in the plating procedure and gives the highest and presumably the best estimate of bacterial population numbers (Hobbie et al., 1977). Biomass estimates were made by utilizing the bacterial numbers obtained and multiplying the (microscopically) estimated average cell volume. This number was then multiplied by a carbon to mass correction value obtained from the literature (Watson et al., 1977).

Enumeration of specific physiological bacterial types was made by use of various selective media described elsewhere (Hollaway et al., 1980). Sulfur utilizing and oil degrading bacteria were examined in detail. The oil MPN tube procedure, which in our hands was the most reproducible technique and produced the least number of false positive tests, was used to enumerate oil degrading microorganisms. Sulfur metabolizing bacteria were enumerated on a variety of media (Postgate, 1967) to ensure that the many different types of sulfur oxidizing and reducing bacteria could be detected.

Bacterial taxonomy was performed using isolates collected from either the non-selective medium (MSWYE) used for enumeration or from selective media (e.g., oil MPN tubes) used to characterize specific types of bacteria. Strains were identified using a simple dichotomous key. Some strains, especially those isolated from fish, were subjected to identification utilizing API 20E strips (Analytab Products, Plainview, New York). These strips are designed for clinical isolates and therefore could not identify all the marine strains. Where the strips did not provide a positive identification the test data from the strips were utilized in a dichotomous key.

Bacteria from the study area were examined for physiological diversity and the ability to tolerate discharge products from the production platforms. Bacteria from the platform and control sites as well as laboratory strains (*Escherichia coli* and *Pseudomonas aeruginosa*) were tested for their ability to grow, attach to glass surfaces, and to exhibit normal chemotaxis in the presence of various concentrations of platform discharge. The concentrations used varied from full strength brine discharge to diluted discharge such as would be found in the area around the platform. Some brine discharge samples suspected to contain biocide were also used.

This study estimated the capacity of bacteria from the study area to utilize hydrocarbon discharge from the platforms. These experiments utilized laboratory conditions which were desigened to simulate optimal environmental conditions for oil degradation (i.e., optimal NO_3 + PO_4 concentrations). The hydrocarbon utilizing capacity of individual cultures and mixed cultures from the platform area were compared to isolates from the uncontaminated control site. During the first two years of study the technique and equipment used only permitted the examination of n-alkane degradation. In the last year of the study model petroleum (Walker and Colwell, 1974) was utilized as a substrate in place of Buccaneer Field petroleum. Model petroleum contains very few components relative to the number of components in crude oil and is thus different from the petroleum discharge actually found in the field. However, the utilization of model petroleum as a substrate for biodegradation permitted the identifaction and quantification of a variety of hydrocarbon components (e.g., aromatic compounds) with existing equipment. This was not possible and/or practical utilizing Buccaneer Field brine discharge or crude oil.

RESULTS AND DISCUSSION

Bacterial numbers and biomass tended to be slightly higher at the control site than at platform 296-B (see Hollaway et al., 1980). For example, during the 1978-1979 study period bacterial counts in the water column around platform 296-B averaged 465 bacteria per ml whereas the water in the control site averaged 564 bacteria per ml (as determined by plating). Epifluorescent bacterial counts were similar to total viable counts (plating) but

TABLE 1. TAXONOMIC COMPOSITION OF ISOLATES.

Taxon	Cruise I: Aug. Platform	Cruise I: Aug. Control	Cruise II: Nov. Platform	Cruise II: Nov. Control	Cruise III: Feb. Platform	Cruise III: Feb. Control	Cruise IV: May Platform	Cruise IV: May Control
Aeromonas	8	9	12	12	6	21	8	12
Acinetobacter	2	1	0	1	0	4	0	2
Acinetobacter iwoffi	6	11	3	11	6	3	0	2
Bacillus	0	1	1	0	1	0	0	0
Cryneforms	5	3	3	1	1	0	1	0
Cytophaga	5	4	3	6	1	0	0	0
Enterobacteriaceae	2	1	1	0	1	0	0	0
Flavobacterium	3	4	2	5	0	0	0	0
Flexibacter	3	2	3	4	3	1	0	0
Lactobacillus	0	0	1	0	0	0	0	0
Lucibacterium	0	0	0	0	0	0	2	3
Micrococcus	0	2	1	0	0	0	0	0
Moraxella	3	3	2	6	3	1	2	0
Planococcus	0	0	1	0	0	0	0	0
Pseudomonas	4	3	2	6	8	7	0	3
Pseudomonas Gp 2	0	4	0	0	5	0	1	0
Pseudomonas Gp 3	8	11	12	8	8	7	8	7
Vibrio	12	4	23	5	23	24	45	40
Diversity (H'=)	2.45	2.44	2.14	2.23	2.01	1.63	1.11	1.33

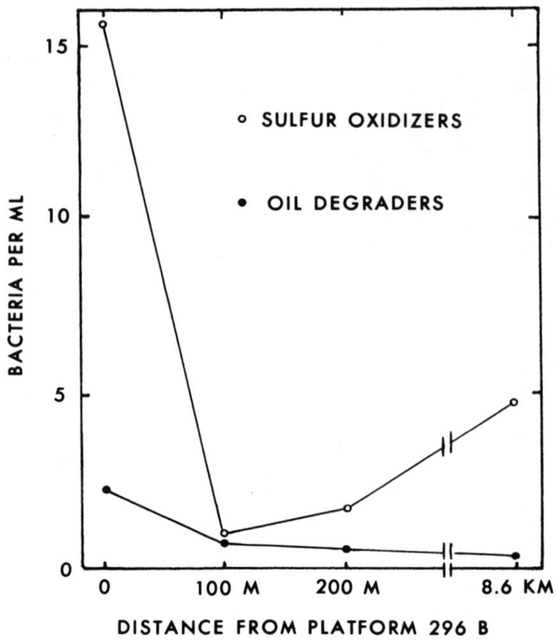

Figure 1. Average number of oil degrading and sulfur oxidizing bacteria in the water column of 4 sampling sites collected between August 1978 and August 1979. The sample collected at a distance of 8.6 km from the platform was in the control area.

were usually a thousand-fold higher. Bacterial numbers in the sediment were also higher at the control site (4.1×10^6 bacteria per g dry weight sediment) as opposed to sediment under platform 296-B (3.4×10^6 bacteria per g). A seasonal pattern in bacterial numbers was also seen in the 1978-1979 sampling period. The highest bacterial numbers were found in the water of the February sampling cruise (e.g., 1123 bacteria per ml at platform 296-B) and the lowest in the November cruise (e.g., 141 bacteria per ml at platform 296-B). These trends in bacterial numbers are discussed in detail elsewhere (Olsen and Sizemore, 1981). During the 1977-1978 and 1979-1980 sampling periods this trend was constant. During the 1978-1979 period, the pattern was inconsistent and was hidden by the combination of data from platform 288-A and platform 296-B which at the time were assumed to have similar rates of petroleum production discharge (brine). Throughout the latter two years of this study, platform 288-A closely resembled the control site in bacterial numbers and biomass. This resemblance could be due to the cessation of routine discharge of brine from platform 288-A in November, 1978. If only platform 296-B was compared to the control site, differences in bacterial numbers were detectable

TABLE 2. PHYSIOLOGICAL CAPACITY OF TAXONOMIC ISOLATES COLLECTED FROM SEDIMENT SAMPLES BETWEEN AUGUST 1978 AND AUGUST 1979.

Cruise	% Oil degraders	% Sulfur oxidizers	% Sulfate reducers
		Platform 296-B	
I	0	7	18
II	0	4	21
III	9	4	18
IV	9	6	36
		Control site	
I	0	0	3
II	0	0	8
III	1	3	6
IV	1	1	13

by both the plating technique and the epifluorescent counts. The difference in numbers between the two sites however was small, usually less than two-fold. Variations between sampling dates or different water depths at the same date and site were more dramatic (often 10 or 100-fold) than between sites. This general phenomenon was also true for the taxonomic makeup of the bacterial population. No taxonomic differences were noted in the bacterial population of the platform vs. the control site in either the water column (Hollaway et al., 1980; Sizemore et al., 1979) or the sediment (Sizemore and Olsen, 1980). The taxonomic diversity of both sites changed dramatically seasonally but the taxons and diversity were essentially the same for both sites (Table 1). *Pseudomonas*, *Vibrio*, *Aeromonas*, and *Acinetobacter* were the predominant bacterial genera at all sites.

The similarity between the platform site 296-B and the control site also held for the general physiological profile of the bacterial population (Hollaway et al., 1980). The percentages of proteolytic, chitinoclastic, lipolytic, and salt-requiring bacteria were essentially identical for the platform site (296-B) and the control site. Bacteria from the two sites were also compared for their ability to utilize two specific components of the discharge; oil and sulfur. The number of oil degrading bacteria and sulfur oxidizing bacteria was in general low at both sites. However, the platform site consistently contained a higher number and percentage of both types of bacteria than the control site (Figure 1). Sulfate reducing bacteria were also found to be more prevalent in the sediment under the platform (Table 2). These data suggest that the platform discharge has a very small yet detectable effect on the bacterial communities. Bacterial populations around the platform have adapted to the presence of the platform discharge products but this adaptation has not required a dramatic change in the population.

Brine discharge is continuously being released from the production platforms. At least three components are known to be in this discharge; sulfur, petroleum residue, and a sporadic inclusion of biocides. The biocides are added to the brine discharge to limit bacterial growth within the tanks on the platform. The brine discharge was found to be inhibitory to the growth and functioning of some laboratory cultures. However, diluted brine discharge did not affect the growth of isolates from the platform area. Indeed, some strains grew more rapidly in the dilute discharge than the control medium. Full strength brine discharge did not support the growth of some platform strains but did not appear to be lethal to these strains. However, immediately below the platform the suspended particulate matter did not contain normal bacterial numbers (i.e., less than 1% of the total viable count). In all other water samples examined, most (approximately 90%) bacteria were found to be attached to particles larger than 3 m. In the laboratory, platform isolates were able to attach to particles and could exhibit chemotaxis in dilute brine discharge. The lack of attached bacteria under the discharge point could be due to the effect of the high concentration of the discharge or an artifact of the composition of the discharge water itself. Biocide within the tank may kill the attached bacteria in the discharge before release. In this study no effect of biocide could be detected. However, there was some difficulty in obtaining fresh discharge samples known to contain biocides. These data are supported and explained by a chemical examination of brine discharge which was thought to contain biocide that showed no evidence of appreciable biocide levels (Middleditch et al., 1979).

Two materials released in brine discharge, sulfur and hydrocarbons, were examined for their ability to serve as nutrients for bacterial growth. Sulfur was not utilized in detectable amounts in our experiments. Visual indication of sulfur oxidation was seen but gravimetric analysis did not detect sulfur loss. An examination of the taxonomic composition of the sulfur oxidizing bacteria suggests an explanation for this discovery. Less than 1% of the bacteria capable of attacking elemental sulfur in the study area were obligate autotrophic sulfur oxidizing bacteria (i.e., genus *Thiobacillus*). Heterotrophic *Pseudomonas* sp. which can utilize sulfur as a supplemental nutrient (Tuttle et al., 1974) were more prevalent, comprising more than half of the sulfur

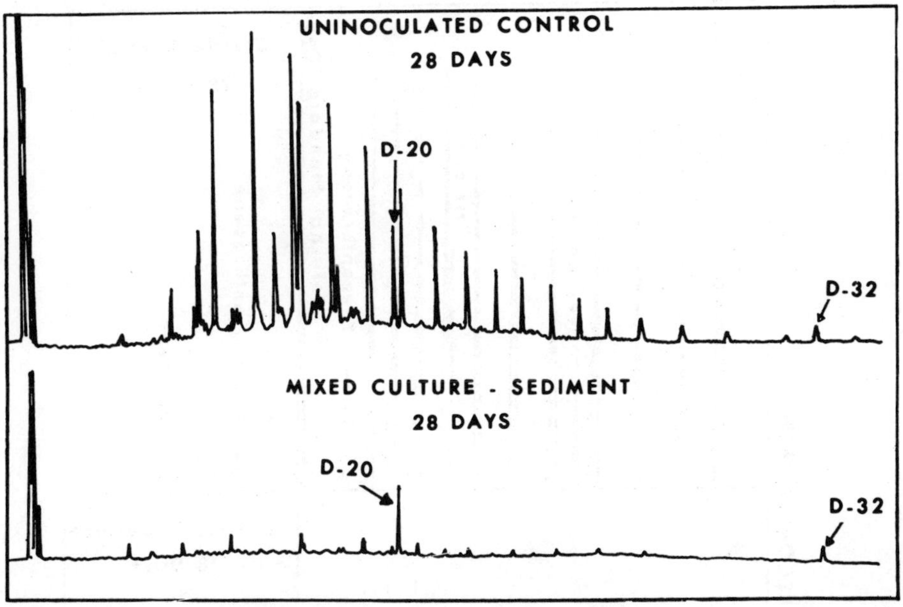

Figure 2. Gas-liquid chromatographic tracing of uninoculated weathered control (upper) vs. biodegraded sample (lower). D-20 and D-32 are deuterated internal standards added at a concentration of 2.5 ppm. (taken from Olsen and Sizemore, 1981)

oxidizing bacterial population. These strains preferentially utilize organic material for growth and would not be expected to utilize large quantities of sulfur. Sulfate reducers, 32% of the bacterial population, were common in the sediment of the platform area but no attempt was made to measure their activity.

Petroleum hydrocarbons are continuously released in small quantities (Middleditch et al., 1979) from the production platforms. Measurements have been made of the potential of microorganisms from the platform area, compared with strains from other sites, to utilize different components of the discharge hydrocarbons. During the first two years of this study (1977-1979), only biodegradation of the n-alkane fraction of the hydrocarbons were studied. This portion of petroleum was readily utilized by a number of bacteria (Faw and Sizemore, unpublished data). In general, mixed cultures prepared from water and sediment samples collected around the platform could utilize hydrocarbons (Faw and Sizemore, unpublished data; Olsen and Sizemore, 1981). The activity of these mixed cultures, combined with abiotic factors, could result in loss of most of the n-alkane hydrocarbons within 7 to 28 days (Figure 2). Individual pure cultures, usually belonging to the genus *Pseudomonas*, were also able to utilize

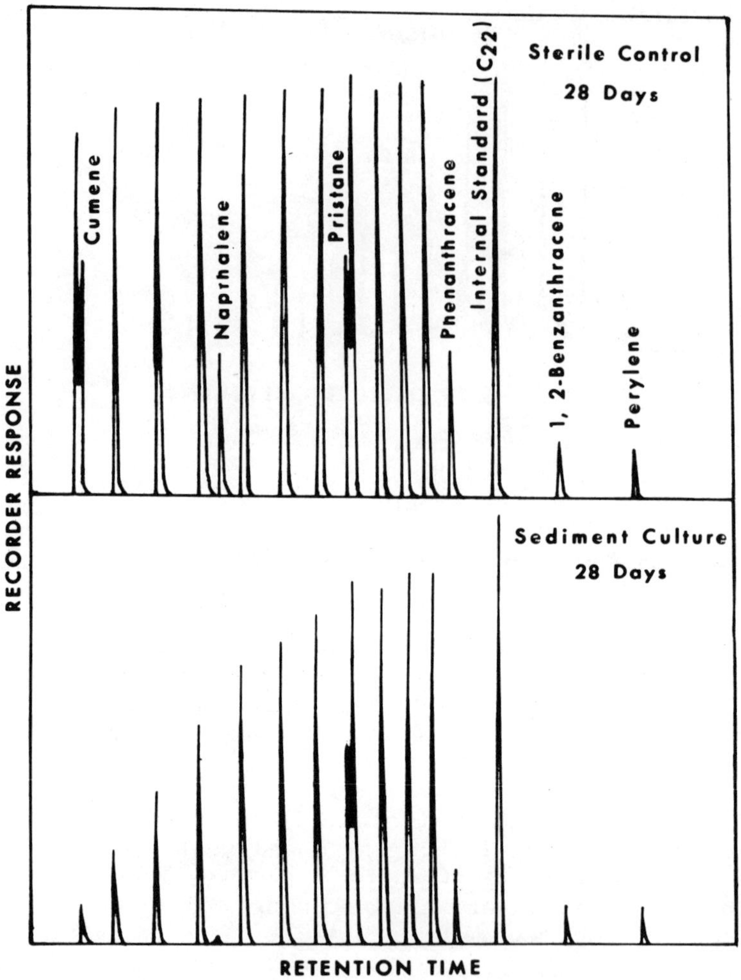

Figure 3. Gas-liquid chromatographic tracing of 1% model petroleum without (upper) and with (lower) inoculum of a mixed sediment culture derived from Buccaneer sediment. Unlabeled peaks are straight-chain alkanes from C_{10} to C_{20}. Peaks in the inoculated culture which are smaller than the control have been biodegraded.

n-alkanes. While most of the pure cultures did not utilize oil as quickly or thoroughly as mixed cultures, some isolates collected from around the platform appeared to be exceptional alkane utilizers. One strain isolated from the sediment under platform 296-B was able to utilize the majority of n-alkanes in the study mixture in a seven day period (Faw and Sizemore, unpublished data).

Alkanes, however, are only a portion of the hydrocarbons found in petroleum. To study the biodegradation of other components of oil, a model or synthetic petroleum was used. Mixed cultures prepared from sediment samples collected from around platform 296-B were particularly effective in degrading the different classes of hydrocarbons in model petroleum. In general, aromatic compounds were degraded less rapidly than the aliphatic alkanes. Of the non-alkane hydrocarbons tested, cumene was the most readily degraded followed by naphthalene, phenanthracene, 1,2-benzanthracene, and perylene in that order. Certain pure cultures were able to utilize only some of the specific components of model petroleum. Figure 3 shows a degradation experiment which utilized an oil field sediment culture which attacks naphthalene.

In a separate study, the genetic basis for adaptation to hydrocarbon utilization was studied (Hada and Sizemore, 1981). Bacterial strains collected from the control site and platform site were examined for the presence of extrachromosomal plasmid DNA. The plasmid DNA represents an efficient way by which bacterial cultures adapt to new stresses or environments. The comparison of the two sites showed that strains from the oil field contained a higher incidence of plasmids than the control site. This may represent a mechanism by which the platform area bacteria adapt to the low level pollutants without exhibiting a large change in bacterial numbers or types. This type of adaptation represents a minimum of metabolic energy expenditure by the populations but permits maximal genetic flexibility.

During the course of this study, Gallaway and co-workers (this Volume) reported an apparent increased incidence of diseased fish around the platforms. A modest effort was made to examine the predominant bacterial flora of the diseased and healthy fish from the study area in an attempt to detect the presence of bacterial pathogens. A number of different members of the genera *Vibrio* and *Aeromonas* were found associated with diseased and apparently healthy fish. One bacterium, *Aeromonas hydrophilia*, a known fish pathogen was found associated with a few diseased fish. It is impossible to conclude from the small number of diseased fish examined that *Aeromonas hydrophila* is the causal agent. However, *Aeromonas hydrophilia* has been linked with fish diseases normally triggered by polluted waters (Mitchell, 1980). This discovery is preliminary and an additional detailed study would be helpful.

ACKNOWLEDGEMENTS

We wish to thank Ms. A. Esqueda and Mr. G. Bevel for their excellent technical assistance. This work is a result of research sponsored by the Environmental Protection Agency and Department of

Commerce, National Oceanic and Atmospheric Administration, National Marine Fisheries Service, Southeast Fisheries Center, Galveston Laboratory under Contract Nos. 03-7-042-35123, FSE43-8-ERP-47, and NA79-GA-C-0040.

REFERENCES

Atlas, R. M., 1975, "Effect of temperature and crude oil composition on petroleum biodegradation," Appl. Microbiol., 30:396.

Hada, H. S., and Sizemore, R. K., 1981, "Survey of the incidence of plasmids in marine *Vibrio* spp.," Appl. Environ. Microbiol., in press.

Hobbie, J. E., Daley, R. J., and Jasper, S., 1977, "Use of nucleopore filters for counting bacteria by fluorescence microscopy," Appl. Environ. Microbiol., 33:1225.

Hollaway, S. L., Faw, G. M., and Sizemore, R. K., 1980, "The bacterial community composition of an active oil field in the northwestern Gulf of Mexico," Mar. Pollut. Bull., 11:153.

Kaper, J. B., Mills, A. L., and Colwell, R. R., 1978, "Evaluation of the accuracy and precision of enumerating aerobic heterotrophs in water samples by the spread plate method," Appl. Environ. Microbiol., 35:756.

Kerr, R. A., 1977, "Oil in the ocean: circumstances controlling its impact," Science, 198:1134.

LaRock, P. A., and Severance, M., "Bacterial treatment of oil spills: some facts considered," in: Stevenson, L. H., and Colwell, R. R. (eds.), "Estuarine Microbial Ecology," University of South Carolina Press, Columbia, South Carolina, 536 pp.

Loughry, J., 1977, "Fate and effects of oil in the ocean," Surveyor, 11:3.

Middleditch, B. S., Basile, B., and Missler, S. R., 1979, "Determine levels, pathways, and bioaccumulation of selected discharge constituents (non-metals) in the marine ecosystem in the oil field/hydrocarbon modeling," in: Jackson, W. B. (ed.), "Environmental Assessment of an Active Oil Field in the Northwestern Gulf of Mexico, 1977-1978, Vol. III: Chemical and Physical Investigations." NOAA, pp. 2.4.1-1 to 2.4.1-302. Available from: NTIS, Springfield, Virginia, Accession No. PB80107899.

Mitchell, R., 1980, "Environmental effects on microbial processes," in: Schlessinger, D. (ed.), "Microbiology-1980." American Society for Microbiology, Washington, D.C.

Olsen, K. D., and Sizemore, R. K., 1981, "Effects of an established offshore oil platform on the autochthonous bacterial community," Dev. Indust. Microbiol., in press.

Postgate, J. R., 1967, "Media for sulfur bacteria," Lab. Practice, 15:1239.

Sizemore, R. K., and Olsen, K., 1980, "Bacterial Communities," in: Jackson, W. B., and Wilkens, E. P. (eds.), "Environmental Assessment of Buccaneer Gas and Oil Field in the Northwestern Gulf of Mexico, 1978-1979, Vol. IV." NOAA Technical Memorandum NMFS-SEFC-38, 32 pp. Available from: NTIS, Springfield, Virginia.

Sizemore, R. K., Hollaway, S. L., and Faw, G. M., 1979, "Determine and compare groups of bacteria from an active oil field and unaltered area," in: Jackson, W. B. (ed.), "Environmental Assessment of an Active Oil Field in the Northwestern Gulf of Mexico, 1977-1978, Vol. II: Data Management and Biological Investigations." NOAA, pp. 2.3.7-1 to 2.3.7-41. Available from: NTIS, Springfield, Virginia, Accession No. PB80165970.

Tuttle, J. H., Holmes, P. E., and Jannasch, H. W., 1974, "Growth rate stimulation of marine Pseudomonads by thiosulfate," Arch. Microbiol., 99:1.

Walker, J. D., and Colwell, R. R., 1974, "Microbial petroleum degradation: use of mixed hydrocarbon substrates," Appl. Microbiol., 27:1053.

Watson, S. W., Novistsky, T. J., Quinby, H. L., and Valois, F. W., 1977, "Determination of bacterial numbers and biomass in the marine environment," Appl. Environ. Microbiol., 33:940.

EFFECTS ON ARTIFICIAL REEF AND DEMERSAL FISH AND MACROCRUSTACEAN COMMUNITIES

Benny J. Gallaway, Larry R. Martin, Randall L. Howard, Gregory S. Boland, and George D. Dennis

LGL Ecological Research Associates, Inc.
1410 Cavitt Street
Bryan, Texas 77801

INTRODUCTION

This chapter provides a synthesis of the observed effects of the Buccaneer Gas and Oil Field (Figure 1) on biological systems and fisheries of the study area. Demersal fishes and macrocrustaceans, the biofouling community, and reef and pelagic fishes were used as indicators of impact. Other biological components (bacteria, plankton, and infaunal benthos) were considered as part of the "environment" and were addressed only to the extent that they directly impacted or limited the primary indicator groups.

INVESTIGATIVE PROGRAM

Project Development and Approach

The first year of the study (1976-1977) was expended in characterizing the regional environment and biota. While most of the work was, by necessity, descriptive (the area had been studied little, if at all), the experimental design was (1) to compare the Buccaneer Gas and Oil Field to control areas located 9.3 km from the field in each of the northeast and southwest directions and (2) to document and describe the "reef effect" of Buccaneer Field platforms. Results of the 1976-1977 investigations are presented in Jackson (1977).

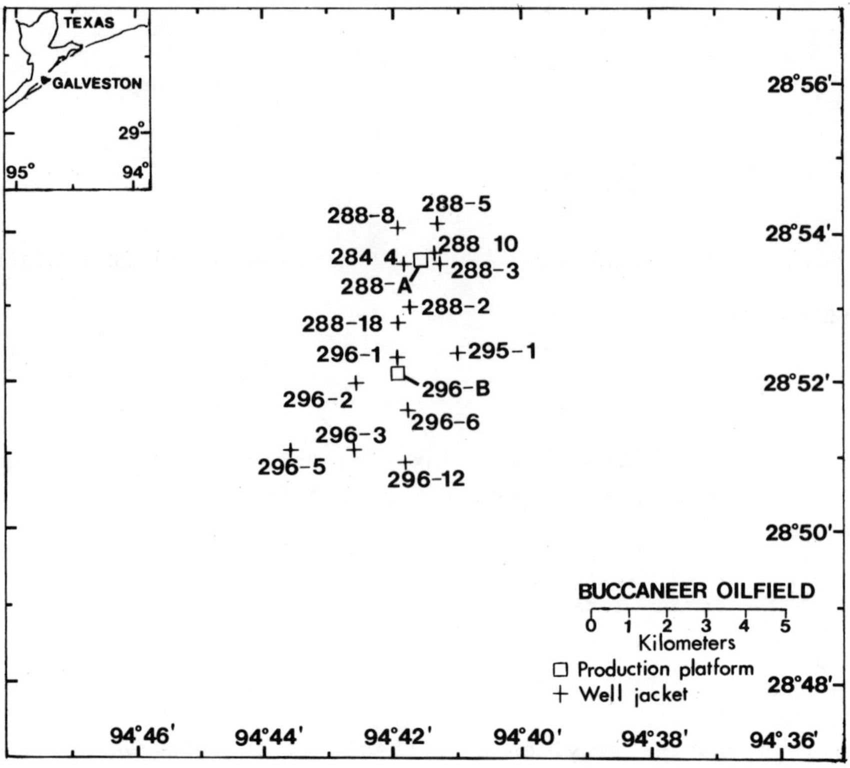

Figure 1. Map showing locations of production platforms and well jackets in the Buccaneer Field.

As the first year's survey was in progress, an initial conceptual model of the ecosystem surrounding the Buccaneer Field was developed (Gallaway et al., 1976) as a means to facilitate program integration, future research planning, and impact assessment. Important ecosystem components and key processes were identified, and a detailed description of the Buccaneer Field in terms of historical and current activities was developed. Indicators of impact from the Buccaneer Field on biological systems were selected based largely upon their perceived importance to man (direct or indirect), and the likelihood of their being subject to impact either from direct exposure to contaminants or indirect exposure through the food chain. Information from other disciplines which was believed necessary in order to be able to make the biological assessment was identified and incorporated as additional objectives for the non-biological disciplines. Additionally, sample needs for other disciplines from the biological work group were identified. The program was thus integrated by means of a matrix of interdisciplinary data needs which identified the kinds of information each discipline

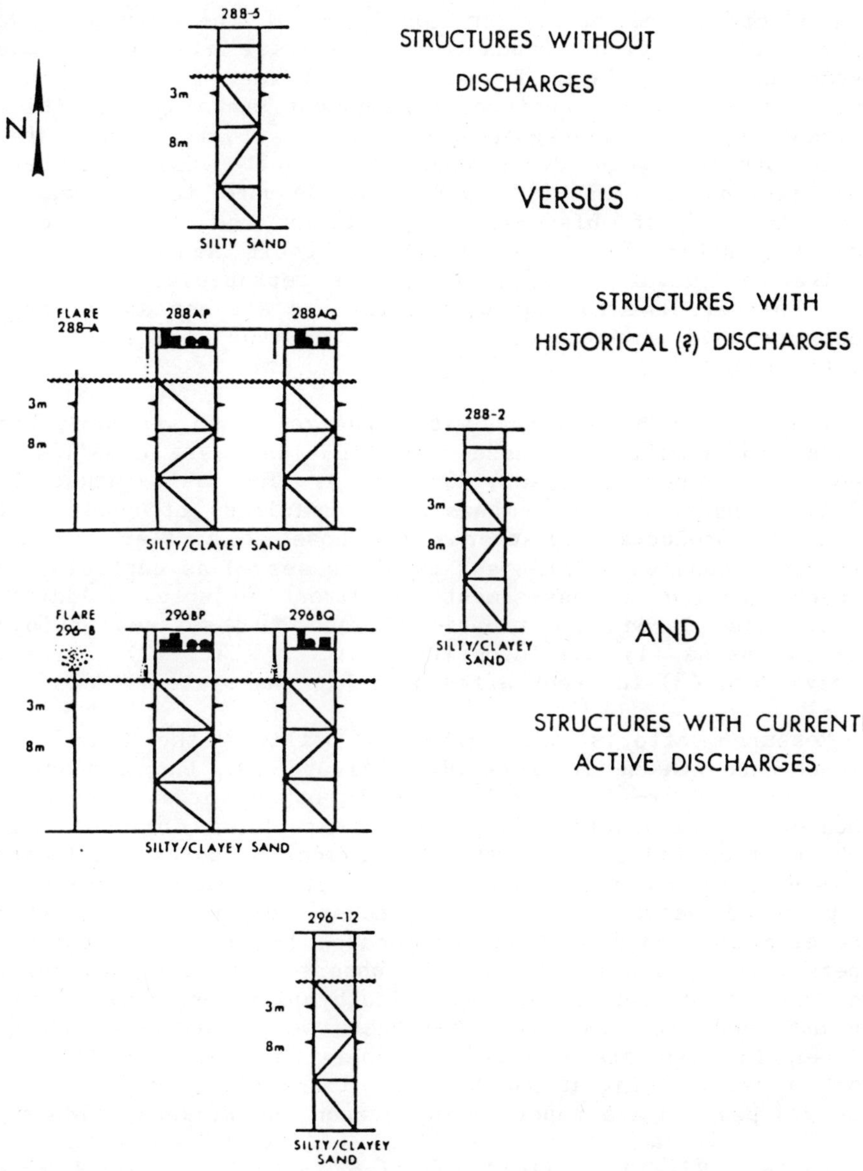

Figure 2. Idealized experimental design.

needed from other disciplines. Care was taken to ensure that data which were not readily appropriate for impact assessment purposes were not included in the study design.

The indicators selected for study of biological impact were (1) standing crop biomass, community structure and composition, and production and health or condition of the biofouling community; (2) relative abundance of demersal fishes and macrocrustaceans; (3) pelagic fishes; and (4) reef fishes. Ichthyoplankton studies were continued to document spawning activities of important species aggregated at the platforms, but infaunal benthic studies were discontinued. Results from the benthic investigations of 1976-1977 appeared adequate to delineate the direct impacts of historical operations and to characterize infaunal benthic faunas in terms of their seasonal abundance relative to the different bottom types represented in the study area. Further, the infauna was sparse and did not appear of much direct trophic importance to most of the indicators being investigated.

The 1977-1978 research was organized into 11 work units. Results and details of these investigations were compiled in a three-volume report (Jackson, 1979a,b,c). The basic thrust of the 1977-1978 investigations emphasized comparisons of conditions at and around production platforms to those at similar depths and distances around satellite jackets which served as controls. This approach enabled an assessment of direct effects. Additional emphasis was placed upon trophic linkages and systems ecology in order to assess (1) the integrity of the system, (2) contaminant pathways, and (3) indirect effects within the system.

Research efforts for 1978-1979 utilized the same general approach undertaken in 1977-1978 (Figure 2), but emphasis was placed upon "near-field" effects. In other words, emphasis was placed upon delineating the spatial extent of observed effects which had been indicated by the comparisons of sites and locations on discharging and non-discharging structures in the proximity of the produced water discharge. Further, major information gaps concerning suspended sediments, plankton biomass, and contaminant dispersal and diffusion (given the observed hydrological regimes) were identified and addressed. Ichthyoplankton data which had been obtained over the first two years were considered adequate, and sampling was discontinued. Otherwise, the distribution of effort remained similar to that of 1977-1978. Results of the 1978-1979 program are reported in Jackson and Wilkens (1980a).

During 1979-1980, field sampling in the biological program was reduced to only that necessary to fill data gaps addressing the remaining uncertainties considered to be of importance (red snapper population levels, trophic linkages, and condition of barnacles). The primary goal of the final year of the program was to produce a milestone report, an integrated synthesis of the major findings of the project (Jackson and Wilkens, 1980b).

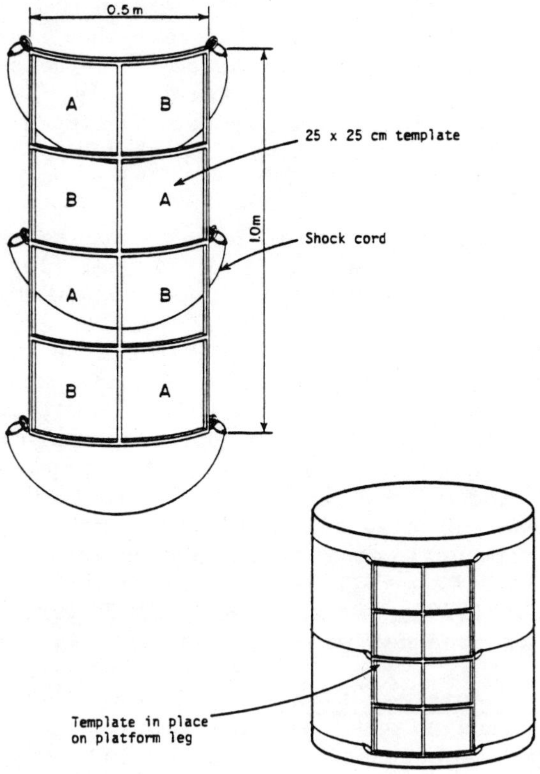

Figure 3. Diagrammatic representation of scraping templates.

Sampling Methods

A precis of the methods that were employed in the study of the selected indicator groups is presented below. Detail as to exact methods can be found in referenced reports of the Principal Investigators. For convenience, the descriptions provided below are grouped under (1) demersal fishes and macrocrustaceans, (2) biofouling, and (3) structure-associated fishes.

Demersal fishes and macrocrustaceans. Field studies of this group were conducted throughout the program by means of otter trawling (Emiliani et al., 1977; Workman and Jones, 1979; Gallaway and Martin, 1980). A 12-m semiballoon otter trawl was used in

1976-1977, a 6.1-m trawl in 1977-1978, and a 12-m trawl was again utilized in 1978-1979. Time of trawl hauls and number of replicates differed among investigators. Studies performed in 1978-1979 consisted of triplicated trawl hauls, each of which was of 10-min duration. These samples were believed to provide the best estimates of seasonal and spatial abundance and are heavily relied upon herein.

Biofouling community. Initial biofouling studies in the Buccaneer Field were conducted by Fotheringham (1977) and utilized scraping and photogrammetric techniques to characterize fouling community composition and structure, particularly in terms of relative abundance and coverage of platform substrate. During the remainder of the program, biofouling efforts were conducted by Gallaway et al. (1979a), Howard et al. (1980), and this study. These studies relied largely upon scraping techniques utilizing templates to obtain replicated quadrat samples (Figure 3). Standing crop biomass and rates of recolonization by the biofouling community of cleaned areas were emphasized and supplemented by experimental studies of production and condition of community dominants. *In situ* respirometry investigations (Gallaway et al., 1979a; Howard et al., 1980) were used to further define effects of produced water on the biofouling community.

Structure-associated fishes. Pelagic, reef, and other structure-associated fishes were investigated using a variety of field sampling techniques, including (1) diver observation (Fotheringham, 1977; Workman and Jones, 1979; Gallaway et al., 1979a; Gallaway and Martin, 1980; this study), photography (Workman and Jones, 1979; this study), trolling, gill-netting (Trent, 1977), hook-and-line (Fotheringham, 1977; Emiliani et al., 1977; Trent, 1977; Workman and Jones, 1979; Gallaway et al., 1979a; Gallaway and Martin, 1980; this study), air lift and other diver-operated devices (Fotheringham, 1977; Workman and Jones, 1979; Gallaway et al., 1979a; Gallaway and Martin, 1980; this study), quantitative diver census of sedentary reef fish (Workman and Jones, 1979; Gallaway and Martin, 1980), and mark-recapture experiments (Workman and Jones, 1979; Gallaway et al., 1979a; Gallaway and Martin, 1980; this study). Field sampling was designed to yield (1) qualitative and quantitative estimates of population levels and structure, (2) descriptions of trophic ecology, (3) an assessment of the health and condition of species being investigated, and (3) characterizations of the Buccaneer Field recreational fisheries.

Sample and Data Analysis

Samples returned to the laboratory for analysis were analyzed following published protocols for the respective disciplines and

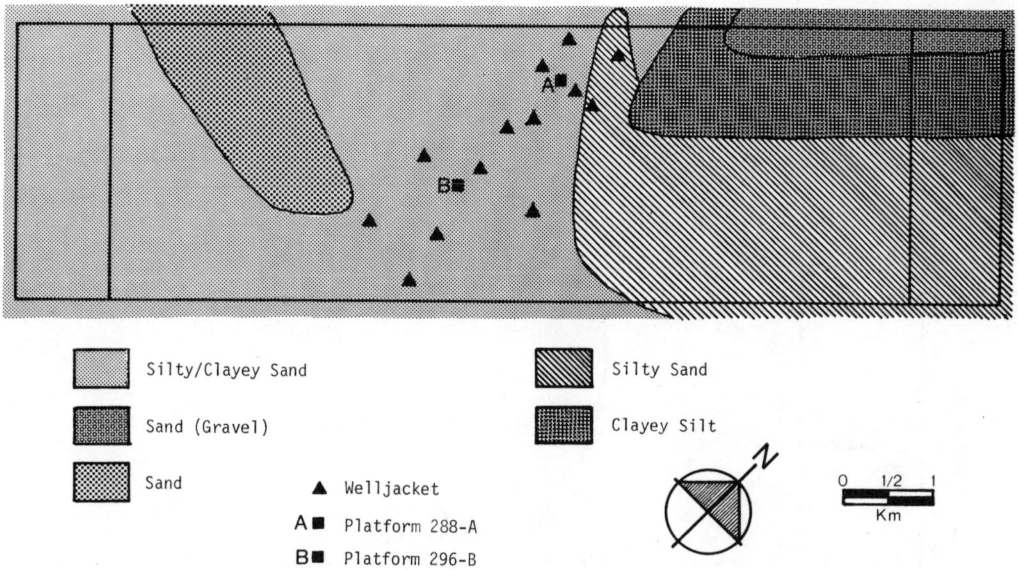

Figure 4. Sediment distribution in the Buccaneer Field.

all data were provided to project data management. In general, collection data were summarized using information theory species diversity indices and cluster analysis (see Gallaway et al., 1979a). Where data were adequate, statistical analyses were made using factorial analysis of variance supplemented by either orthogonal contrasts or Duncan's Multiple Range Tests (see Howard et al., 1980) to compare significance of differences among means.

Seasonal population estimates were made using both single- and multiple-census techniques. (Gallaway and Martin, 1980) and, for small cryptic species, by replicated quadrat counts (Workman and Jones, 1979; Gallaway and Martin, 1980). Food habitat investigations were based upon both qualitative (Workman and Jones, 1979) and quantitative gravimetric methods (Gallaway et al., 1979a; Gallaway and Martin, 1980; this study), with feeding periodicity evaluated using index of fullness (IF) values (Gallaway et al., 1979a).

Health and condition evaluations for fishes were based upon analysis of covariance of length-weight regressions (Workman and Jones, 1979; Gallaway et al., 1979a; Gallaway and Martin, 1980) and by direct histopathological (Gallaway and Martin, 1980) and microbial examination (Gallaway and Martin, 1980; Sizemore and Olsen, 1980). For barnacles, health and condition indices were determined using regressions of cavity volume on meat weight (Boland, 1980).

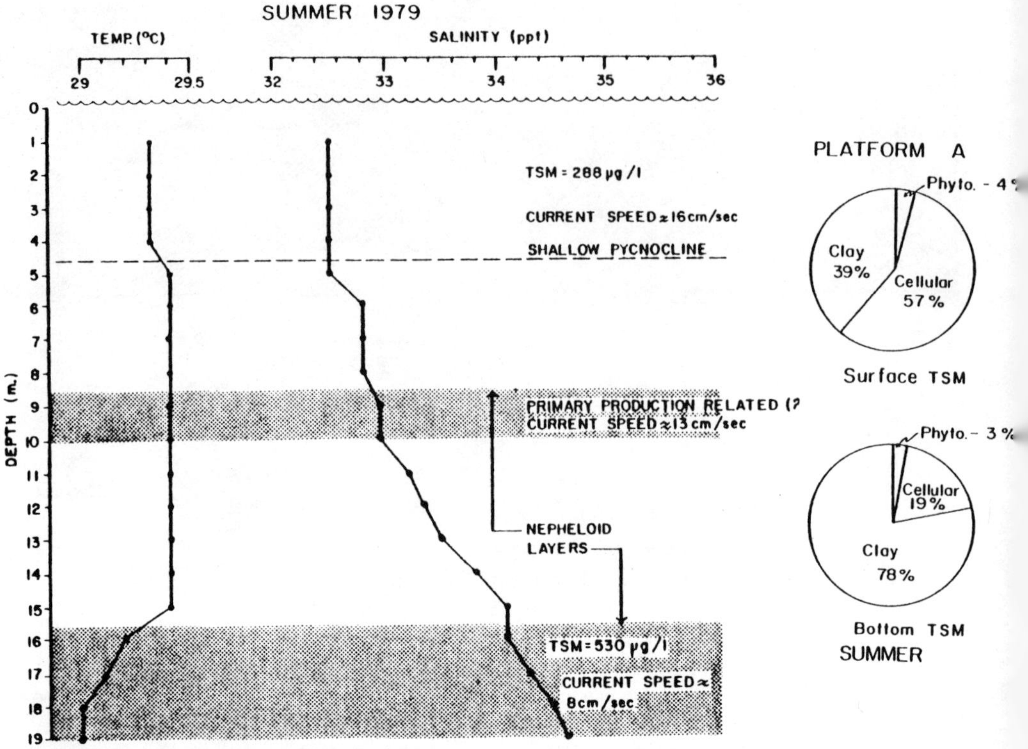

Figure 5. Characteristic environmental structure observed in the Buccaneer Field during summer, 1979.

ENVIRONMENTAL SETTING AND ALTERATIONS

Setting

Water depths in the Buccaneer Field range from 17 to 22 m and overlay a rather diverse bottom (Anderson et al., 1979; Figure 4). The majority of the Buccaneer Field structures are sited over silty/clayey sand, but two (including one of the controls, well jacket 288-5) were located on silty sand. The consensus reached from examination of all available information (Anderson et al., 1977, 1979; Behrens, 1977; Middleditch et al., 1977, 1979; Middleditch and West, 1980; Brooks and Estes, 1980; Martin, 1977; Armstrong, 1979, 1980) was that the bottom sediments of the area were scoured; i.e., sedimentation rates were characteristically low due to resuspension and transport out of the area by bottom currents.

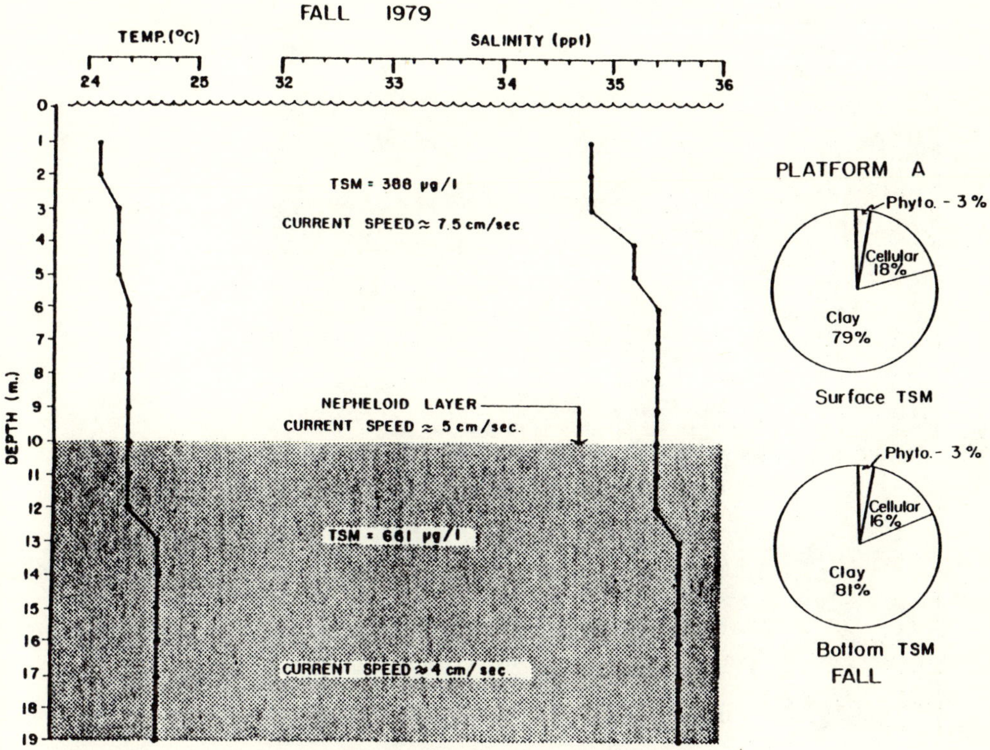

Figure 6. Characteristic environmental structure observed in the Buccaneer Field during fall, 1979.

Currents in the Buccaneer Field were found to be aligned principally in longshore directions, reversing seasonally from upcoast toward the northeast in summer (May-August) to downcoast toward southwest for October-April (Armstrong, 1979, 1980). Transitional conditions appeared to rule in September and April. Current meter records showed layering of contrasting flows during some seasons. Local winds were apparently the main driving force for the circulation. Flow was typically with the wind but was deflected by the coastline such that there was compensating offshore transport with offshore winds, or to the right of the winds due to Ekman transport. Distinct departures from local wind-driven circulation develop during spring, when it seems high river discharge may establish a downcoast, geostrophic current which, from current meter records, may account for the layered currents of summer. Also, during early fall, currents of the area do not appear to relate to local winds, but may be responding to larger-scale atmospheric alterations. Spectral analyses of current meter records indicate that tidal currents and wind shifts account for most of the variability in flow dynamics, with

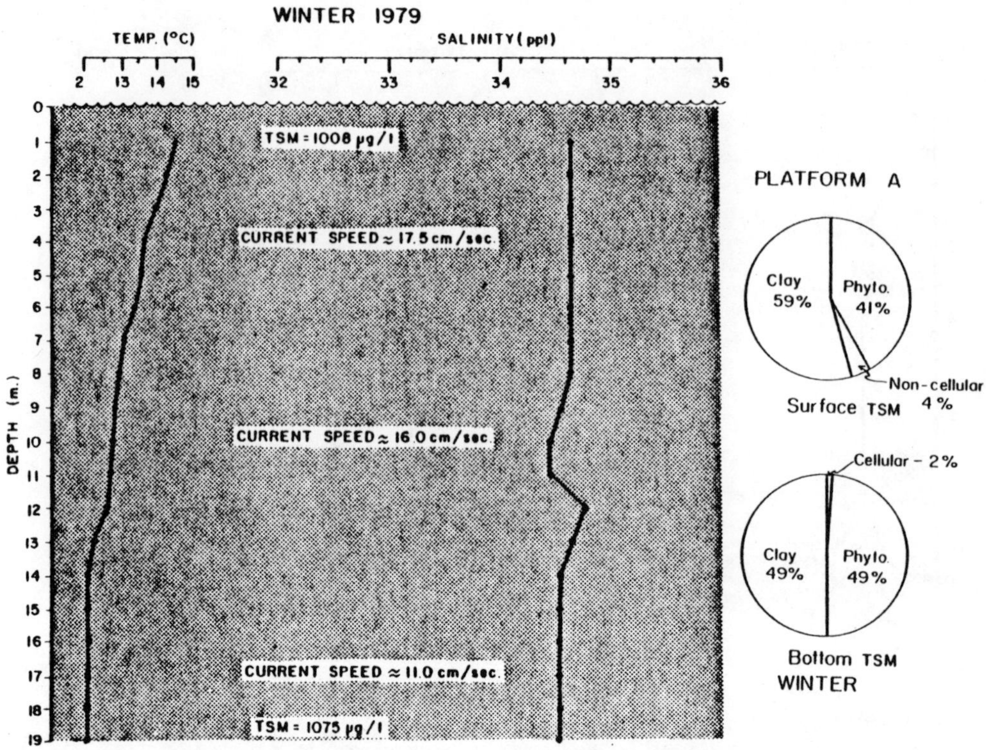

Figure 7. Characteristic environmental structure observed in the Buccaneer Field during winter, 1979.

dominant periods perhaps associated with passage of continental air masses in winter and fall, and longer-period maritime air mass development during summer.

Water column structure. As might be deduced from the above, the general structure of the Buccaneer Field water column underwent marked seasonal changes which are characterized below using 1979-1980 observations (Figures 5-8). In general, during periods of vertical stratification, salinity stratification was more pronounced than temperature, and the distribution of turbid, or nepheloid, layers corresponded with pycnoclines. In summer 1979, two nepheloid layers were present under the conditions depicted by Figure 5. The level of total suspended particulate matter (TSM) in surface waters (0.288 mg/l) was lower than that of bottom waters (0.538 mg/l) and also differed in composition. On a relative basis, zooplankton ("cellular") particulates dominated in the upper part of the water column whereas clays were the dominant particulate in near-bottom water (Brooks et al., 1980).

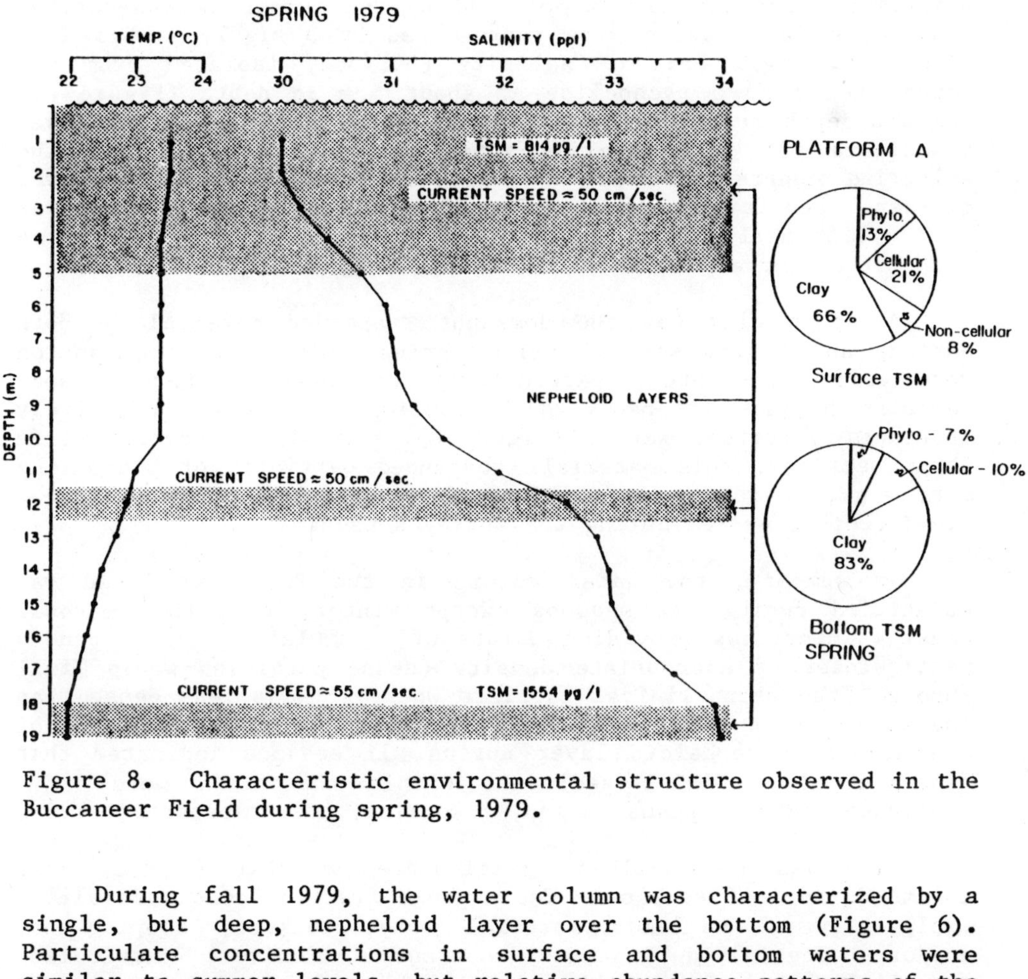

Figure 8. Characteristic environmental structure observed in the Buccaneer Field during spring, 1979.

During fall 1979, the water column was characterized by a single, but deep, nepheloid layer over the bottom (Figure 6). Particulate concentrations in surface and bottom waters were similar to summer levels, but relative abundance patterns of the components differed greatly - clay was the dominant particulate in each water column mass (Figure 6). During winter 1979-1980, the entire water column was highly turbid (TSM >0.1 mg/l) and temperature-salinity stratification was weak or absent (Figure 7). Clay and phytoplankton were the codominant particulate types on a relative basis, and zooplankton was scarcely represented in the samples. Surface waters contained measurable levels of non-cellular organic material during winter - probably representing winter-blooming mat organisms sloughed from the biofouling community.

The water column during spring 1980 was characterized by the presence of three turbid layers and marked salinity stratification. During this period of transition, the water column was changing from a vertically mixed system to the characteristic stratified condition as exemplified by the summer data described above. Surface waters were represented by a highly turbid lens (TSM = 0.8 mg/l) of low salinity ($30^{o}/oo$), isolated from the waters below by a pycnocline at about 5 m in depth (Figure 8). The mid-depth nepheloid layer was at about 12 m and, again, was associated with a pycnocline which probably accounted for the selective congregation of sinking particulates at the observed mid-depth density interface. The dramatic bottom nepheloid layer (TSM = 1.55 mg/l) was attributable to turbulent mixing associated with bottom currents.

Although clay was the dominant suspended material in both surface and bottom waters during spring, phyto- and zooplankton were well represented, particularly in surface waters. Noncellular biomass (probably of biofouling origin) was relatively abundant in surface waters (Figure 8). As will be described in later sections, this material (suspended particles of biofouling mat; e.g., hydroids) is particularly important to the Buccaneer Field trophic system during the spring season.

In summary, the water column in the Buccaneer Field was stratified during all seasons except winter, with the seasonal density interfaces providing points of accumulation for suspended particulates of appropriate density (dense particles would "fall through" the pycnoclinical barriers whereas those less dense than the waters below would settle at the interface). The presence of a near-bottom nepheloid layer during all seasons indicated that fine-grained surficial sediments within the field were in a continued state of resuspension, reworking, and transport.

Clay was the dominant particulate material in the water column during all seasons. The organic fraction consisted almost exclusively of cellular material (phytoplankton, zooplankton, and/or bacteria) during most seasons. However, particulate noncellular carbon, probably of biofouling origin, comprised 4-8% of the particulate load in surface waters during late winter and spring, respectively. The winter season was characterized by the highest levels of organic nutrients and an associated phytoplankton bloom.

Bacteria. Bacterial populations of the Buccaneer Field were of marine origin (94% of those enumerated required salt for growth) with biomass estimated to range between 5.2 and 44.0 x 10^{-6} g C/l (Sizemore and Olsen, 1980). The dominant genera represented included *Vibrio*, *Pseudomonas*, *Aeromonas*, *Acinetobacter*, and *Moraxella*. Bacterial diversity changed with season and was lowest in spring. Ninety percent of the bacteria in the water column were found to have been attached to particles greater than three microns in diameter.

A number of potential fish pathogens were well represented as part of the typical bacterial assemblage found in the water and on suspended sediments, surficial sediments, and fish (Sizemore and Olsen, 1980; Gallaway and Martin, 1980). These included several hemolytic species of *Vibrio* as well as species of *Aeromonas*. These opportunistic pathogens were implicated as potential agents for the spadefish disease epidemics observed duing the winter seasons.

Benthos. The benthic macroinfauna of the Buccaneer Field was diverse (estimated between 400 and 420 species) and abundant (Harper et al., 1976; Harper, 1977). During summer 1976, mean density of benthic organisms in the field was approximately 8,000 individuals/m^2, but declined from this level to approximately 4,500/m^2 in January 1977. By spring 1977, populations had increased to an average of about 6,000 individuals/m^2. Although the seasonal trends observed in the Buccaneer Field were similar to those observed for a nearby, more inshore, area off Freeport, Texas, the densities of macroinfauna in the Buccaneer Field were an order of magnitude larger than the densities observed off Freeport. Polychaetes (65-70%) and amphipods (10-20%) dominated the fauna. Biomass levels of benthos were not measured in the study, but were considered low. Attempts to obtain a large enough (5-10 g wet weight) sample of infauna for chemical analyses were seldom successful due to the small size of the average infaunal organism.

Alterations

Types and amounts of contaminants. The most obvious environmental alteration in the Buccaneer Field was the addition of an estimated 16,000 m^2 of hard substrate habitat extending from the bottom to some 21 m above the water's surface. The only discharges from these structures of any consequence were produced water from the production platforms and treated sewage effluent from the quarters platforms. Produced water discharge, although variable, was estimated to have averaged about 2.5 l/sec. Discharge of sewage effluent was intermittent, but was estimated to have been approximately 30 l/h (0.006 l/sec).

Produced waters were characterized in terms of alkanes, aromatics, volatiles, sulfur, and biocides by Middleditch and West (1980). They estimated the daily discharge of alkanes to have been, on the average, 382 g which represented 19% of the estimated 2 kg/day total of oil discharged from production platforms in produced waters. The light aromatic fraction of hydrocarbons in produced water was represented by some 68 different compounds having an average total concentration of 104.2 ppb. Twelve normal, branched, and cyclic alkanes were characterized in the

TABLE 1. TOXICITY OF PRODUCED WATERS TO BUCCANEER FIELD ORGANISMS

Organism	96-h LC50 (ppm)	48-h LC50 (ppm)	Concentration (%)
Larval brown shrimp	-	9500	0.95
Subadult white shrimp	68,000	-	6.8
Adult white shrimp	70,000	-	7.0
Barnacle	83,000	-	8.3
Subadult brown shrimp	100,000	-	10.0
Adult brown shrimp	116,000	-	11.0
Crested blenny	269,000	-	26.9

analysis of produced water volatiles. Three aromatics, comprising 64% of the volatile components measured, were identified as benzene, toluene, and ethylbenzene.

In contrast to the low levels of hydrocarbons being discharged in produced water, some 207 kg of sulfur was believed to have been discharged daily from production platforms. As sulfur has a specific gravity of about two and is insoluble in water, it may serve as the major transporter of oil through the water column and into sediments - if hydrocarbons can be adsorbed on sulfur particles. The acrolein biocide used to control microbially aggravated corrosion of pipes and vessels was not detected in produced water discharge samples. This biocide, while highly toxic, is quite labile.

Tillery (1980) found produced waters to have been enriched in barium and strontium, but characterized the levels of other trace metals in the discharge as being extremely low. These findings confirmed the previous work of Anderson et al. (1979), also performed in the Buccaneer Field.

Produced water was found to have been toxic to marine organisms, with crustaceans more sensitive than fish (Zein-Eldin, 1979; Rose and Ward, this volume) see Table 1. The most sensitive organism among those tested was the larval brown shrimp, which had a mean 96-h LC50 of about 9,500 ppm (1% produced water in seawater).

<u>Water column</u>. We used two methods to project the concentrations of produced water in the receiving seawater. The first method was used by Smedes et al. (1980) to establish an initial concentration for use in a hydrodynamic model. Based upon dye studies performed by Workman and Jones (1979), the produced water

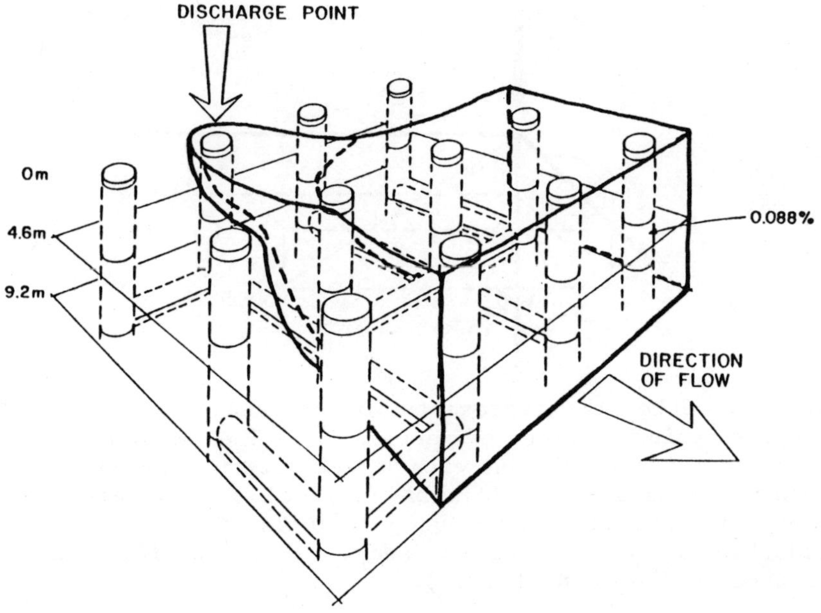

Figure 9. Hypothetical distribution of produced water in the water column.

(due to turbulent mixing beneath the platform) was assumed to completely mix in a volume of water approximately 1/8 of the volume occupied by the platform (Figure 9). [The effects of pycnoclines on dye dispersion can be seen by comparing Figures 5 and 9.] The relative concentration of 0.088% of produced water in this volume beneath the platform was determined by letting

Q = rate of discharge (m^3/sec)
U = ambient current (m/sec)
L = length of the platform (m)
Z = water depth (m)

and then estimating the time (t) for the pollutant to be swept past the platform using:

$$t = L/U$$

The volume (V) of pollutant discharged during this time was calculated using:

$$V = Qt = QL/U$$

This volume of pollutant was considered to have been initially mixed into a volume (V_o) of ambient water approximately equal to:

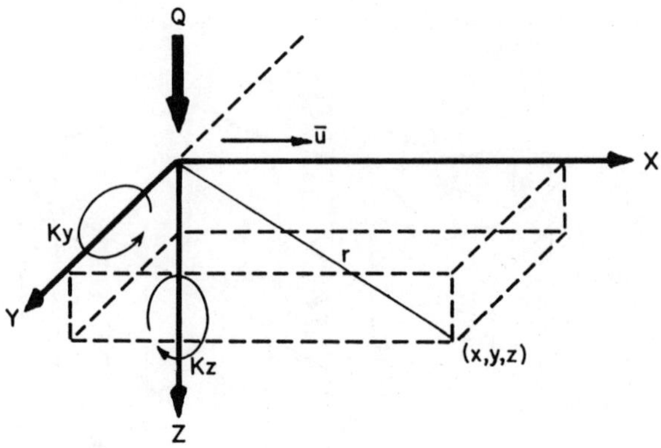

Figure 10. Schematic representation of idealized dispersion from continuous point source of effluent. The ambient current is flowing in the x direction with a velocity $\bar{u}$ (m/sec). K_y and K_z are eddy diffusion coefficients (m^2/sec) in the horizontal and vertical directions respectively. Q is the rate of effluent introduction at the source (m^3/sec). r is the distance (m) from the source to the point (x,y,z).

$$V_o = 0.5\ L^2Z$$

leading to the relative concentration (X_o) of 0.088% calculated:

$$X_o = V/V_o = 8Q/ULZ$$

for the conditions where

$U = 0.05$ m/sec
$L = 50$ m
$Z = 20$ m
$Q = 0.055\ m^3/sec$

The relative contaminant concentration projected by this method agrees well with the pollution concentrations observed in waters beneath the platforms (see Middleditch and West, 1980; Tillery, 1980) given the initial levels of contaminants in the produced water being discharged.

The second method employed was to use an analytic steady state approximation for diffusion from a continuous point source discharge having a mean advective component (ambient current) perpendicular to the dispersion (Figure 10). The mean concentration ($\bar{C}$) at any point was determined using:

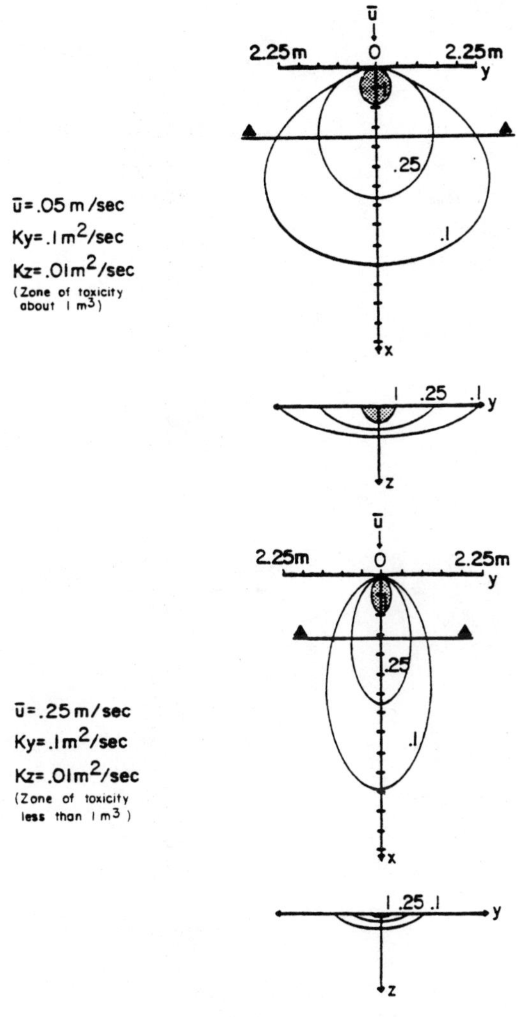

NOTE: all contour numbers expressed in % of produced water.

Figure 11. Distribution of produced water based upon analytic steady state approximation, conditions 1 and 2.

$$\bar{C} = \frac{Q}{4\pi r(k_y k_z)^{1/2}} \cdot e^{\left\{-\frac{\bar{u}}{4x}\left(\frac{y^2}{k_y} + \frac{z^2}{k_z}\right)\right\}}$$

Explanation of the symbols is provided in Figure 10. Although this model can rightfully be criticized (mixing beneath the

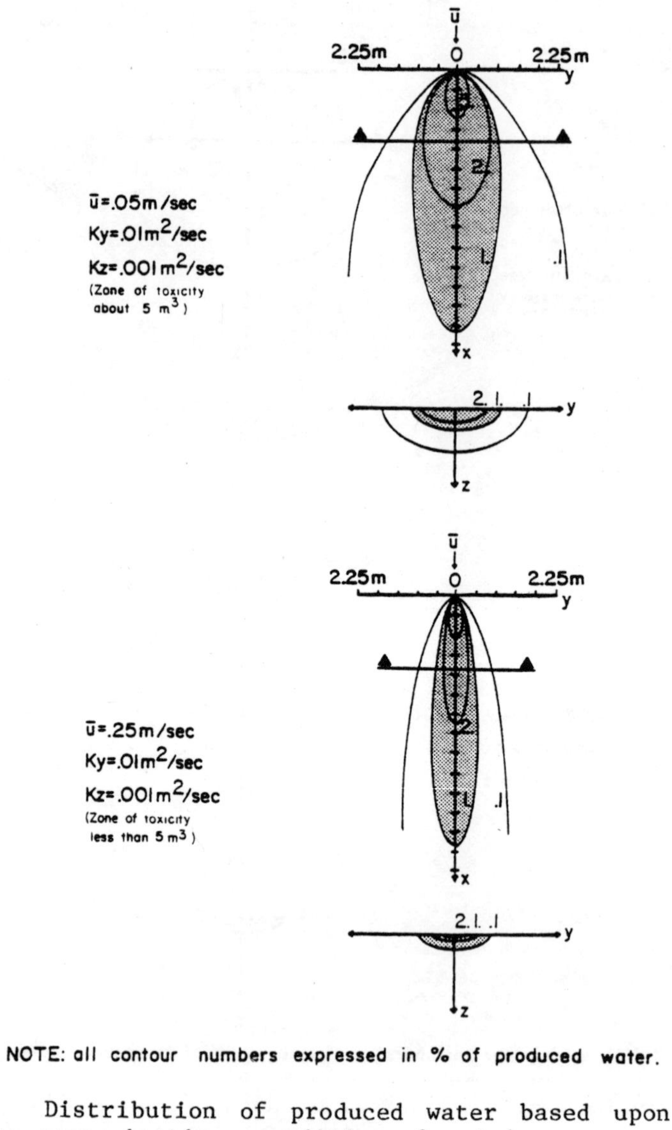

Figure 12. Distribution of produced water based upon analytic steady state approximation, conditions 3 and 4.

platform is far more complex and effective than simple eddy diffusion), we believe it could be used to estimate maximum zones of toxicity around the point source under "worst case" conditions.

Eddy diffusion coefficients (K_y = 0.1 m^2/sec; K_z = 0.01) were selected based upon Nichoul (1975) and, although considered typical for the mixed layer, represent the greatest uncertainty associated with the above model. Typical current velocities in the Buccaneer Field range between 0.05 and 0.25 m/sec (Danek and Tomlinson, 1980) and were used in the above model in conjunction with the average loading value at the point source (2.5 l/sec of produced water) to calculate the results shown by Figure 11. Under the conditions depicted, the maximum zone of toxicity (assuming a 1% concentration of produced water in seawater to be toxic) was about 1 m^3. Decreasing the diffusion rates each by an order of magnitude resulted in the "potentially toxic" volume increasing to up to 5 m^3, mostly in the direction of current flow (Figure 12). Increasing the diffusion rates resulted in a decrease in the potentially toxic volume and, in the limiting case, approximated results from method one described above.

Sedimentary regimes. Particulate and sedimentary regimes of the Buccaneer Field were highly dynamic. The combination of wave and current energy served to resuspend and transport particulates out of the area following only intermittent and brief periods of static conditions. Buccaneer Field operations appeared to alter sedimentary regimes in several ways. The structures themselves not only contributed to turbulent mixing, but yielded a "rain" of metal ranging in size from microscopic flakes (Anderson et al., 1977) to large pieces of grating and batteries. We believe that the source of the small metal flakes which were very abundant in the water and sediment samples taken beneath the platforms was corrosion of the metal gratings which comprised the decks of the platforms. In any case, the observed metal flakes probably were a major source of the trace metal contamination of sediments reported for samples taken adjacent to the platforms.

As mentioned above, particulate sulfur was a major component in the produced water and, along with the metal flakes, may have served as a transporter of hydrocarbons to the bottom. Hydrocarbon levels in sediments beneath the production platforms, although highly variable, were typically higher than levels in sediments in control areas. Another possible transporter of hydrocarbons to the bottom was oily sand which was sometimes present in the skim tank and discharged overboard.

The presence of the structures allows development of a reef community which contributes particulates of biogenic origin ranging from parts of colonial organisms to fish scales to fecal pellets to whole barnacles. The latter break off the platforms during storms and have, in effect, formed a shell-rubble pad beneath the structures in the field. The observed decreasing gradient in organic and inorganic carbon away from the platforms in the Buccaneer Field was probably attributable to the contribution from the reef communities which develop on and around the structures.

With the exception of the large pieces of metal and the barnacle shells, the residence time of particulates beneath the platforms was presumably very short due to resuspension (waves) and transport out of the area (currents). The direction of sediment transport appeared to be controlled by seasonal current patterns. Dilution and/or biodegradation appears to reduce levels of contaminants to that of background conditions within very short distances (<50 m) from the platforms.

Bacteria. The bacteria data provide evidence that the degree of hydrocarbon contamination emenating from the Buccaneer Field was, indeed, minimal. Bacterial diversity and density levels in the Buccaneer Field were markedly similar to those in control areas. Although numerical densities and taxa represented in collections from the two areas were the same, the relative abundance of taxa was different between the Buccaneer Field and control areas. Buccaneer Field samples contained relatively more oil-degrading, sulfur-oxidizing, and sulfate-reducing bacteria than did samples from control areas outside the field. These data indicate that chronic, low-level pollution was occurring, but not to the extent that population levels were significantly increased.

Produced water inhibited or retarded the growth of laboratory cultures of bacteria, but appeared to have no effect, or a stimulatory effect, on isolates obtained from the Buccaneer Field. Both pure and mixed cultures of bacteria from the Buccaneer Field exhibited the ability to degrade significant portions of the n-alkanes in Buccaneer crude oil.

Presumptive coliform microorganisms were more commonly encountered in samples taken near quarters platforms than in samples obtained from control areas, but no fecal coliforms were observed in any samples. Based upon the evidence provided by bacteria data, the Buccaneer Field was an environmentally clean operation during the period of study.

Benthos. Numerical (cluster) analysis of the benthic data based upon 67 abundant species yielded four distinct site groups. Site group I consisted of the majority of the stations in the study, all of which exhibited a high degree of ecological similarity. Site group II was a group of five stations, mostly associated with clay substrates, characterized by reduced populations. Site group III was comprised of stations in the vicinity of the production platforms and was also considered to have had reduced populations. The last site group delineated was one consisting of only two stations. Both were markedly dissimilar from all the other platform groups as well as from each other. The reduced populations near production platforms are believed by us to have been attributable more to sediment differences than to contaminant levels.

The benthic collections around platforms were sometimes characterized by high densities of intact shells of a planktonic pteropod. We believe, based upon results of our Atlantic spadefish trophic investigations described below, that these accumulations were attributable to predation by Atlantic spadefish. On occasions during summer months, the guts of spadefish would be literally packed to maximum capacity with pteropods, many of which were probably passed with little or no digestive deterioration.

MAJOR ECOSYSTEM COMPONENTS AND EFFECTS

Demersal Fishes and Macrocrustaceans

Characterization. Demersal nekton communities in the vicinity of the Buccaneer Field were diverse and abundant. During 1976-1977, Emiliani et al. (1977) trawled 97 species of finfish and macrocrustaceans represented among 47,530 specimens. A much reduced trawl program conducted in 1977-1978 yielded 61 species (Workman and Jones, 1979), and 104 species and 49,481 specimens were taken in 1978-1979 by Gallaway and Martin (1980). One undescribed species of jawfish was taken in the 1978-1979 program and the Buccaneer Gas and Oil Field records will be included in a forthcoming revision of the jawfishes by William F. Smith-Vaniz of the Academy of Sciences of Philadelphia. Specimens were deposited with the Academy.

As exemplified by Figure 13, demersal nekton were basically represented by two seasonal assemblages (summer and winter), separated by periods of transition in the spring and fall. The summer assemblage was characterized by low abundances, few species, and strong dominance by one species, the longspine porgy (*Stenotomus caprinus*). In contrast, the winter assemblage was characterized by high abundance, many species, and dominance by a macrocrustacean - namely sugar shrimp (*Trachypenaeus similis*). Commercially important shrimps (*Penaeus*) were never a collection dominant although brown shrimp were relatively abundant in collections taken during the fall periods of migration. The commercial shrimping fleet was seldom observed fishing within the Buccaneer Field but was active during summer and fall in the silty-sand area to the east of the Field (see Figure 4).

Most species of the demersal nektonic community were benthic feeders and they, in turn, were found to serve as an important source of food for some benthic reef species such as red snapper. The demersal nekton community was a major contributor to the ichthyoplankton - engraulids, sciaenids, and bothids were the three most abundant taxa of fish larvae represented in the

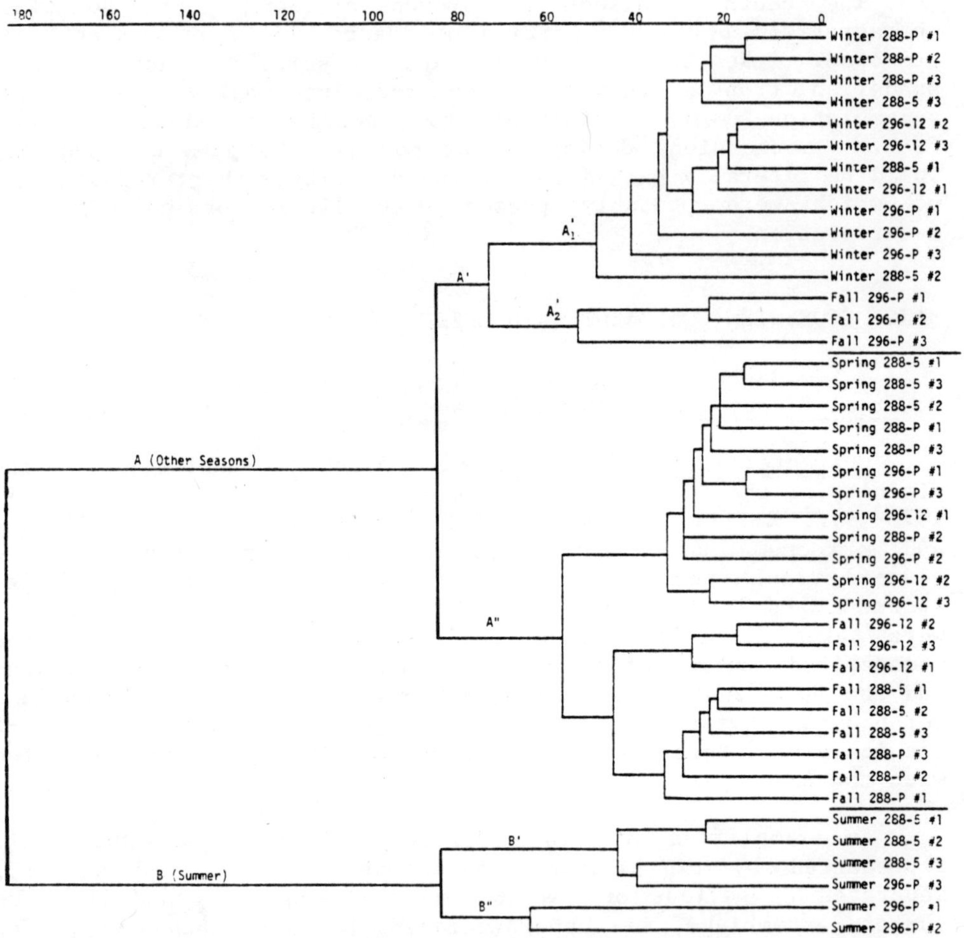

Figure 13. Normal cluster analysis dendrogram for trawl collections made during the 1978-1979 Buccaneer Gas and Oil Field Study.

Buccaneer Field (Finucane et al., 1979). Surprisingly, based upon egg abundance the Buccaneer Field is a spawning ground for anguilliforms, callionymids, clupeids, sciaenids, scombrids, and soleids, but eggs and ichthyoplankton of reef fish were not abundant. Finucane et al. (1979) did not detect any effects of contaminant discharges on ichthyoplankton.

<u>Effects</u>. The results reported by Emiliani et al. (1977) suggest that demersal nekton communities within the Buccaneer Field were more diverse and abundant than those associated with the control areas. These differences may have been partly attributable to the greater habitat diversity associated with the

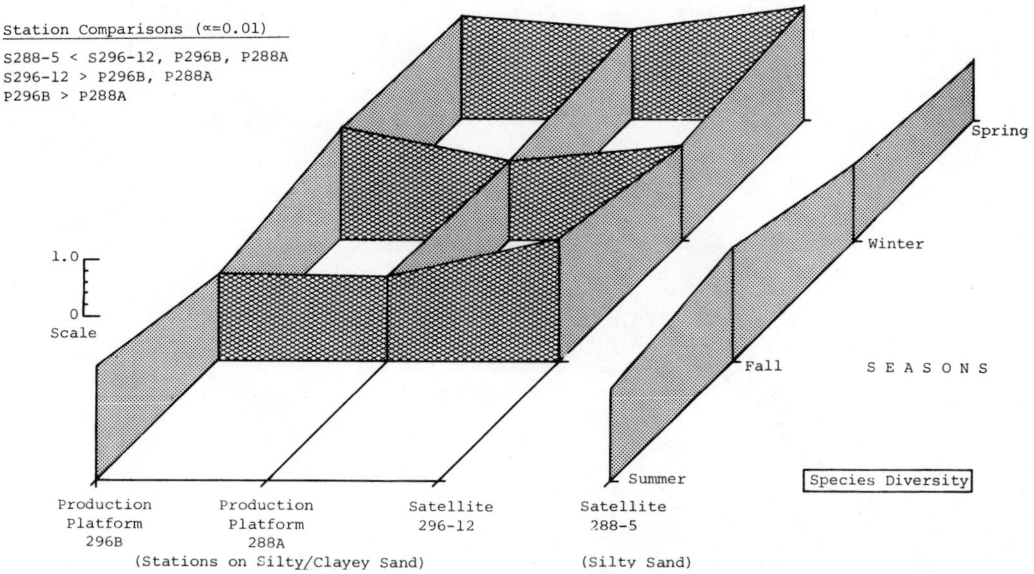

Figure 14. Comparisons of mean species diversity levels at study area stations or platforms by season.

field, particularly the presence of structures and exposed pipelines.

Comparisons of levels of species diversity and the abundances of dominant species within the Buccaneer Field were made by Gallaway and Martin (1980). Collections taken at a structure over a silty-sand substrate had significantly lower diversity than collections taken at structures over silty-clayey-sand substrates (Figure 14). Within the latter substrate type, species diversity was significantly higher at the control well jacket structure (no discharge) than at the production platforms (with contaminant discharge). This difference in species diversity was not associated with marked differences in the number of taxa represented in collections taken at two types of habitats, but appeared to result from the greater abundance of a few of the seasonally dominant species at production platforms than at well jackets (e.g., sugar shrimp, Figure 15; chevron shrimp, Figure 16). Most of the abundant species represented in the Buccaneer Field collections exhibited neither an attraction to, nor an avoidance of, discharging structures as compared to non-discharging structures. We believe that most of the observed differences in diversity and abundance were more attributable to substrate than to any other factor.

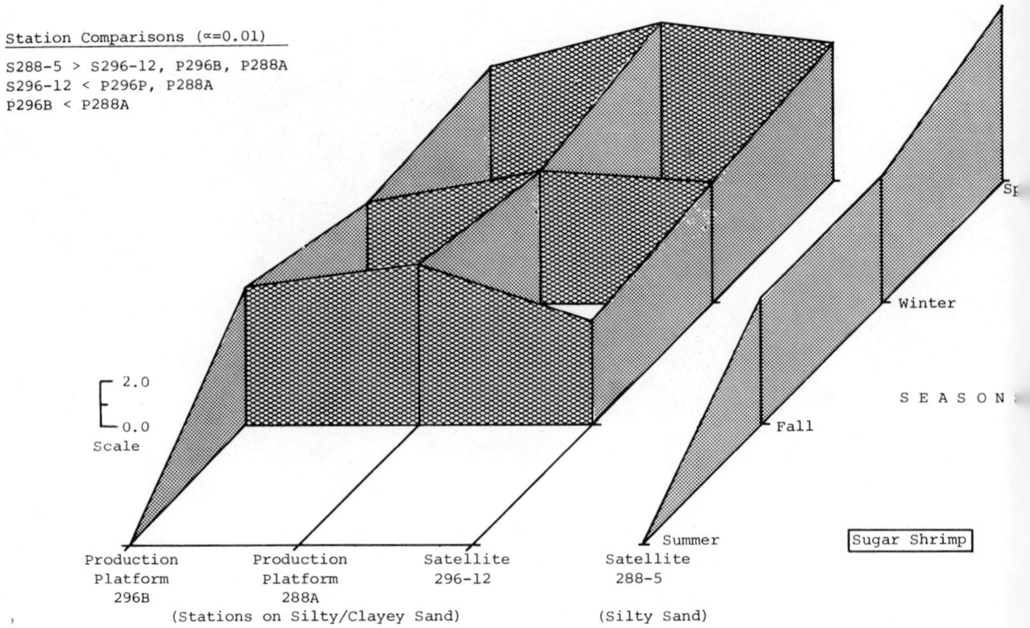

Figure 15. Comparisons of mean abundance levels of sugar shrimp (*Trachypenaeus similis*) at study area stations or platforms by season.

Brown shrimp were used as indicators of the health of the demersal nekton community. No histopathological anomalies were found and no evidence of any disease was observed. Bacterial flora on brown shrimp collected at production platforms was highly similar to the flora of specimens collected at control structures.

Biofouling Community

Characterization. The biofouling community colonizing Buccaneer Field structures was diverse and abundant. Fotheringham (1977) identified 16 algae and 101 species of invertebrates. The community consisted of two main components - shelled organisms which comprise and shape the overall habitat, and an encrusting "mat community" which provides additional habitat. Each of the above is characterized by a closely associated cryptic fauna.

The large Mediterranean barnacle, *Balanus tintinnabulum*, was the perennial dominant of the biofouling community and was estimated to occupy some 77% of the original platform substrate. Individuals of this barnacle were observed to obtain basal diameters of 3-4 cm and heights of 6-8 cm. They characteristi-

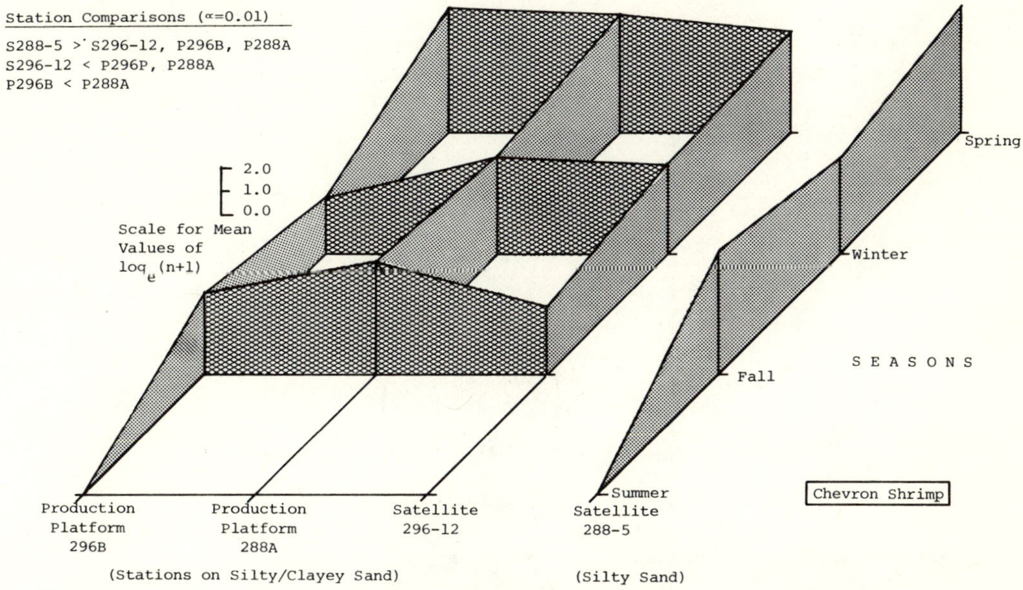

Figure 16. Comparisons of mean abundance levels of chevron shrimp (*Sicyonia dorsalis*) at study area stations or platforms by season.

cally grew in clusters forming a three-dimensional habitat some 10-15 cm thick.

The dominance of biofouling communities on Buccaneer Field structures by the Mediterranean barnacle represents a major zoogeographic finding of the program. It has been reported as an incidental species in the Gulf for some 20 years, and remains so on most platforms we have examined in the Western Gulf of Mexico off Louisiana. Our observations indicate that this species may be the dominant barnacle on platforms from the Buccaneer Field to areas off West Cameron, but it is seldom abundant on structures further east.

The Mediterranean barnacle feeds mostly during the night (Figure 17) and is a filter feeder on particulates and plankton. It spawns during the late spring or early summer and in fall, usually somewhat later than the competing acorn barnacles which are seasonally abundant in the Buccaneer Field. The combination of the rapid growth and large size of the Mediterranean barnacle enable it to settle on and overgrow the smaller acorn barnacles.

The principal predators of the barnacles were sheepshead and triggerfish and, to a much lesser degree, stone crabs. Another major source of natural mortality was large clusters breaking off

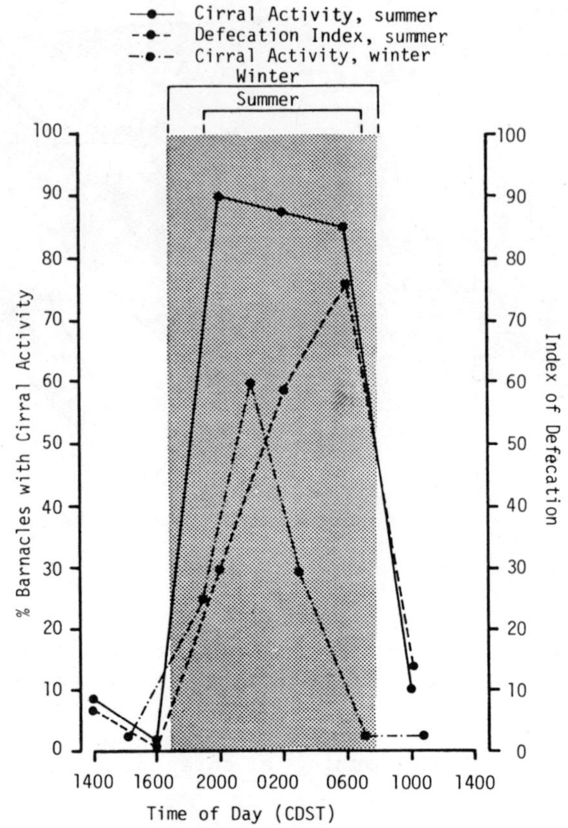

Figure 17. Diurnal activity of the Mediterranean barnacle (*Balanus tintinnabulum*), 18-19 August 1978.

and sinking to the bottom during periods of high waves and/or currents. Barnacle shells were the major component of the rubble pads observed on the bottom beneath the structures.

The barnacle community provided critical habitat for cryptic species such as pistol shrimp, stone crab, and blennies, as well as clean surface for colonization by the mat community. Clean surface was provided not only by shell growth but also by scars left where clusters had broken off and fallen to the bottom.

The mat portion of the biofouling community was characterized by a virtually inseparable interspersion of macroalgae, sponges, bryozoans, and hydroids. The macroalgae (mostly green and red algae) represented a relatively small percentage of the standing crop biomass, and were more abundant in summer than in winter. The faunal component of the mat, however, bloomed during winter seasons (particularly the stalked bryozoan *Bugula neritina* and the

Figure 18. Artistic rendition of the spring decline of a hydroid colony at 8-m depth on quarters platform 296-Q, as indicated by time-lapse photography.

hydroid *Tubularia crocea*), but declined markedly over short periods during spring (Figure 18) resulting in the characteristic low levels of mat observed for summer and fall.

The faunal components of the mat community are also filter feeders on particulates and plankton, and were utilized for food by sheepshead, triggerfish, and small reef fishes. During periods of the spring decay, hydroid stalks suspended in the water column provided an important food for Atlantic spadefish. The "bushy" hydroids and bryozoans were used as habitat by small micro-crustacean species (amphipods and copepods) as well as by brittle stars.

Although the numerical dominants of the cryptic assemblage associated with the biofouling community were represented by microcrustaceans, it was found that blennies, stone crab, pistol shrimp, polychaetes, and brittle stars dominated from a biomass standpoint.

Cryptic species which were dependent upon bushy hydroids and bryozoans as cover and/or food (microcrustaceans and brittle stars) bloomed in winter and declined during warm seasons. In contrast, cryptic species dependent upon barnacles for habitat, and which did not outgrow the cover provided (e.g., pistol shrimp, polychaetes, and blennies), were characterized by rather stable seasonal population levels. Other cryptic species were apparently recruited to the structures from the plankton, then flourished and grew until they exceeded a size allowing use of the habitat as cover. They were then either harvested by predators or left in the area prior to reaching a reproducing size (e.g., stone crabs).

Principal predators on microcryptic species were sheepshead, triggerfish, blennies, and small reef fishes; sheepshead and triggerfish were principal predators on larger cryptic species. Almaco jacks showed a marked preference for blennies.

A species of blenny new to the Gulf of Mexico was discovered during the final year of the Buccaneer Gas and Oil Field study. The species represented (*Hypsoblennius invemar*) is currently being described by Dr. Smith-Vaniz and, prior to our discovery, was known only from St. Bartholemy Island (Lesser Antilles), Venezuela (precise locality unknown), and the Gulf of Uraba, Columbia.

The biomass dynamics of the biofouling community by depth and biological season with respect to characteristic water column conditions are summarized by Figure 19. During all seasons, biomass levels near the bottom were markedly lower than biomass levels in the upper water column. The depth of this biomass discontinuity appears to coincide with the distribution of the year-round bottom nepheloid layer, and was mainly attributable to the absence of barnacles. As shown, biomass levels in winter were significantly higher than summer levels. Most of the observed seasonal change was attributable to the blooms of the mat community. High dissolved nutrient levels and phytoplankton blooms were also characteristic of the early winter season, and may have contributed to the increased biofouling levels.

Effects. The discharge of produced water had detrimental effects on the biomass levels and production rates of the biofouling community but, using the 5% level to determine differences, significant alterations of the community were restricted to a vertical distance of about 1 m and a horizontal distance of less than 10 m (Figures 20 and 21). These results, which were obtained in situ, agree well with the projected zones of toxicity described for worst-case conditions above. The near-

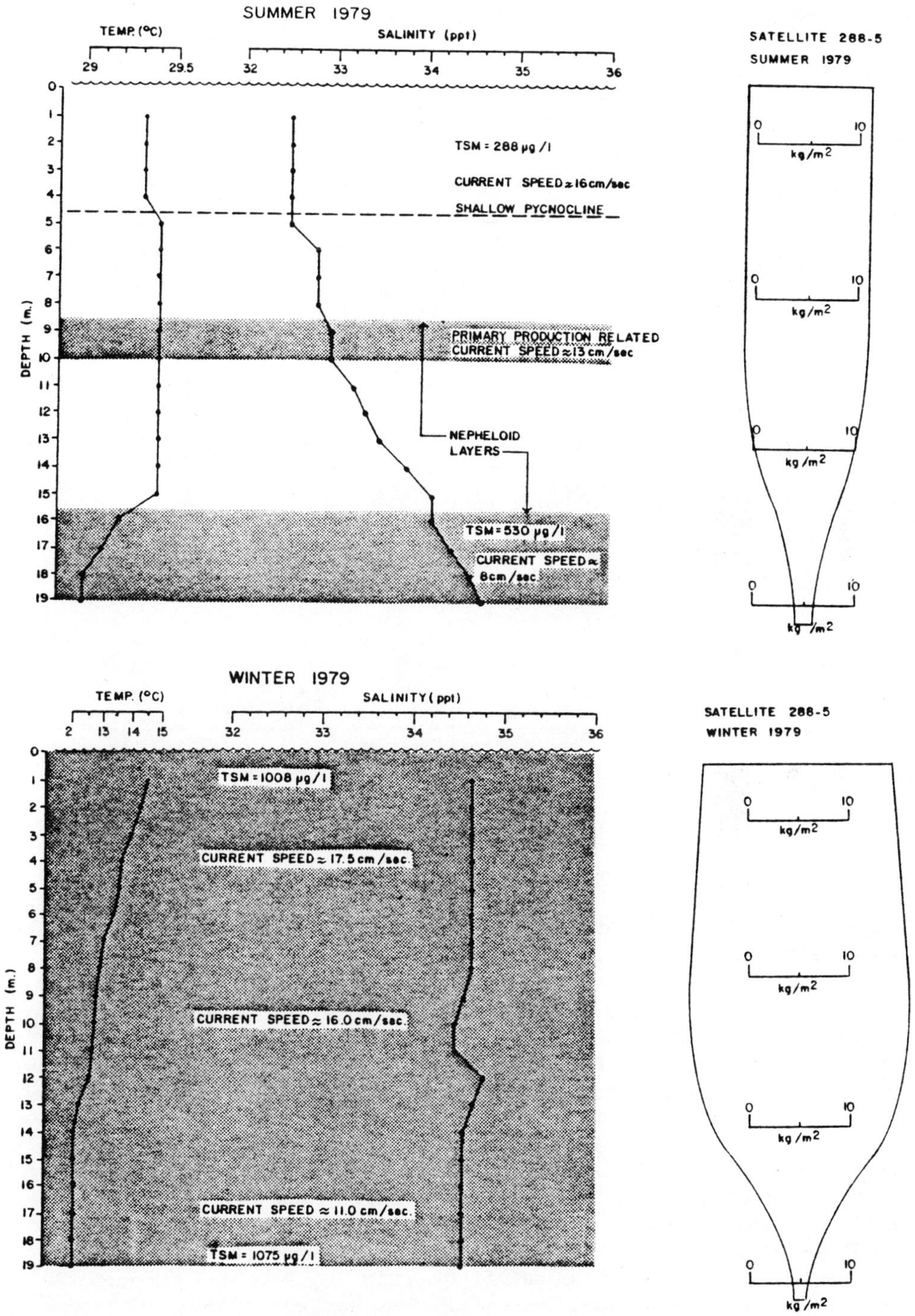

Figure 19. Seasonal levels of biofouling biomass and characteristic hydrographic conditions.

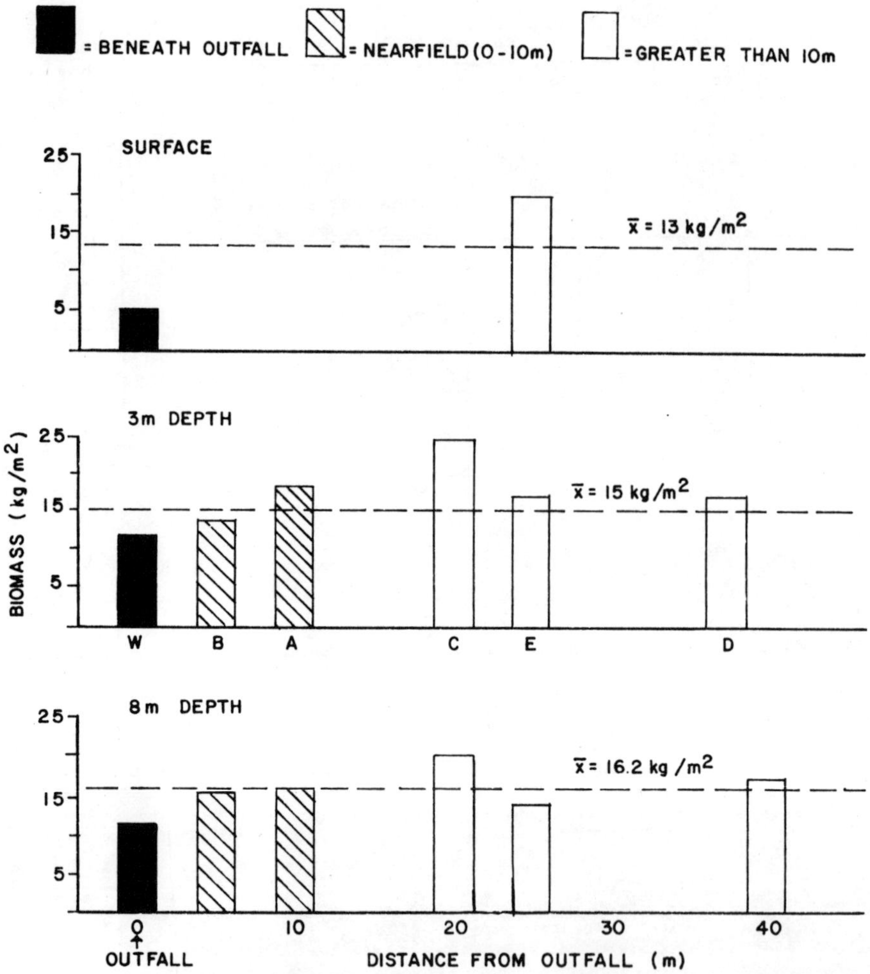

Figure 20. Standing crop biomass of biofouling community in the Buccaneer Field, summer 1980.

surface zones in the immediate vicinity of the outfall were characterized by the virtual absence of any living large barnacle, but small (usually dead) barnacles were sometimes obtained in the collections taken there. Organisms colonizing this area may do quite well until worst-case hydrographic conditions occur. In addition, organisms colonizing this zone were probably periodically subjected to nearly 100% concentrations of produced water when they are exposed in the troughs of the waves. Based upon recolonization information (Figure 21), worst-case conditions were apparently encountered more frequently in spring through fall periods than during other seasons. The surface effect of produced water on recolonization rates for spring to summer and summer to

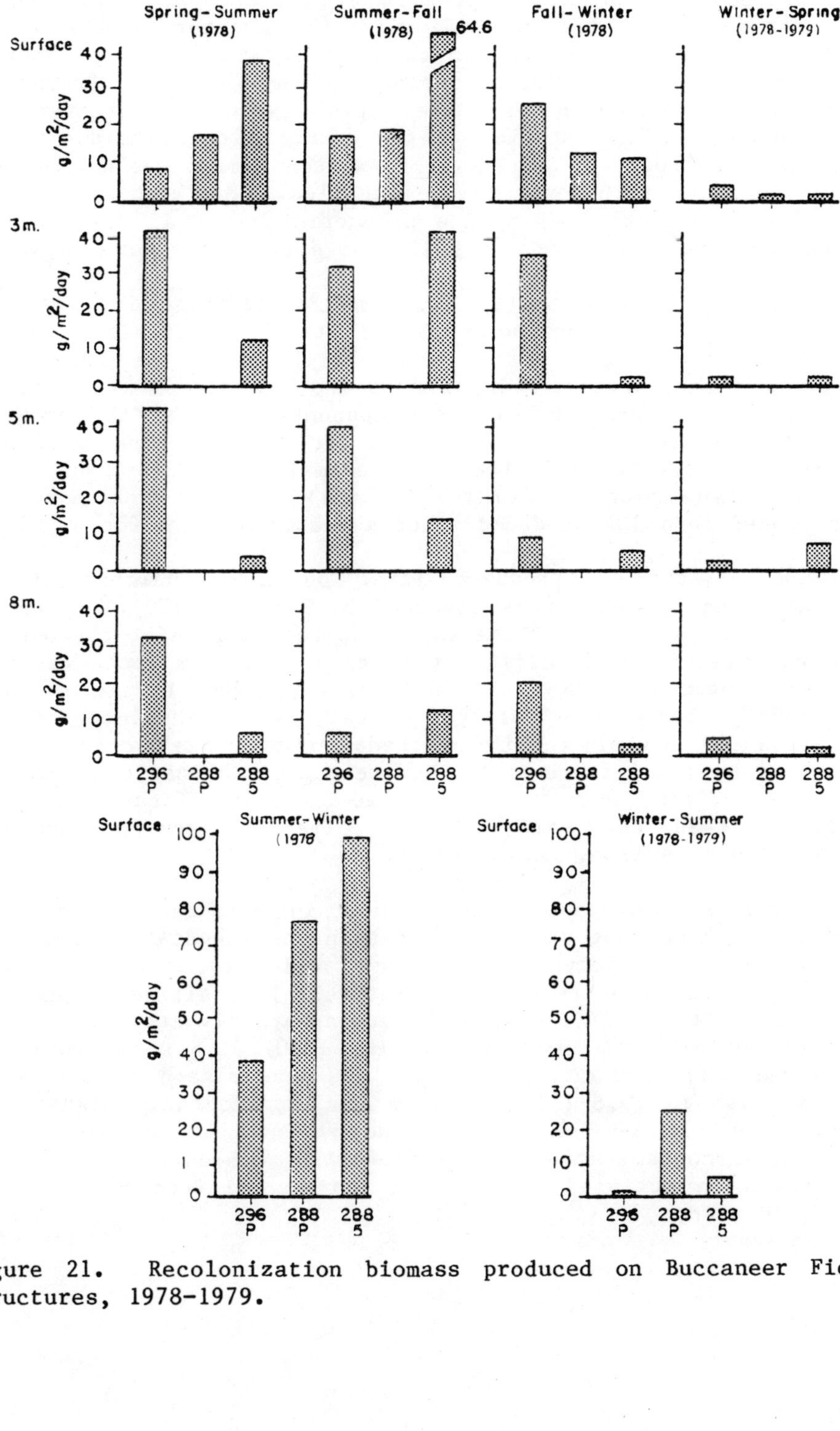

Figure 21. Recolonization biomass produced on Buccaneer Field structures, 1978-1979.

fall periods is readily apparent in Figure 21. However, production rates beneath the outfall at depths greater than 1 m were typically equal to, or greater than, production rates on control structures at the same depths during the same periods. Production rates at the surface beneath the outfall were even greater than rates at the surface station on the control structure during the fall to winter period. The fall to winter period was one of high energy in which turbulent mixing prevailed. The winter to spring season was characterized by low production rates at all stations throughout the field and no significant differences were apparent.

Results of respirometry experiments indicated low rates of biofouling primary production and that a stress response (increased oxygen uptake) had been elicited from the communities subjected to treatment. In retrospect, the stress response was attributable to the fact that the concentrations of the produced water to seawaters (10-25%) exceeded the 96-h LC50 value of most of the organisms being tested. For example, a common amphipod of the biofouling community (*Jassa falcata*) suffered 100% mortality when placed in a 10% produced water/seawater mixture for 48 h.

The effects of produced water on the condition of the Mediterranean barnacle were reported by Boland (1980). Barnacles taken from locations as close as 1 m to the surface at the outfall were not significantly different in condition from those taken at control stations. He did find, however, that barnacles taken immediately below the sewage outfall were characterized by significantly higher condition than barnacles taken in control areas. He also found that Mediterranean barnacles from the Buccaneer Field were characterized by significantly higher condition than the same species collected from a structure offshore West Cameron, Louisiana.

Barnacles were not found to contain measurable amounts of petroleum alkanes but the fouling mat in the immediate vicinity of the outfall was observed to have been oiled by direct exposure. The cryptic blennies which were relatively insensitive to produced water (96-h LC50 = 27%) and which were apparently attracted to the area of outfall (Gallaway and Martin, 1980) also showed evidence of marked petroleum contamination. The mean alkane concentration in this fish in 1978-1979 was 6.79 ppm, considerably higher than levels found in any other fish. No evidence of any significant trace metal contamination of the biofouling community attributable to production activities was found during the Buccaneer Gas and Oil Field study.

Structure Associated Fishes

Natural and artificial structures, including petroleum platforms, in the marine environment serve as aggregation points for large numbers of fishes representing many species. The mechanism of attraction (increased food, thigmotropism, etc.) and degree of permanency varies depending upon the ecological role of the species in question, as well as on the time of year and related hydrographic conditions encountered.

In the most general sense, the structure-associated fish fauna of the Buccaneer Field can be classified as either seasonal transients or residents. The most important, from at least a fisheries standpoint, of the seasonal transients are the warm-season pelagic predators and their plankton-particulate feeding prey. The predatory species representing this group in the Buccaneer Field included king mackerel, cobia, bluefish, little tunny, dolphin, sharks, blue runner, sharksuckers, and jack crevalle. Prey species included spanish sardine, scaled sardine, and rough scad. The attraction of the seasonal-transient assemblage of fishes appears to be the structures per se, but residence times at the structures for most of the species were believed to have been short.

Klima and Wickham (1971) and Wickham et al. (1973) documented the effectiveness of artificial structures and floating objects in attracting pelagic predators and their prey. Although large variations in daily numbers of fish were observed, as many as 10,000 pelagic fish were estimated around floating structures one day after they had been positioned. The congregations of fishes were observed to be highly transient in nature, with different schools constantly moving into and away from structures.

The predator and prey species maintained different spatial relationships with the structures. Prey species were normally in the upper half of the water column either around the structure or up-current from it. Predators stayed either at the level of the structure or below it, seldom swimming above. Feeding was observed among the prey species but never among predators. Although large predators were infrequently observed directly, considerable evidence of their presence and feeding was noted in the form of mutilated fish. The authors interpreted their data as evidence that the initial attraction of fishes to structures is probably the result of a visual stimulus provided by the structures in the optical void of the pelagic environment.

Wickham et al. (1973) reported that the attraction of the game fish species involved species-specific behavioral mechanisms. King mackerel and little tunny were seldom observed unless baitfish were present, but dolphin, cobia, and great barracuda were attracted to the structures per se. They presented evidence that baitfish were able to use artificial structures for predator avoidance in that the competing visual stimuli of structures disrupted the predator's visual fix on the prey which was required for a succesful attack.

Studies of the seasonal transient fishes in the Buccaneer Field were directed towards bluefish (*Pomatomus saltatrix*) which, because of their tendency towards remaining in the vicinity of platforms for longer times and being more visible than other species, seemed most susceptible to investigative effort. In contrast to most pelagic predators, bluefish were typically abundant during cool seasons (fall, winter, and spring) and rare during summer (except during 1978-1979 when they were abundant during all seasons). Although no tagged fish were observed in census efforts, visual estimates made by divers indicated as many as 3,000 to 5,000 bluefish might be associated with a structure at a given time during the periods of seasonal abundance. Most specimens of Buccaneer Field bluefish were about 45-50 cm in fork length. We received no tag returns from bluefish by sportfishermen.

Bluefish fed mainly on fish during the fall and winter, but relied heavily upon demersal macrocrustaceans during spring (Figure 22). Although some pelagic prey species were represented in the "unidentified fish" category, most of the food contained in the stomachs of specimens collected were representative of the demersal fish community. Bluefish, when present in the Buccaneer Field, were characterized by healthy populations and no evidence of any impacts from petroleum operations were observed for this species.

The resident species in the Buccaneer Field were found to include (1) fishes which were indicated to be directly dependent upon the biofouling community for both food and cover, and (2) those which appeared to be attracted to the structures mainly for cover alone (they exhibited little or no trophic dependency on the biofouling community). In the former category, we have included sheepshead (biomass dominant), blennies (numerical dominant), triggerfish and amberjacks as well as small pelagic (damselfishes, butterfly fishes, angelfishes, small sea basses, etc.) and demersal (cubbyu, wrasses) "reef" fish. In the latter category, we have included pelagic reef forms such as spadefish (usually the numerical and biomass dominant of the entire fish community) and tomtate as well as benthic reef species such as red snapper and groupers. Sheepshead and blennies were studied as representatives of the non-reef-trophically-dependent segment of the population.

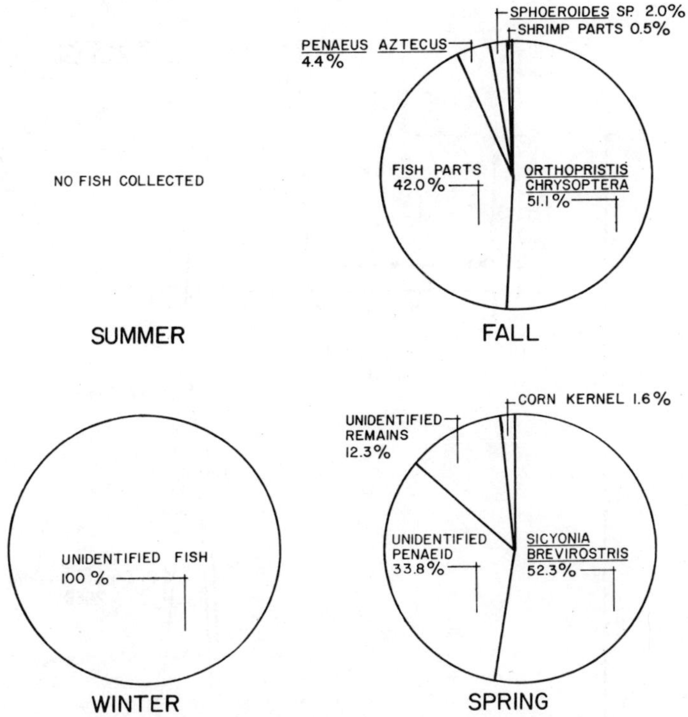

Figure 22. Seasonal food habits of bluefish (*Pomatomus saltatrix*) in the Buccaneer Field, 1978-1979.

Sheepshead. The sheepshead (*Archosargus probatocephalus*) is a common inshore sportfish of the Atlantic and Gulf coasts. Along the coast, the sheephead frequents pilings, jetties, and oyster reefs and sometimes moves up rivers into fresh waters (Pew, 1971). In coastal habitats, the sheepshead feeds on barnacles and small fish. Young sheepshead are collected in spring and early summer along the beach and in the marsh (Dahlberg, 1975), and we have observed young sheepshead particularly abundant in high-salinity grassbeds in Texas during spring and summer.

Very little work has been done with respect to the biology of offshore sheepshead populations. Gallaway et al. (1979b) reported sheepshead as one of the dominant species around petroleum platforms off Louisiana seaward to about the 37-m depth contour. At this point, however, the relative abundance of sheepshead began to decline with distance offshore, or depth, and they were not observed at platforms investigated which were located seaward of the 64-m depth contour. Sheepshead investigations were conducted during the 1978-1979 research year by Gallaway and Martin (1980) and the results are summarized below.

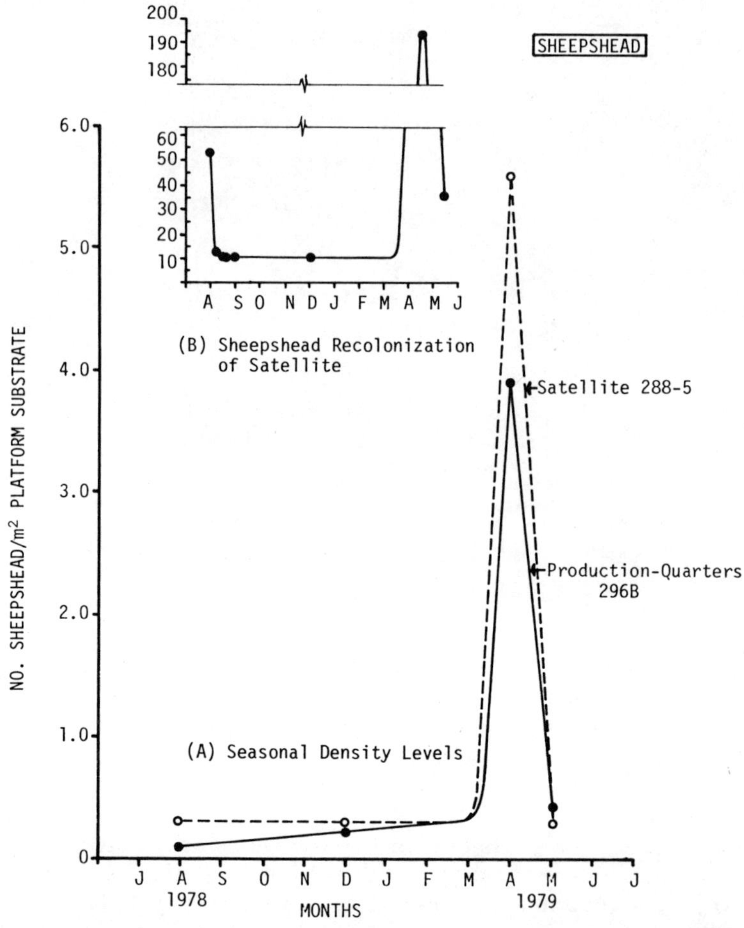

Figure 23. Seasonal densities of sheepshead (*Archosargus probatocephalus*) at Buccaneer Field structures, 1978-1979, and population levels following harvest of specimens at well jacket 288-2.

Characterization - With the exception of early April 1979, populations of sheepshead at Buccaneer Field structures were relatively stable during the research year 1978-1979 (Figure 23). The observed population levels in April represented 17- to 19-fold increases over the population sizes estimated for each structure the previous quarter. We believe that the observed concentrations represented a spawning congregation as the fish were mostly running ripe, and many were exhibiting what we interpreted as courtship behavior. Similar congregations were observed at all the structures examined in the Buccaneer Field during early April 1979. We do not know exactly how long the

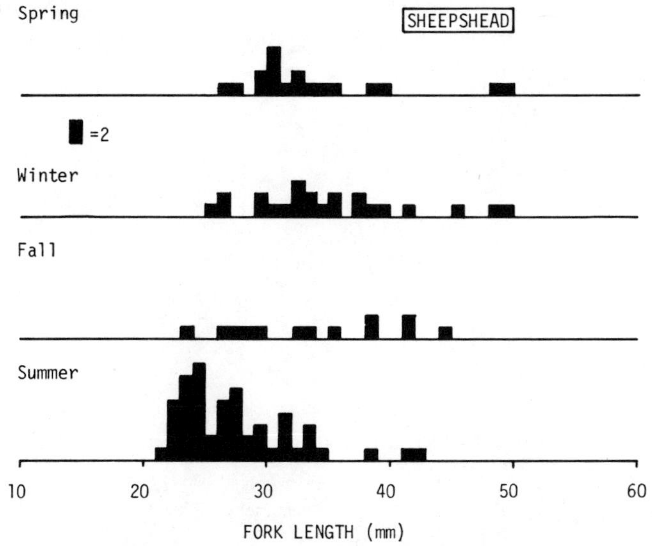

Figure 24. Size distribution of sheepshead (*Archosargus probatocephalus*) in the Buccaneer Field, summer 1978 - spring 1979.

congregations persisted; populations had returned to the normally observed ranges by mid-May. We believe that these observations represent the first evidence of spawning migration and aggregations for this species.

Recruitment of sheepshead to Buccaneer Field structures appears to be an annual event related to spring spawning aggregations at this habitat (Figure 23). During August 1978, we were able to harvest all but about 10 of the sheephead observed at well jacket 288-2. The observable population remained at about 10 fish until April 1979, after which it was estimated at about 36 individuals (69% of the pre-harvest level: insert B, Figure 23). Densities of sheephead at the two censused habitats (platform 296-B, well jacket 288-5) were remarkably stable during each season, with the obvious exceptions of April at both structures and summer at platform 296-B (Figure 23). With the exception of movement between the adjacent production-quarters structures, sheepshead appeared to be habitat-faithful. No marked fish were seen at a structure other than where they had been marked. As indicated by Figure 23, the density of sheepshead was typically slightly higher at the well jackets than at the production-quarters platforms.

Sheepshead in the Buccaneer Field during 1978-1979 ranged from about 22- to 50-cm fork length (Figure 24). Fish between 22 and 35 cm usually dominated the collections, particularly during

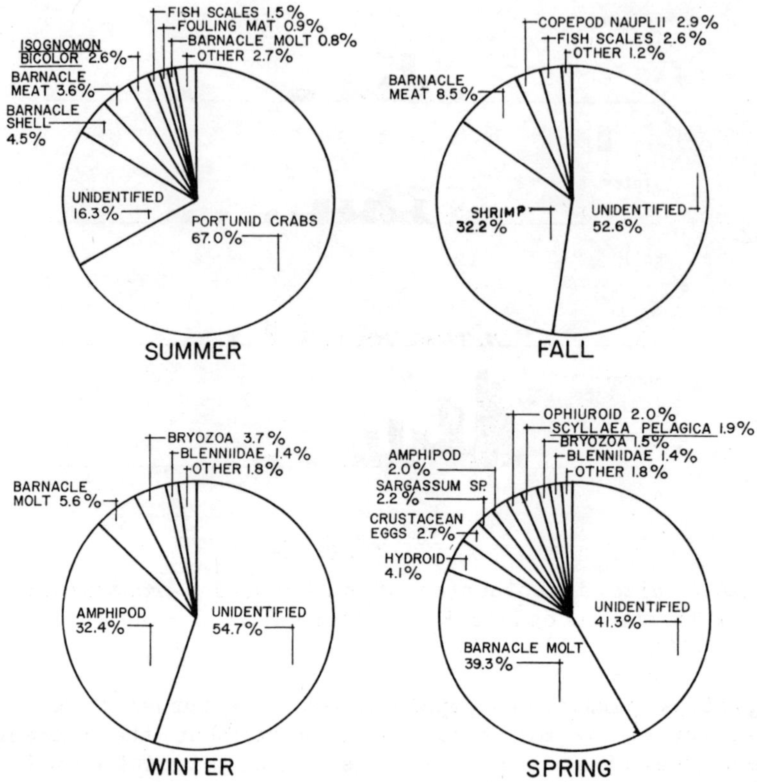

Figure 25. Seasonal food habits of sheepshead (*Archosargus probatocephalus*) in the Buccaneer Field, 1978-1979.

summer. In the latter case, nearly all of the specimens at well jacket 288-2 were harvested and comprised a rather complete sample of the resident population.

The seasonal length-weight relationships for sheepshead were characterized by equal slopes, and significant differences (5% level) were indicated among the seasonal levels of condition (F = 2.99 at 3 and 90 d.f.). Sheepshead collected in winter and summer were significantly heavier at an adjusted mean length than fish at the same length in fall and spring. Fall fish were not significantly different from summer and spring fish. The greatest difference in condition was observed between April and May sheepshead; the former group were in spawning condition and, at a given length, averaged 6.3% heavier than fish collected after the spawning activity.

The food habits of sheepshead varied with season (Figure 25). During summer 1978, portunid crabs comprised 67% (by weight) of the diet and were supplemented by biofouling organisms. During each of the fall, winter, and spring seasons, the biofouling community comprised the majority of the diet. Most of the material in the "unidentified" category was believed to have been of biofouling origin. The presence of sargassum in stomachs of spring specimens supports visual observations of sheepshead grazing on rafts of this material as well as on the organisms utilizing the sargassum as habitat.

Sheepshead were characterized by higher index of fullness values in winter (26.0) and spring (24.0) than in summer (19.0) and fall (14.0), but the data showed an almost uniform feeding periodicity over the 24-h cycle during each of the four seasons. Index of fullness values for the periods 2401-0600, 0601-1200, 1201-1800, and 1801-2400 h were 19.0, 19.3, 19.0, and 12.6, respectively. This species was heavily dependent upon the biofouling community for food but, as described below, also obtained food from other sources (including food scraps from the platforms).

Effects - The bacterial flora of sheepshead was similar to that observed for other fish species collected from the Buccaneer Field. Species of *Vibrio* were represented each season and were usually the dominant form. Of interest, hemolytic *Vibrio* sp., typically abundant, were not isolated from sheepshead collected during fall 1979. *Aeromonas* sp. was one of the dominant taxa on sheepshead collected at each of the two sampled habitats in summer, and was also represented on fish taken at each structure in fall. This potential fish pathogen was not isolated from winter specimens and was represented only on specimens from platform 296-B in spring. The two categories of structure types (discharging and non-discharging) did not show marked differences in terms of sheepshead bacterial flora.

The most notable histopathological finding with respect to sheepshead was the virtual absence of any anomalous condition in tissue samples taken from specimens collected during the brief period of the spawning aggregation observed in the Buccaneer Field in April. Typically, sheepshead collected during other seasons exhibited five to seven different tissue anomalies, with each condition represented in 20-100% of the specimens collected. With the exception of gill hyperplasia which was characteristic of all specimens collected during the summer and four of five specimens collected at platform 296-B in fall, most of the anomalies in the tissues examined were lesions in association with the presence of, or attributable to, a parasite (e.g., nematodes). If the fish collected during April were indeed representative of a migrant population, it would appear from these data that resident sheepshead are characterized by a higher degree of histopatho-

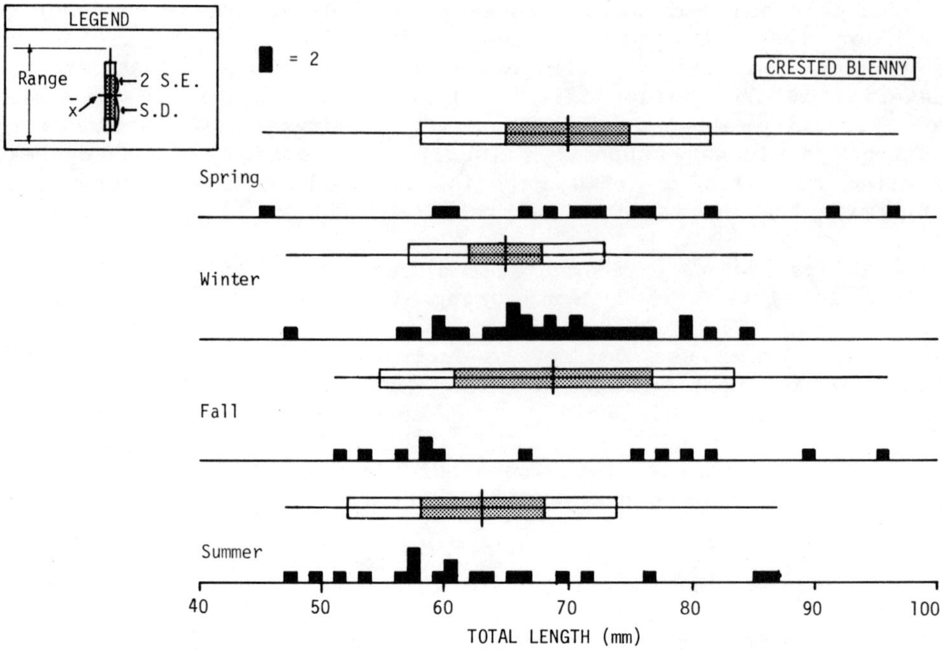

Figure 26. Seasonal size distribution of crested blenny (*Hypleurochilus geminatus*) in the Buccaneer Field, 1978-1979.

logical anomalies (or parasitism) than are sheepshead which migrate in and out of the study area for spawning purposes.

Comparisons of condition of sheepshead at the treatment and control structures were based upon specimens subsequently submitted for histopathological and bacterial flora analysis. The data set was reduced to the December 1978 and May 1979 collections, as the sheepshead represented during April were not considered resident fish, and weights were not obtained for the specimens analyzed from August. The length-weight regressions for fish from the two habitats had equal slopes (F = 1.75 at 1 and 16 d.f.) and, although fish from platform 296-B were 10.6% heavier than fish from well jacket 288-5, the differences were nonsignificant (F = 3.79 at 1 and 17 d.f.).

Sheepshead were characterized by the presence of petroleum alkane contaminants in both liver (6.08 ppm) and muscle tissues (4.57 ppm). These levels were lower than those observed for blennies, but higher than those observed for fishes not trophically dependent upon the biofouling community for food. No significant trace metal contamination related to Buccaneer Field operations was demonstrated (Tillery, 1980).

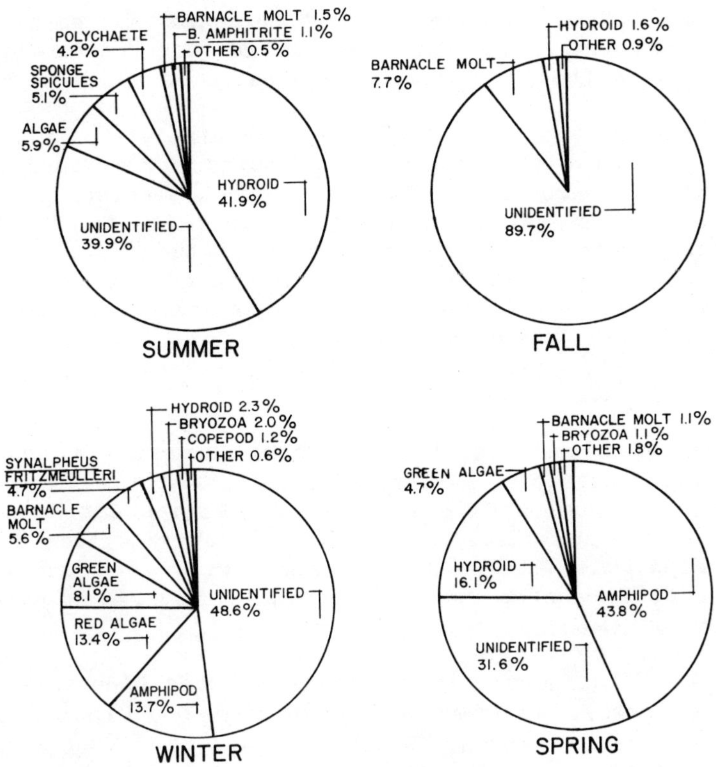

Figure 27. Seasonal food habits of crested blenny (*Hypleurochilus geminatus*) in the Buccaneer Field, 1978-1979.

Crested Blenny. The blennies are small, "personable" fishes that live in and around rocks, reefs, and other hard substrates, particularly barnacle shells. Populations in the Buccaneer Field consisted of at least four species - crested blenny (*Hypleurochilus geminatus*, dominant), seaweed blenny (*Blennius marmoreus*), molly miller (*Blennius cristatus*), and the new species (*Hypsoblennius invemar*) noted above. The crested blenny is the most common form on Texas jetties and petroleum platforms of the shallow Gulf.

Characterization - Average density of blennies on Buccaneer Field structures dominated by the Mediterranean barnacle ranged from 12 to over $50/m^2$ (Workman and Jones, 1979; Gallaway and Martin, 1980). This compares with a range of $8-16/m^2$ on platforms off Louisiana dominated by smaller acorn barnacles (Gallaway et al., 1979b). Densities of blennies was significantly higher during summer periods than during other seasons, presumably due to recruitment. Spawning of the crested blenny extended from spring to at least August, and the entire life cycle is completed at the structures. Eggs and larvae are brooded in empty barnacle shells. Seasonal size distribution of crested blenny in the Buccaneer Field (based upon 1978-1979 data) is shown by Figure 26.

The crested blenny relied almost entirely on the biofouling community as food (Figure 27). Hydroids, bryozoans, and algae were commonly ingested and small, cryptic species (e.g., amphipods, polychaetes) were also important food items. During summer 1979, sponge spicules represented 5.1% of the diet. Much of the above "fouling mat" material is probably ingested as the blennies take discrete, cryptic organisms. Barnacles provide not only critical habitat for blennies but also food, mainly through the work of sheepshead who leave bits of barnacles in the crushed shells as a result of feeding.

Based upon the identifiable food contents in the stomachs from samples obtained during 1978-1979, hydroids and barnacle molts were the dominant food items of blennies during summer and fall; amphipods and algae were important during winter and, during spring, amphipods, hydroids, and algae were the dominant foods. The unidentified category is believed to have been primarily of biofouling origin.

Index of fullness values for the crested blenny were highest in winter (52.5), intermediate in spring (36.0), and lowest in summer and fall (29.6 and 28.2, respectively). On a daily basis, stomach contents of specimens collected between 0601 and 1200 h were lower (IF = 27.0) than contents from specimens collected during other periods (IF $\overline{x}$ = 49.0 for period of 1201-1800 h; 49.5 for 1801-2400 h; and 42.0 for 1201-0600 h).

Effects - Blennies exhibited an apparent attraction to the produced water discharge. Highest densities were observed at production platforms and significantly higher densities were observed near the outfall than elsewhere. Based upon in situ observations, the effluent had no significant effects on recolonization rates of areas harvested of blennies nor were there any significant effects on condition of blennies.

The apparent attraction to the outfall area may have been attributable to the combination of a greater level of habitat availability due to higher densities of both live and dead barnacles in these areas (Howard et al., 1980) and the apparent lack of sensitivity of the crested blenny to produced water (96-h LC50 was 269,000 ppm or about 27% produced water in seawater).

Results of simple bioassays performed in situ by Workman and Jones (1979) confirmed that the crested blenny was tolerant of produced discharge. The experiment was performed by suspending caged blennies just beneath the surface in the area of the outfall and beneath a well jacket for an approximate 48-h period. Three of 20 and 3 of 22 crested blennies suspended in each of the respective areas died during the experiment. In contrast, all 20 of the seaweed blennies in the cage suspended beneath the effluent died. None of 17 in the control cage were known to have died although 5 specimens were missing (believed to have escaped during a transfer of fish into the cage).

The crested blenny differed little from other Buccaneer Field fishes in terms of its external bacterial flora. Species of *Vibrio* were the most common taxa during each season; hemolytic *Vibrio* were not isolated from fall specimens. *Moraxella* sp. was apparently a codominant with *Vibrio* sp. during spring. There was no marked difference in the bacterial flora of blennies taken from the production platform as compared to those collected at well jacket habitats. No diseased blennies were noted in any of the areas sampled.

The crested blenny was a "clean" fish in terms of histopathological anomalies. Other than a light infestation of microsporidean parasites, no significant histopathological anomalies were detected in the specimens which were examined.

However, the average alkane concentration in this fish was 6.79 ppm, higher than the mean levels observed for any other fish from the Buccaneer Field (Middleditch and West, 1980). Trace metal contamination of blennies attributable to Buccaneer Field operations was not indicated.

<u>Spadefish</u>. The spadefish (*Chaetodipterus faber*) represents another common coastal food and game fish with little known about its offshore biology. In the United States, its range extends along the Atlantic coast from Cape Cod to Florida and throughout the Gulf of Mexico and into the Caribbean. In Texas, large numbers of small spadefish show up in the surf along the beach during the spring of each year and remain in nearshore habitats until about fall. Spadefish are generally found in schools and, along the coast, they congregate around jetties, wrecks, pilings, and bridges where the average size seldom exceeds 454 g. Offshore, schools are sometimes observed in open water, but spadefish characteristically congregate around structures, particularly petroleum platforms which differ from other reefs in that they extend from the bottom to the surface. Around production platforms, large fish are common, some up to 9 kg. The offshore distribution of spadefish in the Gulf was found by Gallaway (1979b) to be similar to that described for sheepshead above - they were not observed at deep-water platforms. Spadefish investigations were conducted during 1977-1978 (Gallaway et al., 1979a) and 1978-1979 (Gallaway and Martin, 1980).

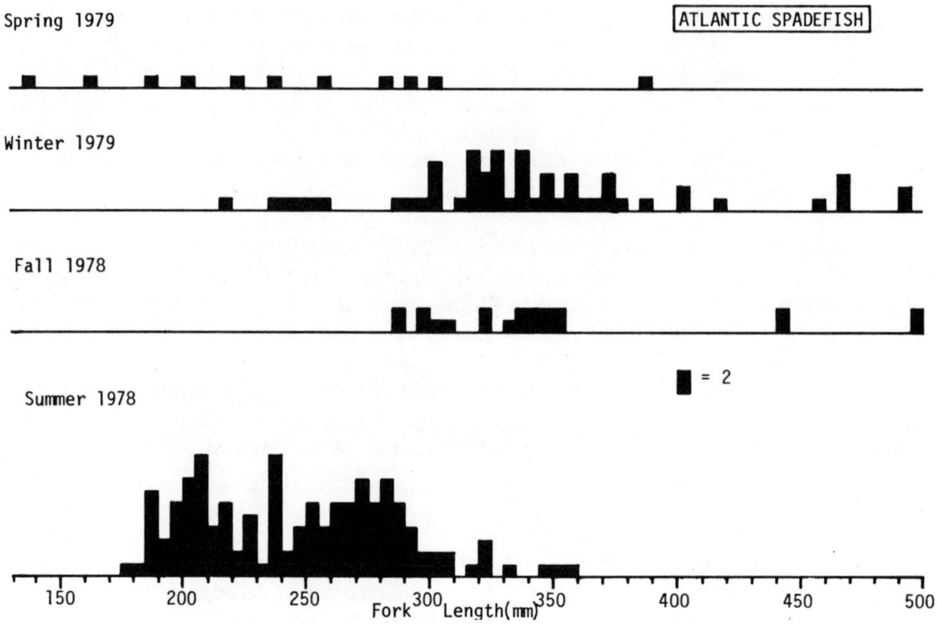

Figure 28. Size distribution of Atlantic spadefish (*Chaetodipterus faber*) in the Buccaneer Field, summer 1978 - spring 1979.

Characterization - Based upon results of the mark-recapture studies, spadefish were observed to be habitat faithful; i.e., there was little exchange among populations associated with different structures except for the closely-allied quarters and production platforms. We believe that the best estimates of seasonal densities of spadefish around Buccaneer Field structures were about 0.11-0.15 fish/m^3 in summer, 0.20 fish/m^3 in fall, 0.22 fish/m^3 in winter, and 0.16 fish/m^3 in spring. The seasonal size distribution of spadefish populations appeared to differ by season. During winter and fall, length of individuals ranged from about 210 to 500 mm, with fish greater than 400 mm rather common. These large individuals were scarce or absent around the structures during spring and summer seasons (Figure 28). In spring, the observed length range of spadefish was about 135-385 mm, with no apparent size group dominant. Summer populations ranged in length from 175 to 360 mm with at least two different size groups represented (Figure 28). We believe that relatively high densities observed during the fall and winter are attributable to recruitment during fall and to the influx of large fish during winter which were absent (spawning?) during spring and summer.

Recently-spawned spadefish (5-30 mm) were not observed in the Buccaneer Field by us during any season. Additionally, results

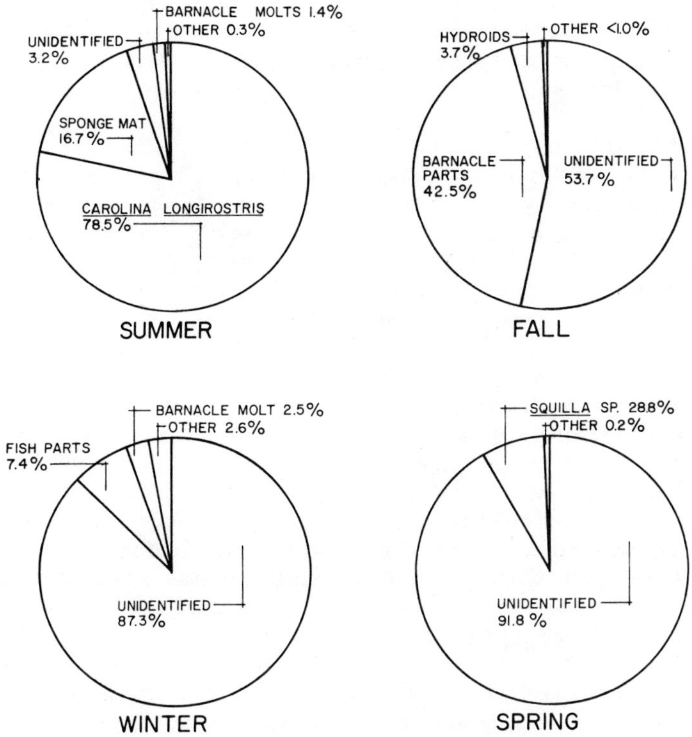

Figure 29. Seasonal food habits of Atlantic spadefish (*Chaetodipterus faber*) in the Buccaneer Field, 1978-1979.

from the ichthyoplankton sampling program of 1976-1978 (Finucane et al., 1979) indicated that larval spadefish were not abundant during any season. As described above, young spadefish are abundant in the surf zone of Galveston Island in late May and June. We suspect that the absence of large fish during spring and summer and the relatively large size of the smallest recruits indicates that spawning of this species generally occurs elsewhere.

Food habits of Atlantic spadefish varied seasonally (Figure 29). During summer, the diet of this species was dominated by a planktonic pteropod, *Carolina longirostris*. During each of the fall, winter, and spring seasons, we were unable to identify most of the material in the stomachs, but suspect that it is primarily of biofouling origin. During winter of the 1979-1980 research year, we were able to obtain a series of stomach contents grading progressively from intact hydroid stalks to an unidentifiable mass. Food habitat data for spadefish indicate that, when plankton are unavailable, spadefish will utilize the biofouling community as a food source. This appeared especially true during the spring when biofouling organisms were being sloughed from the substrate and, as suspended particulates, were harvested by the plankton-feeding spadefish.

Based upon comparisons of index of fullness values (Gallaway et al., 1979a) determined from specimens collected by spear, daily feeding periodicity of Atlantic spadefish was not markedly different among seasons. Feeding appeared to have been greatest during the period from mid-morning (1000 h) to early evening (2000 h), particularly during late afternoon (1600-1700 h). Although patterns of daily feeding were similar among seasons on a relative basis, the magnitude of the IF values varied greatly among seasons. The respective average IF values for specimens speared during summer, fall, winter, and spring were 25.7, 3.3, 1.0, and 20.5. The high summer value was associated with plankton-particulate feeding in the upper water column, whereas the low fall and winter levels were believed associated with near-bottom grazing of the biofouling community. As described above, most of the food in the spring samples was also considered to have been of biofouling origin (probably *Tubularia crocea*) but was believed to have been harvested from the water column as opposed to having been grazed from the structures.

During winter of 1977-1978, we observed that spadefish populations in the Buccaneer Field experienced a disease epidemic characterized by large, external lesions and varying degrees of fin rot. Based upon bacterial isolates cultured from diseased specimens collected at that time, the fish pathogen *Vibrio* was abundantly represented. Badly diseased fish (those with large lesions and advanced cases of fin rot) were more in evidence at platform 296-B (61% of the sample) than at well jacket 288-5 (25%). The spadefish disease epidemic was also in evidence during the winter period of 1978-1979. In contrast to the previous year, there was little difference among structures (73 and 74% of the spadefish at the production platform and well jacket, respectively, were badly diseased). In early March practically no diseased spadefish were observed at the V.A. Fogg Liberty Ship Reef which was used as a control against which to compare Buccaneer Field populations.

In addition to gross examination, spadefish tissues were collected each season for microscopic examination. Histopathological anomalies in spadefish tissues (gills, intestine, liver, kidney, skin) determined by microscopic examination varied among seasons and habitats. Although larger sample sizes would have been needed to make definitive comparisons, gill hyperplasia was observed to be prevalent during summer, particularly at platform 296-B, where each of the five specimens collected evidenced the anomaly. Fatty infiltration of the liver was pronounced in spadefish at all sites sampled in the winter and fall, less prevalent during summer, and present in only one of ten fish collected during spring. Fatty infiltration of the liver of spadefish colonizing offshore Louisiana platforms during summer was also observed by BLM investigators (C.A. Bedinger, Southwest Research Institute, personal communication). Lesions in skin and fin tissues of Buccaneer Field spadefish were restricted to winter samples.

Bacteriological analysis of fish tissues yielded results similar to those of Sizemore and Olsen (1980) in that *Vibrio* sp. was a predominant genera in all samples, particularly during winter. Potential fish pathogens (*Vibrio* sp., hemolytic; *Seromonas* sp.) were also represented during all seasons. Of these, Sizemore and Olsen (1980) found only *Aeromonas hydrophila* associated with the four diseased fish he examined.

Spadefish disease epidemics appear best explained as the result of the actions of opportunistic pathogens during a period of natural seasonal stress for the host. Much of the seasonal stress is presumably attributable to the observed combination of high density, change in habitat (spadefish move from surface waters to the bottom during winter), reduction in apparent feeding efficiency, and the change in food habits from plankton to suspended particulates of biofouling origin (Figure 30). The observed fatty infiltration of the liver may represent results of a nutritional deficiency of this alternate food source utilized during winter. During the 1978 research year, Atlantic spadefish were characterized by significantly ($\alpha = 0.05$) better condition in summer than in winter (Gallaway et al., 1979a); however, no significant seasonal differences were observed during the 1979 research year (Gallaway and Martin, 1980).

Effects - Spadefish showed no evidence of either petroleum or trace metal contamination attributable to Buccaneer Field operations, and were characterized by the lowest levels of total alkanes of any fish tested (Middleditch and West, 1980; Tillery, 1980). Density levels were found to have been equivalent among the various structure types in the field. The observed disease epidemics seem best explained by natural phenomena. Whereas condition of spadefish was significantly lower during winter than during other seasons, as might have been expected, condition did not differ significantly among the habitats sampled during winter.

However, the possibility remains, although considered by us to be slight, that the winter disease epidemics may have been related to the chronic, low-level discharge of contaminants.

Minchew and Yarbrough (1977) found that 96% of the mullet, *Mugil cephalus*, held in ponds subjected to a low-level oil spill

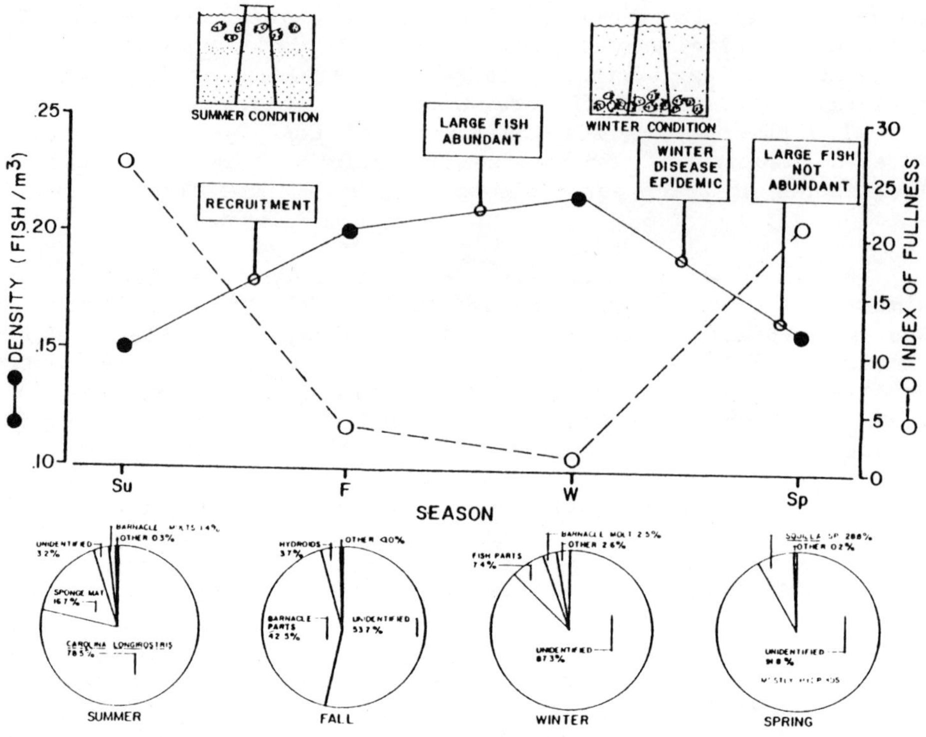

Figure 30. Seasonal distribution, abundance, foods, and feeding patterns of Atlantic spadefish (*Chaetodipterus faber*) in the Buccaneer Field.

(4-5 ppm) suffered fin rot, whereas only 6% in a control pond developed eroded fins. The primary pathogen considered responsible for the fin erosion was a species of *Vibrio*. Subsequent laboratory work by Gilles et al. (1978) confirmed the above results and showed that chronic, low-level exposure of mullet to oil significantly altered the bacteria on the fish, allowing for a large population of potentially pathogenic *Vibrio*. They also suggested that the *Vibrio*, through utilization of the oil, may have acquired an enhanced virulence. Our field studies agree with the findings of the above pond and laboratory experiments in that fish exposed to chronic low levels of hydrocarbons in discharges developed external lesions and fin rot which may have been attributable to a *Vibrio* sp.

Red snapper. The red snapper (*Lutjanus campechanus*), a highly prized sport fish and commercially valuable offshore species, occurs in association with hard-bottom habitats throughout the Gulf of Mexico (where it is probably most abundant) as well as in the Western Atlantic and Carribean Sea. It is believed

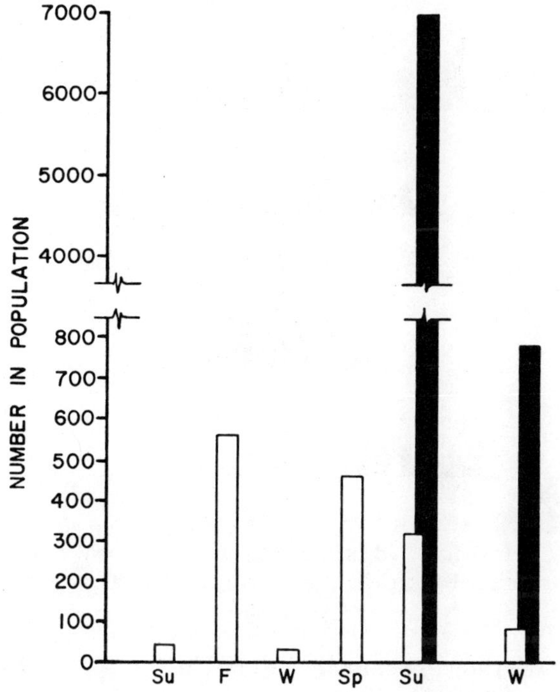

Figure 31. Seasonal population estimates for red snapper (*Lutjanus campechanus*) at Buccaneer Field platform 296-B and West Cameron platform 333-A, 1978-1979.

that red snapper tend to be associated with deeper water (30-65 m) during winter but that, during warmer periods, there is a general movement from offshore to inshore reefs (20-30 m). Spawning occurs between June and October, and juveniles utilize open sand and other soft substrates in water 10 to 30 m in depth as nursery grounds. Red snapper are recruited from the demersal nekton to the reef community at about the end of their first year, at which time they have attained lengths of about 140-250 mm.

Red snapper investigations in the Buccaneer Field were conducted by Workman and Jones (1979), Gallaway and Martin (1980), and during the final year of the Buccaneer Gas and Oil Field Study, reported herein.

Characterization - All of the tag return data indicated that red snapper were structure-faithful. No tag was returned or taken during a census other than at the location where the fish had been tagged. Population levels of "reef" (>180 mm fork length) red snapper at platform 296-B in the Buccaneer Field (heavily fished)

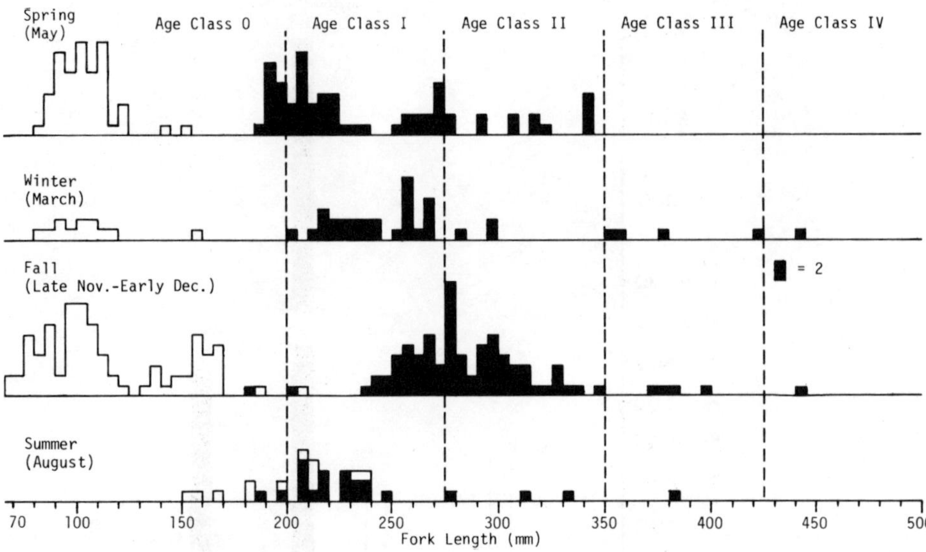

Figure 32. Seasonal distribution of red snapper (*Lutjanus campechanus*) in the Buccaneer Field, 1978-1979.

during the 1978-1980 research years and at a relatively little-fished structure in the West Cameron 333 BLM lease block during research year 1980 are shown by Figure 31. Population levels of reef-sized snapper in the Buccaneer Field were highest during the fall and spring seasons, and summer levels typically exceeded those observed during winter, except in 1978. Although some of the population fluctuations of red snapper in the Buccaneer Field were undoubtedly associated with seasonal movements, we believe that most of the reductions were directly attributable to sportfishing. The difference in population size between the heavily fished Buccaneer Field structure and the relatively unfished petroleum platform in an area with difficult access is apparent (Figure 31). We received a high rate of tag returns from sportfishermen and directly observed fishing pressure to have been heavy, particularly during the warm season. We believe that most of the annual recruitment of reef red snapper to the Buccaneer Field is harvested by sportfishermen as elaborated under "Recreational Fisheries" below.

The typical seasonal size distribution of red snapper (including small specimens taken in Buccaneer Field trawls) based upon 1978-1979 data is shown by Figure 32. Within Age Class 0 (after Bradley and Bryan, 1976) two size groups were represented in fall, winter, and spring - only the larger of these was represented in summer collections. Moseley (1966) stated that spawning off the Texas coast extended from early June to mid-

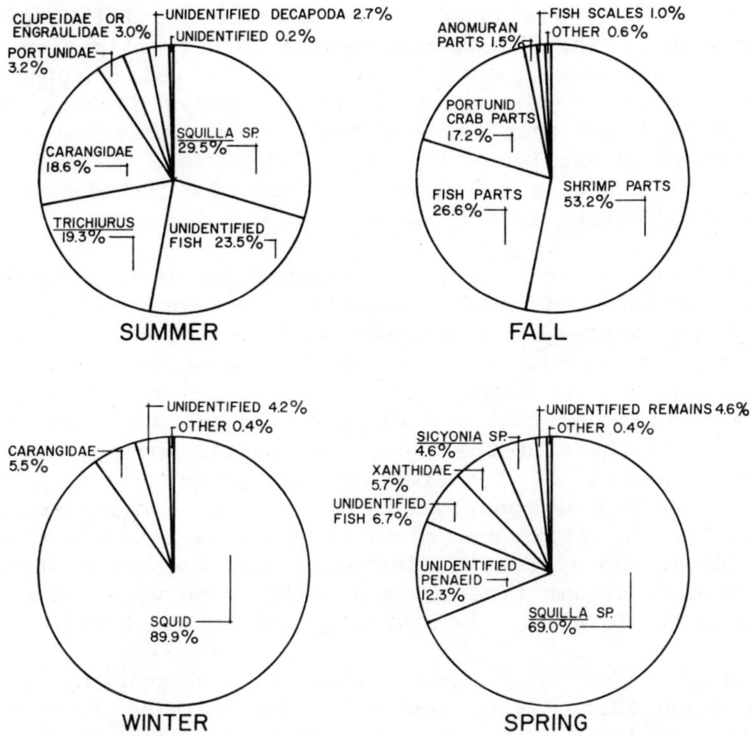

Figure 33. Seasonal food habits of red snapper (*Lutjanus campechanus*) in the Buccaneer Field, 1978-1979.

September; Bradley and Bryan (1976) provided evidence extending the spawning period from April to as late as November. The latter authors also reported that snapper attain fork lengths of about 200 mm during their first year and grow at a rate of about 75 mm per year after age or year class I. Using the above data as criteria, Age Class 0 fish in the study area during summer of 1978 were probably those of the previous summer's spawn. The small size group of age 0 fish represented in fall and winter collections were probably spawned in spring of 1978, grew little during the period December 1978 - March 1979, and were probably represented by the two specimens between 140 and 155 mm trawled in spring 1979 (Figure 32). The larger size group of Age 0 fish in fall and winter 1978 collections were probably spawned in early spring of 1978 and were represented by the large Age 0 and small Age I specimens represented in spring 1979 collections. The small size group of Age 0 fish in spring were probably fish spawned the preceding month.

Snapper enter the hook-and-line fishery at about 200 mm (Bradley and Bryan, 1976). The red snapper fishery in the

Buccaneer Field is dominated by Age Classes I and II, and no fish older than Age Class IV was represented in our collections (Figure 32). The dominance of relatively young red snapper in the Buccaneer Field fishery also indicates heavy fishing pressure. Fable (1977) found that heavily fished structures and reefs were characterized by smaller and younger (mostly Age Classes I and II) red snapper than were taken at less heavily fished banks, where fish up to Age Class V were taken.

The diet of red snapper varied with season, with winter appearing to be the most dissimilar in terms of diet (Figure 33). During winter, red snapper were indicated to feed almost exclusively on squid, although small carangids (probably scad) were also taken. We suspect that the squid was provided by us in the form of bait and that red snapper rely primarily on other fish as food during the winter season. The mantis shrimp, *Squilla*, was a major component of the diet of red snapper during both the summer and spring seasons, with fish also well represented during summer. Shrimp, fish, and swimming crabs were the most abundant food items of red snapper collected in the Buccaneer Field during fall. Results of our findings generally agree with those of other investigators (Moseley, 1966; Bradley and Bryan, 1976).

Average index of fullness values of red snapper captured by angling were 68.7, 29.0, and 18.0 for morning (0600-1200 h), afternoon (1201-1800 h), and early evening hours (1801-2000 h), respecively. No specimens were obtained for the period 2000 to 0600 h. These data, in combination with the above food habit data, could be interpreted to indicate that red snapper feed during the night or early morning over soft bottom away from the platforms. Hastings et al. (1976) obtained similar results for lutjanids in the northeastern Gulf and the Gulf of Mexico Fishery Management Council (1980) reported that most of the benthic prey "consumed by red snapper are not obligate reef or rock dwellers and therefore the inference can be made that the species feeds away from these areas." Red snapper obtained in our study exhibited very little, if any, dependence upon the biofouling community as food.

Effects - Of the 34 red snapper examined for histopathological anomalies, 62% were characterized by gill hyperplasia and 47% by intestinal parasites, usually accompanied by intestinal inflamation, fibrosis, and lesions. Gill parasites were believed to be largely responsible for the observed hyperplasia. No marked difference in the frequency of the various anomalies was observed for production platforms vs. well jacket populations or among seasons. Bacterial flora of red snapper varied seasonally, with *Vibrio* sp. usually the dominant form on fish from each of the two habitats sampled. Hemolytic *Vibrio* sp., which include representatives of potential fish pathogens, were well represented on

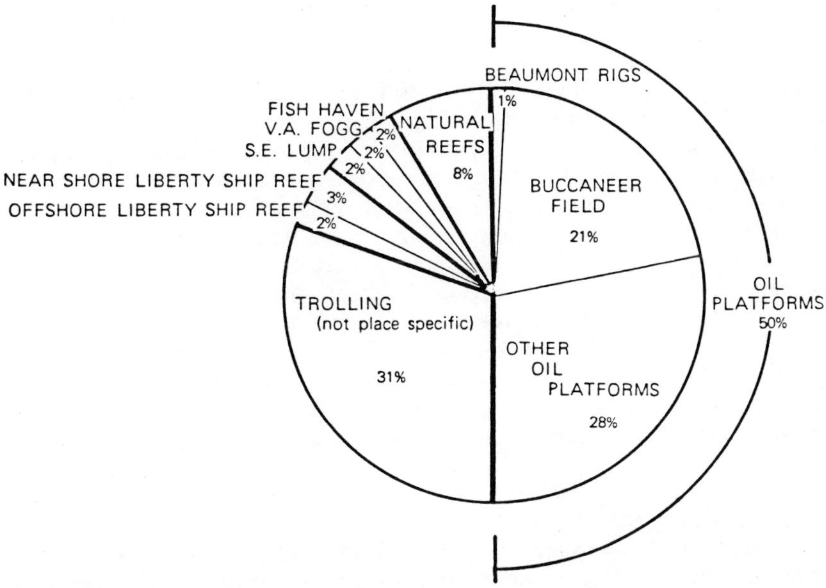

Figure 34. Distribution of fishing effort off the Texas coast from Freeport to Beaumont/Port Arthur.

specimens from each structure during each season except fall 1978, when none was isolated from any sample. *Aeromonas* sp., which also contains fish pathogens, were isolated from specimens taken at platform 296-B in summer 1978 (27% of the total 26 colonies isolated from snapper tissue were *Aeromonas* sp.) and in spring 1979 (15% of the total 47 isolated colonies). *Aeromonas* sp. were not isolated from red snapper specimens taken at well jacket 288-5 during any season. No evidence of disease or red snapper in poor condition was observed at any location or during any season. Hydrocarbon contamination was variable but typically low (Middleditch and West, 1980); no significant trace metal contamination was observed (Tillery, 1980).

Recreational Fisheries

Information provided by Bob Ditton of Texas A&M University (personal communication) shows that 50% of all offshore sportfishing effort along the upper Texas coast is directed at "oil" platforms and that the Buccaneer Field receives an estimated 21% of the effort (Figure 34). Data for the Buccaneer Field obtained by Trent et al. (1977) indicated that, weather permitting, the average number of fishing boats in the Buccaneer Field ranged between 1 and 6 on weekdays and from 5 to 16 boats on weekend days. Fishing pressure was highest during late summer and early fall.

Of the total number of boats categorized as to type of effort, 77% were bottom fishing, 17% were trolling, and 6% were being used as a diving platform. Bottom fishing boats were concentrated around the large production platforms; divers mostly utilized the well jackets, perhaps to avoid the bottom fishing. The most fish per fisherman hour were caught by bottom fishing. Of all species reported caught by bottom fishing, red snapper comprised 80% of the catch, Atlantic spadefish 7%, and bluefish comprised about 4%. The average bottom fishing boat contained 3.2 people, who averaged fishing about 4 h. The number of red snapper caught per fisherman hour ranged from 0.2 (July) to 6.8 (December). Red snapper were also taken in both the trolling and diving efforts. In the former type of effort, the king mackerel was dominant.

We developed a model to simulate the Buccaneer Field recreational red snapper fishery based upon the above information and the biology of the species. The purpose of the model was to evaluate the observed effort indicated at the Buccaneer Field in terms of impact on the commercial fishery and snapper stocks in the Gulf of Mexico. The results of the modeling effort produced projections of marked red snapper stock declines directly attributable to the recreational fisheries associated with petroleum platforms if as few as 100 of these platforms were to receive the same fishing pressure as a Buccaneer Field production platform. The Gulf of Mexico Fishery Management Council (1980) had published data confirming that red snapper stocks are being overfished, and that this condition is directly attributable to the recreational fishery as opposed to the commercial fishery.

Biological Systems Overview

As depicted in Figure 35, we conceptualize the Buccaneer Field as a large sink in the marine environment. The physical presence of the platforms is the major factor controlling, or accounting for, the three general biotic assemblages which aggregate at these sites - the biofouling, pelagic, and benthic reef fish communities. Of these, the biofouling community is the most complex. The diversity and biomass level of the biofouling community that develops is controlled by the type of perennial shelled animal (in the case of the Buccaneer Field, barnacle) that dominates. Barnacles (or other shelled organisms) provide habitat diversity, space, and food for other organisms. Barnacles are preyed upon largely by sheepshead and triggerfish, species capable of crushing their protective shells. In their grazing, sheepshead do not always consume all the barnacle and the remains are

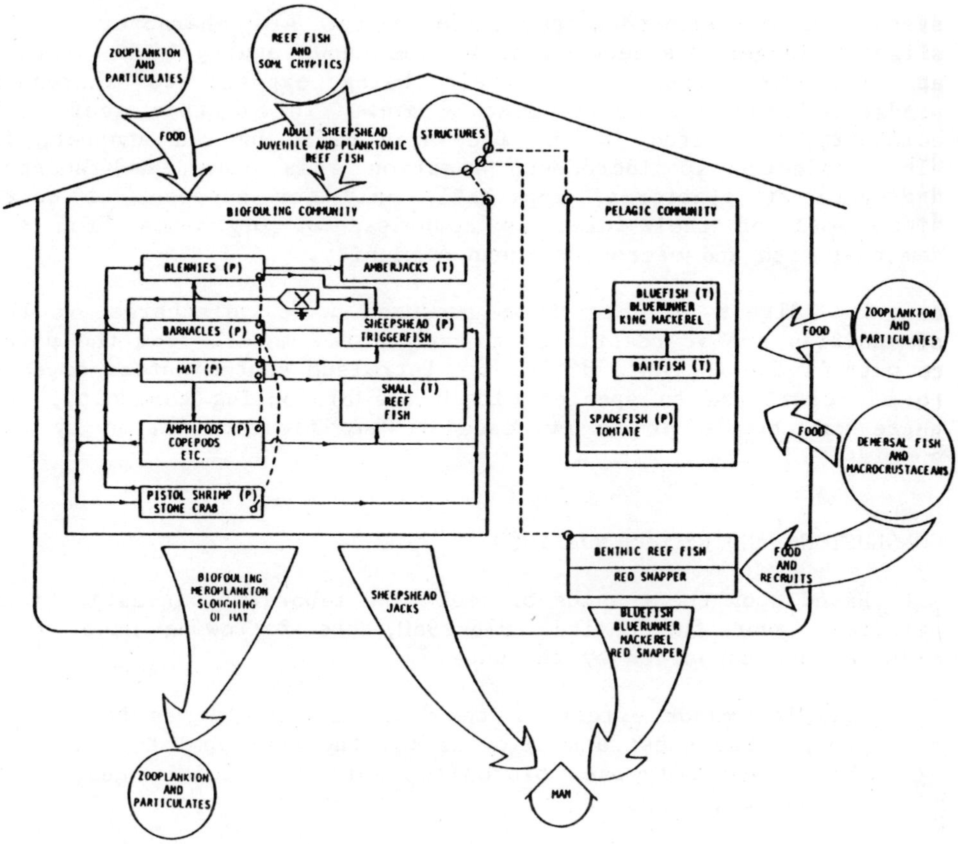

Figure 35. Conceptual model of a Buccaneer Field platform community.

available to smaller predators such as blennies. Blennies, in turn, are sometimes taken by sheepshead as well as other fish predators, particularly amberjack or almaco jack. Even though cycling is undoubtedly high (Gallaway and Margraf, 1979), the biofouling community probably obtains most of its food from the environment in the form of plankton and particulates flowing through the system from the environment. Some species are also recruited to the biofouling community from the environment (e.g., stone crab larvae, adult sheepshead). Losses from the biofouling community to the environment include those from reproduction, breaking off, and sloughing, as well as to man (mainly sheepshead, triggerfish, amberjacks).

With the exception of a few plankton-particulate feeding pelagic species which apparently are platform residents, the pelagic predators (e.g., king mackerel, blue runner) and their prey (e.g., scaled sardine), essentially "drift" through the system as do suspended particulates but are characterized by slightly longer residence times. The known aggregative habitats at structures results in their being exposed to increased predation from the environment - man. The benthic reef fish community, dominated in the Buccaneer Field by red snapper, is also subjected to increased predation from man. Red snapper aggregate at platforms, apparently only for purposes of cover since most of their diet is comprised of organisms from the demersal fish and macrocrustacean community.

The direct effects from produced water discharge on the communities appear restricted to only a few meters from the point of outfall due to rapid dilution. Petroleum contamination appears to be restricted to species within the biofouling community, and there was little or no evidence of any bioaccumulation by any species.

CONCLUSIONS AND RECOMMENDATIONS

Based upon the results of field and laboratory investigations performed over the period 1976-1980, the following major conclusions are supported by the data:

(1) The major effect of the Buccaneer Field has been to provide substrate allowing for the development of a rich and diverse biofouling or artificial reef community;

(2) The structures and reefs aggregated nektonic species preferring these habitats, as well as those who preyed upon them - particularly man;

(3) Produced waters were toxic but the direct effects on the biota were restricted to within only a few meters of the outfall;

(4) Indirect effects of chronic, low-level contaminant discharge, although not known with certainty, were not indicated to have been significant;

(5) Measurable uptake of contaminants seemed minimal and restricted to those species in the biofouling food chain, and there was little evidence of bioaccumulation; and

(6) More concern seems to be in order relative to the effects of the recreational fishery on the fish rather than concern about the effects of the field on either the fish or the recreational fishery.

Based upon subjective evaluations of the certainty of the conclusions, two areas should receive additional investigation. Laboratory and field studies should be directed towards examining the relationship of chronic, low-level contaminant discharges and seasonal fish disease epidemics, and population dynamics studies of red snapper at natural and artificial reefs should be determined as a basis for enlightened management of an apparently dwindling resource.

ACKNOWLEDGEMENTS

This work is a result of research sponsored by the Environmental Protection Agency and Department of Commerce, National Marine Fisheries Service, Southeast Fisheries Center, Galveston Laboratory under Contract No. NA79-GA-C-00026.

REFERENCES

Anderson, J. B., Schwarzer, R. R., and Clark, H. C., 1977, "Sedimentology, geochemistry, and trace metal analysis," in: Jackson, W. B. (ed.), "Environmental Assessment of an Active Oil Field in the Northwestern Gulf of Mexico, 1976-1977." NOAA, pp. 89-174. Available from: NTIS, Springfield, Virginia, Accession No. PB283890.

Anderson, J. B., Schwarzer, R. R., Clark, H. C., Wheeler, R. B., Hokanson, C. L., Anderson, J. E., and Okafor, J., 1979, "Describe the fine sediments and nepheloid layer of the oil field, focusing upon their relationship to heavy metal adsorption/determine levels, pathways, and bioaccumulation of heavy metals in the marine ecosystem in the oil field," in: Jackson, W. B. (ed.), "Environmental Assessment of an Active Oil Field in the Northwestern Gulf of Mexico, 1977-1978. Volume III: Chemical and Physical Investigations." NOAA, pp. 2.3.2/2.4.2-1 to 2.3.2/2.4.2-118. Available from: NTIS, Springfield, Virginia, Accession No. PB80107899.

Armstrong, R. S., 1980, "Current patterns and hydrography," in: Jackson, W. B., and Wilkens E. P. (eds.), "Environmental Assessment of the Buccaneer Gas and Oil Field in the Northwestern Gulf of Mexico, 1975-1980, Vol. IV." NOAA Technical Memorandum NMFS-SEFC-50, 41 pp. Available from: NTIS, Springfield, Virginia.

Armstrong, R. S., and Hamilton, R. C., 1979, "Describe seasonal circulation patterns in the oil field," in: Jackson, W. B. (ed.), "Environmental Assessment of an Active Oil Field in the Northwestern Gulf of Mexico, 1977-1978. Volume III: Chemical and Physical investigations." NOAA, pp. 2.3.9-1 to 2.3.9-250. Available from: NTIS, Springfield, Virginia, Accession No. PB80107899.

Behrens, E. W., 1977, "Total organic carbon and carbon isotopes of sediments," in: Jackson, W. B. (ed.), "Environmental Assessment of an Active Oil Field in the Northwestern Gulf of Mexico, 1976-1977." NOAA, pp. 737-759. Available from: NTIS, Springfield, Virginia, Accession No. PB283890.

Boland, G. S., 1980, "Morphological Parameters of the Barnacle, *Balanus tintinnabulum antillensis*, as Indicators of Physiological and Environmental Conditions." M.S. Thesis, Texas A&M University, College Station, Texas, 69 pp.

Bradley, E., and Bryan, C. E., 1976, "Life history and fishery of the red snapper (*Lutjanus campechanus*) in the Northwestern Gulf of Mexico," in: "Proceedings: Colloquium of Snapper-Grouper Fishery Resources of the Western Central Atlantic Ocean." Rep. Florida Sea Grant Program (No. 17), 77 pp.

Brooks, J. M., Estes, E. L., and Huang, W. H., 1980, "Investigations of surficial sediments and suspended particulates at Buccaneer field," in: Jackson, W. B., and Wilkens E. P. (eds.), "Environmental Assessment of the Buccaneer Gas and Oil Field in the Northwestern Gulf of Mexico, 1975-1980, Vol. II." NOAA Technical Memorandum NMFS-SEFC-36, 261 pp. Available from: NTIS, Springfield, Virginia.

Dahlberg, M. D., 1975, "Guide to Coastal Fishes of Georgia and Nearby States." University of Georgia Press, Athens, Georgia.

Danek, L. J., and Tomlinson, M. S., 1980, "Currents and hydrography of the Buccaneer field and adjacent waters," in: Jackson, W. B., and Wilkens E. P. (eds.), "Environmental Assessment of the Buccaneer Gas and Oil Field in the Northwestern Gulf of Mexico, 1975-1980, Vol. VI." NOAA Technical Memorandum NMFS-SEFC-40, 33 pp. Available from: NTIS, Springfield, Virginia.

Emiliani, D., Baxter, K. N., Jackson, W. B., Morgan, J., Zamora, G., and Lansford, L., 1977, "Demersal finfish and macrocrustaceans collected in the vicinity of Buccaneer oil field during 1976-1977," in: Jackson, W. B. (ed.), "Environmental Assessment of an Active Oil Field in the Northwestern Gulf of Mexico, 1976-1977." NOAA, pp. 275-310. Available from: NTIS, Springfield, Virginia, Accession No. PB283890.

Fable, W. A., Jr., 1977, "Snapper/Grouper Study, Environmental Studies of the South Texas Outer Continental Shelf." Report to the Bureau of Land Management, National Marine Fisheries Service, Panama City, Florida, 235 pp.

Finucane, J. H., Collins, L. A., and Barger, L. E., 1979, "Determine the effects of discharges on seasonal abundance, distribution, and composition of ichthyoplankton in the oil field," in: Jackson, W. B. (ed.), "Environmental Assessment of an Active Oil Field in the Northwestern Gulf of Mexico, 1977-1978. Volume II: Data Management and Biological Investigations." NOAA, pp. 2.3.6-1 to 2.3.6-157. Available from: NTIS, Springfield, Virginia, Accession No. PB80165970.

Fotheringham, N., 1977, "Effects of offshore oil field structures on their biotic environment: benthos and plankton," in Jackson, W. B. (ed.), "Environmental Assessment of an Active Oil Field in the Northwestern Gulf of Mexico, 1976-1977." NOAA, pp. 487-549. Available from: NTIS, Springfield, Virginia, Accession No. PB283890.

Gallaway, B. J., and Margraf, F. J., 1979, "Simulation modeling of biological communities associated with a production platform in the Buccaneer oil and gas field," in: Jackson, W. B. (ed.), "Environmental Assessment of an Active Oil Field in the Northwestern Gulf of Mexico, 1977-1978. Volume II: Data Management and Biological Investigations." NOAA, pp. 2.5.1-1 to 2.5.1-67. Available from: NTIS, Springfield, Virginia, Accession No. PB80165970.

Gallaway, B. J., and Martin, L. R., 1980, "Effect of gas and oil field structures on pelagic and reef fishes, and demersal fishes and macrocrustaceans," in: Jackson, W. B., and Wilkens E. P. (eds.), "Environmental Assessment of the Buccaneer Gas and Oil Field in the Northwestern Gulf of Mexico, 1978-1979, Vol. III." NOAA Technical Memorandum NMFS-SEFC-37, 49 pp. Available from: NTIS, Springfield, Virginia.

Gallaway, B. J., Howard, R. L., Green, K., and Martin, L., 1976, "A Study Plan for the Buccaneer Oil Field Assessment Program," Report to NMFS, Galveston, Texas from LGL Limited-U.S., Inc., Bryan, Texas, 69 pp.

Gallaway, B. J., Johnson, M. F., Howard, R. L., Martin, L. R., and Boland, G. S., 1979a, "A study of the effects of Buccaneer Oil Field structures and associated effluents on biofouling communities and the Atlantic spadefish (*Chaetodipterus faber*)," in: Jackson, W. B. (ed.), "Environmental Assessment of an Active Oil Field in the Northwestern Gulf of Mexico, 1977-1978. Volume II: Data Management and Biological Investigations." NOAA, pp. 2.3.8-1 to 2.3.8-126. Available from: NTIS, Springfield, Virginia, Accession No. PB80165970.

Gallaway, B. J., Johnson, M. F., Boland, G. S., Lewbel, G. S., Martin, L. R., Margraf, F. J., and Howard R. L., 1979b, "The artificial reef studies," in: "Ecological Investigations of Petroleum Production Platforms in the Central Gulf of Mexico." Final Report to Southwest Research Institute for the Bureau of Land Management, Washington, D.C., 108 pp. plus appendices.

Gilles, R. C., Brown, L. R., and Minchew, C. D., 1978, "Bacteriological aspects of fin erosion in mullet exposed to crude oil," J. Fish Biol., 13:113.

Gulf of Mexico Fishery Management Council, 1980, "Draft Environmental Impact Statement and Fishery Management Plan and Regulatory Analysis and Proposed Regulations, Reef Fish Resources of the Gulf of Mexico." Report to Environmental Protection Agency, Washington, D.C., 147 pp.

Harper, D. E., Jr., 1977, "Distribution and abundance of macrobenthic and meiobenthic organisms," in: Jackson, W. B. (ed.), "Environmental Assessment of an Active Oil Field in the Northwestern Gulf of Mexico, 1976-1977." NOAA, pp. 175-273. Available from: NTIS, Springfield, Virginia, Accession No. PB283890.

Harper, D. E., Jr., Scrudato, R. J., and Giam, C. S., 1976, "Pilot Study of the Buccaneer Oil Field (Benthos and Sediments). A Preliminary Environmental Assessment of the Buccaneer Oil/Gas Field." Report to NMFS, Galveston, Texas from Texas A&M University, College Station, Texas, 63 pp. plus appendices.

Hastings, R. W., Ogren, L. H., and Mabry, M. T., 1976, "Observations on the fish fauna associated with offshore platforms in the northwestern Gulf of Mexico," Fish. Bull. U.S. Fish Wildlife Serv., 74:387.

Howard, R. L., Boland, G. S., Gallaway, B. J., and Dennis, G. D., 1980, "Effects of gas and oil field structures and effluents on fouling community production and function," in: Jackson, W. B., and Wilkens E. P. (eds.), "Environmental Assessment of the Buccaneer Gas and Oil Field in the Northwestern Gulf of Mexico, 1978-1979, Vol V." NOAA Technical Memorandum NMFS-SEFC-39, 60 pp. Available from: NTIS, Springfield, Virginia.

Jackson, W. B. (ed.), 1977, "Environmental Assessment of an Active Oil Field in the Northwestern Gulf of Mexico, 1976-1977." NOAA, 759 pp. Available from: NTIS, Springfield, Virginia, Accession No. PB283890.

Jackson, W. B. (ed.), 1979a, "Environmental Assessment of an Active Oil Field in the Northwestern Gulf of Mexico, 1977-1978. Volume I: Synopsis." NOAA, 62 pp. Available from: NTIS, Springfield, Virginia, Accession No. PB301339/AS.

Jackson, W. B. (ed.), 1979b, "Environmental Assessment of an Active Oil Field in the Northwestern Gulf of Mexico, 1977-1978. Volume II: Data Management and Biological Investigations." NOAA, 799 pp. Available from: NTIS, Springfield, Virginia, Accession No. PB80165970.

Jackson, W. B. (ed.), 1979c, "Environmental Assessment of an Active Oil Field in the Northwestern Gulf of Mexico, 1977-1978. Volume III: Chemical and Physical Investigations." NOAA, 722 pp. Available from: NTIS, Springfield, Virginia, Accession No. PB80107899.

Jackson, W. B., and Wilkens, E. P. (eds.), 1980a, "Environmental Assessment of an Active Oil Field in the Northwestern Gulf of Mexico, 1978-1979." NOAA. Available from: NTIS, Springfield, Virginia.

Jackson, W. B., and Wilkens, E. P. (eds.), 1980b, "Environmental Assessment of an Active Oil Field in the Northwestern Gulf of Mexico, 1975-1980." NOAA. Available from: NTIS, Springfield, Virginia.

Klima, E. F., and Wickham, D. A., 1971, "Attraction of coastal pelagic fishes with artificial structures," Trans. Am. Fish. Soc., 100:86.

Martin, L. A., 1977, "Hydrography of the Buccaneer Oil Field and adjacent waters," in: Jackson, W. B. (ed.), "Environmental Assessment of an Active Oil Field in the Northwestern Gulf of Mexico, 1976-1977." NOAA, pp. 575-632. Available from: NTIS, Springfield, Virginia, Accession No. PB283890.

Middleditch, B. S, and West, D., 1980, "Hydrocarbons, biocides, and sulfur," in Jackson, W. B. and Wilkens, E. P. (eds.), "Environmental Assessment of the Buccaneer Gas and Oil Field in the Northwestern Gulf of Mexico, 1978-1979, Vol 8." NOAA Technical Memorandum NMFS-SEFC-41, 119 pp. Available from NTIS, Springfield, Virginia.

Middleditch, B. S., Basile, B., and Chang, E. S., 1977, "Hydrocarbon levels and pathways in the major components of a marine ecosystem," in: Jackson, W. B. (ed.), "Environmental Assessment of an Active Oil Field in the Northwestern Gulf of Mexico, 1976-1977." NOAA, pp. 633-735. Available from: NTIS, Springfield, Virginia, Accession No. PB283890.

Middleditch, B. S., Basile, B., and Missler, S. R., 1979, "Determine levels, pathways, and bioaccumulation of selected discharge constituents (non-metals) in the marine ecosystem in the oil field/hydrocarbon modeling," in: Jackson, W. B. (ed.), "Environmental Assessment of an Active Oil Field in the Northwestern Gulf of Mexico, 1977-1978. Volume III: Chemical and Physical Investigations." NOAA, pp. 2.4.1-1 to 2.4.1-302. Available from: NTIS, Springfield, Virginia, Accession No. PB80107899.

Minchew, C. D., and Yarbrough, J. D., 1977, "The occurrence of fin rot in mullet (*Mugul cephalus*) associated with crude oil contamination of an estuarine pond-ecosystem," J. Fish. Biol., 10:319.

Moseley, F. N., 1966, "Biology of the red snapper, *Lutjanus aya* Bloch, of the Northwestern Gulf of Mexico," Publs. Inst. Mar. Sci. Univ. Texas, 11:90.

Nichoul, J. C. (ed.), 1975, "Modeling of Marine Systems." Elsevier Scientific, New York.

Pew, P., 1971, "Food and Game Fishes of the Texas Coast." Bulletin No. 33, Texas Parks and Wildlife Dept., Austin, Texas.

Sizemore, R. K., and Olsen, K., 1980, "Bacterial communities," in: Jackson, W. B., and Wilkens E. P. (eds.), "Environmental Assessment of the Buccaneer Gas and Oil Field in the Northwestern Gulf of Mexico, 1978-1979, Vol. IV." NOAA Technical Memorandum NMFS-SEFC-38, 32 pp. Available from: NTIS, Springfield, Virginia.

Smedes, G., Calman, J., and Beebe, J., 1980, "Hydrodynamic modeling," in: Jackson, W. B., and Wilkens E. P. (eds.), "Environmental Assessment of the Buccaneer Gas and Oil Field in the Northwestern Gulf of Mexico, 1978-1979, Vol. X." NOAA Technical Memorandum NMFS-SEFC-44, 57 pp. Available from: NTIS, Springfield, Virginia.

Tillery, J. B., 1980, "Trace metals," in: Jackson, W. B., and Wilkens E. P. (eds.), "Environmental Assessment of the Buccaneer Gas and Oil Field in the Northwestern Gulf of Mexico, 1975-1980, Vol. VI." NOAA Technical Memorandum NMFS-SEFC-52, 39 pp. Available from: NTIS, Springfield, Virginia.

Trent, L., Workman, I. K., Dime, S., and Jones, C., 1977, "Recreational fisheries and the distribution of predatory pelagic fishes," in: Jackson, W. B. (ed.), "Environmental Assessment of an Active Oil Field in the Northwestern Gulf of Mexico, 1976-1977." NOAA, pp. 311-337. Available from: NTIS, Springfield, Virginia, Accession No. PB283890.

Wickham, D. A., Watson, J. W., Jr., and Ogren, L. H., 1973, "The efficiency of midwater artificial structures for attracting pelagic sport fish," Trans. Amer. Fish. Soc., 102:563.

Workman, I. K., and Jones, C. E., 1979, "Determine effects of oil field discharges on species composition and abundance of pelagic fishes and demersal fishes and macro-crustaceans in the oil field," in: Jackson, W. B. (ed.), "Environmental Assessment of an Active Oil Field in the Northwestern Gulf of Mexico, 1977-1978. Volume II: Data Management and Biological Investigations." NOAA, pp. 2.3.5-1 to 2.3.5-149. Available from: NTIS, Springfield, Virginia, Accession No. PB80165970.

Zein-Eldin, Z. P., and Keney, P. M., 1979, "Bioassay of Buccaneer oil field effluents with penaeid shrimp," in: Jackson, W. B. (ed.), "Environmental Assessment of an Active Oil Field in the Northwestern Gulf of Mexico, 1977-1978. Volume II: Data Management and Biological Investigations." NOAA, pp. 2.3.4-1 to 2.3.4-25. Available from: NTIS, Springfield, Virginia, Accession No. PB80165970.

ACUTE TOXICITY AND AQUATIC HAZARD ASSOCIATED WITH DISCHARGED FORMATION WATER

Curt D. Rose and Timothy J. Ward

Environmental Sciences Division
Energy Resources Company Inc.
185 Alewife Brook Parkway
Cambridge, Massachusetts 02138

INTRODUCTION

The primary objectives of the investigation described in this chapter were to evaluate acute toxicity of formation water discharged from platform 296-B of the Buccaneer Field (Figure 1) and assess potential aquatic hazard related to the discharged water. The aquatic hazard evaluations presented address just toxicological aspects of hazard. Additional factors must be considered in a comprehensive hazard assessment. These factors include the potential for bioaccumulation of constituents of discharged formation water and impact of the discharged water on other uses of the ocean.

Organisms employed in this investigation were larval, subadult, and adult brown shrimp (*Penaeus aztecus*); subadult and adult white shrimp (*P. setiferus*); a barnacle (*Balanus tintinnabulum*); and the crested blenny (*Hypleurochilus geminatus*). With the exception of the white shrimp, which generally inhabits nearshore waters (Kutkuhn, 1962), all organisms are commonly found in the Buccaneer Field. The barnacle and blenny constitute important components of the biological community attached to oil field platforms and associated structures.

Formation water is discharged from platform 296-B at a rate of 0-232 m^3 (0-1459 bbl) per day (Shell Oil Co., 1978). Discharge occurs through a vertical pipe (10 cm in diameter) that runs approximately 20 m from the well deck of the platform to an

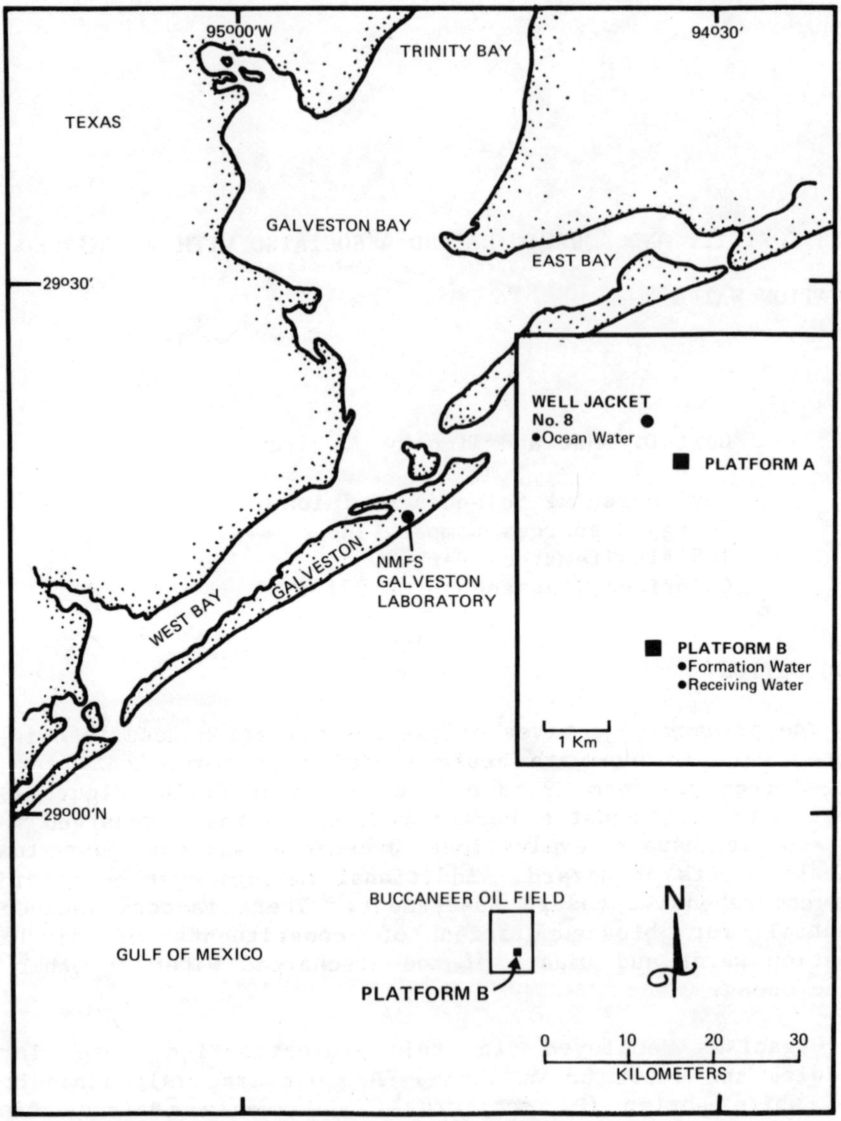

Figure 1. Locations of production platforms 288-A ("A") and 296-B ("B") and well jacket 288-8 ("8") in the Buccaneer Field. Sources of materials tested during the investigation are indicated in inset.

outfall located about 2 m above the surface of the receiving water.

Formation water discharged from platform 296-B contains dissolved mineral salts, sulfur, metals, and hydrocarbons that are not removed by the platform's skimming-scrubbing system. Salinity of formation water is approximately 40-45°/oo. Sulfur content of the material has been reported as 0.06-0.42% (Middleditch and West, 1980). The most abundant metals in formation water (Tillery, 1980) were strontium (58.1-93.6 mg/l), barium (1.03-1.55 mg/l), and iron (0.12-0.38 mg/l). Hydrocarbons reported in the material include alkane, aromatic, and volatile fractions (Middleditch and West, 1980). Total alkane concentration ranged from 0.15 to 8.9 mg/l and total aromatic content (68 constituents) was 0.104 mg/l (one sample of formation water). The most common volatiles (one sample of material) were benzene (30.5% of total volatiles), toluene (27.3%), and ethylbenzene (6.0%). Middleditch and West (1980) concluded that the probability of these latter constituents remaining in the receiving water for "any length of time" is minimized because of their volatility and the "churning" that occurs when formation water impacts the receiving water.

During the period of the investigation, Shell Oil Company (Shell) was treating formation water with acrolein (Magnacide B™), a biocide for controlling sulfate-reducing bacteria in the skimming-scrubbing system of the platform, and ammonium bisulfite, a scavenger employed to neutralize acrolein prior to its discharge to the receiving water. Acrolein and ammonium bisulfite were continuously injected into the formation water at concentrations of approximately 10 and 20 mg/l, respectively, except when maintenance problems (e.g., pump failures) occurred (Shell Oil Co., 1979). Acrolein is highly toxic to marine organisms. For example, Butler (1965) reported a 48-hr median lethal concentration (LC50) of 0.10 ppm for brown shrimp exposed to the material. Although acrolein is very toxic, it is an extremely volatile and reactive substance (a typical half-life in formation water is 3 hr) that is effectively scavenged by ammonium bisulfite (Middleditch and West, 1980). Indeed, several investigators (Magna Corp., 1978a,b; Middleditch and West, 1980) have been unable to detect acrolein residuals in formation water discharged after normal scavenging operations. Middleditch and West (1980) surmised that little, if any, acrolein reaches the receiving water and any that does is rapidly dispersed and inactivated. Ammonium bisulfite, which probably does reach the receiving water, has a substantial oxygen demand that, together with the oxygen-depleted formation water, has the potential to reduce the dissolved oxygen concentration (DO) of the receiving water.

Physical and chemical characteristics of the receiving water at platform 296-B have been monitored seasonally by Danek and Tomlinson (1980). Depth of receiving water is approximately 20 m. Mean minimum and maximum current speeds in subsurface waters (4.5 m deep) were 6.7-20.5 cm/sec and 26.4-36.7 cm/sec, respectively. Temperature varied from 12 to 29°C. Range of chemical variables was: salinity, 28.5 to 35.5‰; total suspended matter, 1.0 to 12.0 mg/l; pH, 8.1 to 8.4; and DO, 4.3 to 9.1 mg/l. Danek and Tomlinson (1980) reported that the discharged formation water, which presumably contained ammonium bisulfite, did not noticeably affect DO of the receiving water.

Toxicity of formation waters has been attributed to their hypersaline nature, which can cause osmotic stress, and low levels of dissolved oxygen (Bureau of Land Management, 1979). A previous evaluation of the toxicity of formation water discharged from platform 296-B of the Buccaneer Field was performed by Zein-Eldin and Keney (1978). These investigators exposed juvenile white shrimp to formation water that had previously been treated with biocide and to untreated formation water. LC50's (96-hr) for treated formation water ranged from 1750 to 6500 ppm and a 96-hr LC50 for untreated water was >100,000 ppm. Neither set of LC50 values reflects toxicity associated with oxygen demand of formation water since DO of test media (various concentrations of formation water in diluent water) was similar to DO of control media. At the time that Zein-Eldin and Keney performed their study, Shell was alternately treating formation water with glutaraldehyde (K-31™) and alkyldimethyl benzyl chloride (KC-14™), which were continuously and intermittently injected into the formation water (Shell Oil Co., 1978). These biocides, which are substantially less volatile and reactive than acrolein, were not being scavenged by Shell. Therefore, LC50's for treated formation water are probably representative of the combined toxicity of formation water and biocide, both of which were being discharged to the receiving water.

METHODS AND MATERIALS

The investigation was performed during the summer (August 21 - September 14) and fall (November 18-28) of 1978 and the winter (March 16-23) and Spring (May 15-30) of 1979. The number of organisms tested each season was dictated by their availability.

Larval brown shrimp (early nauplius to protozoea I) were available only during the spring. Larvae were obtained from a single gravid adult collected by trawling from the RV GUS III approximately 46 km southeast of the Buccaneer Field. The female was captured between 2100 and 2300 on May 24 and induced to spawn in the laboratory (Galveston Laboratory of the National Marine Fisheries Service) at 0200 on May 26.

Subadult (90 to <140 mm total length) and adult (>140 mm) brown shrimp were collected during all seasons by trawling from the research vessel. During the summer, fall, and winter, shrimp were captured approximately 75 km SSW of Galveston Inlet, 12 km north of the Buccaneer Field, and 10 km east of the Field, respectively. In the spring, subadult shrimp were collected about 1 km south of Freeport, Texas (additional animals were purchased from a Galveston bait dealer), and adult shrimp were captured approximately 130 km south of Galveston Inlet.

Subadult white shrimp, which were available in the summer, fall, and winter, were purchased from a bait dealer. Adult white shrimp were obtained during all seasons, either from a bait dealer (winter) or by trawling approximately 2 km south of Galveston Inlet (summer and fall) and 1 km south of Freeport (spring).

Barnacles (1-2 cm in diameter) were collected during all seasons, and crested blennies (25-80 mm total length) were acquired in all seasons except winter. Both organisms were obtained at well jacket 288-8 by SCUBA divers operating from the research vessel. Barnacles were separated from the support legs of the well jacket with a geologist's hammer and blennies were captured with a continuous-action "slurp" gun (Workman, 1978).

After acquisition, organisms were transported to the laboratory (or platform 296-B, where several tests were performed) in aerated fiberglass containers, placed in fiberglass aquaria containing static aerated water, and acclimated to testing temperature over a period of 1-3 days. During acclimation, the maximum temperature change during any one-day period was 3°C.

Organisms were tested with eight samples of formation water. Samples A, B, and C were collected and tested during, respectively, summer, fall, and winter. Samples D-H were obtained and tested in the spring. All samples were collected in an acid-rinsed, 19-l, high-molecular-weight polyethylene bucket attached to the outfall of platform 296-B. Contents of the bucket were transferred by a PVC pump to acid-rinsed, 208-l, high-molecular-weight polyethylene plastic drums (laboratory tests) or by a peristaltic metering pump to mixing chambers of the testing apparatus (platform tests). Formation water employed in laboratory tests was stored in loosely-capped drums, transported to the laboratory on the research vessel, stored at 2-4°C, and tested within nine days after collection. Formation water used in platform tests was tested immediately after being collected. All samples of formation water were analyzed by Energy Resources Company Inc. (ERCO) for salinity and suspended particulate matter (also acrolein in the case of sample H) and by Southwest Research Institute and the University of Houston for metals and hydrocarbons, respectively. Salinity was measured with a refractometer equipped with a salinity scale (American Optical model 1041a). Suspended particulate matter was determined according to methods described by Young (1977), and acrolein was analyzed colorimetrically by the DNPH method (Magna Corp., 1978c), which has a detection limit of 1 ppm acrolein. Methods for analyzing metals are described by Tillery (1980) and procedures used for hydrocarbons by Middleditch and West (1980).

Additional test materials were receiving water and "ocean" water. Receiving water was collected within a 30-m distance from the outfall of platform 296-B. Receiving water employed in laboratory tests was obtained each season from subsurface (~1 m), mid-water (~9 m), and near-bottom (~18 m) depths with a PVC pump operated from the research vessel. The water was mixed aboard the vessel in 208-l plastic drums and, after being transported to the laboratory, expeditiously used in tests or stored at 2-4°C. Water employed in platform tests was acquired in the spring from a subsurface depth with a peristaltic metering pump connected to the platform's fire-fighting system and used immediately in tests. Ocean water was collected within 30 m of well jacket 288-8 according to the procedures described for receiving water employed in the laboratory tests. All water was used in tests within two weeks of collection. At the time of testing, receiving water and ocean water were analyzed by ERCO for salinity, suspended particulate matter, temperature (mercury-filled glass thermometer), DO (YSI model 57 meter), and pH (Orion model 901 meter or Fisher model 150 meter).

Formation water discharged from platform 296-B was tested for acute toxicity in several types of tests. The majority of tests (Test Series No. 1) were static laboratory tests designed to evaluate toxicity of formation water that is not attributable to oxygen demand of the water. Other tests (Test Series No. 2) were static laboratory tests conducted in parallel to Test Series No. 1 to assess toxicity of all components of formation water, including oxygen demand. Another test (Test Series No. 3) was a flow-through laboratory test, performed under otherwise similar experimental conditions as Test Series No. 1, to evaluate the effect of flow-through as compared to static exposure of organisms on estimates of toxicity of formation water. The last series of tests (Test Series No. 4) consisted of flow-through tests performed on platform 296-B. As in the case of Test Series Nos. 1 and 3, Test Series No. 4 did not evaluate toxicity related to oxygen demand of formation water.

Test Series No. 1 (static laboratory tests; oxygen demand of formation water not evaluated) was performed with seven organisms and a broad range of experimental conditions. Larval brown shrimp were tested (48-hr tests) during the spring with samples D-G of formation water. Testing temperature was 28°C, the temperature at which spawning occurred. Other organisms (subadult and adult brown shrimp, subadult and adult white shrimp, the barnacle, and crested blenny) were usually tested (96-hr tests) in the summer, fall, winter, and spring with, respectively, Samples A, B, C, and D of formation water (subadult white shrimp were not available for testing during the spring and crested blennies could not be obtained during the winter). The summer tests were conducted at 25±1°C, the temperature selected by the U.S. Environmental Protection Agency (EPA) and Corps of Engineers (COE) as representing the summer maximum in the northern Gulf of Mexico (U.S. EPA and COE, 1977). The fall, winter, and spring tests were performed at, respectively, 22±1°C, 18±2°C, and 24±1°C, temperatures representing ambient water temperatures at the time of testing. Test were conducted according to protocols recommended by the U.S. EPA and COE (1977) for evaluating the liquid and suspended particulate phases of ocean-dumped material. Testing was performed in an environmental chamber with a 14-hr light and 10-hr dark photoperiod established with cool-white fluorescent bulbs. Receiving water was employed as diluent water, and receiving water and ocean water were used as controls. Randomization procedures were employed to assign animals to test and control containers and to position containers in the environmental chamber. Since discharged formation water did not noticeably affect DO of receiving water (Danek and Tomlinson, 1980), test and control media were aerated to maintain DO above 4 mg/l except in the case of larval tests, which did not require aeration to maintain a minimal DO. Larval shrimp were tested in 100-ml Carolina culture dishes containing 80 ml of test or control medium. Thirty larvae were exposed to each of six test concentrations (1,000 - 320,000 ppm formation water by volume) and each control at a loading rate of 10 animals per culture dish. Subadult and adult shrimp were tested in 38-l aquaria holding 30 l of media. Depending on the availability of animals, 10 or 30 individuals were exposed to each of 5 or 6 test concentrations (5,000 - 560,000 ppm) and each control at a loading rate of 10 animals per aquarium. Barnacles were tested in 19-l aquaria containing 18 l of media. Number of test concentrations (10,000 - 560,000 ppm), number of animals tested per concentration and control, and loading rates were the same as for subadult and adult shrimp. Crested blennies were evaluated in 38-l aquaria holding 30 l of media. Number of test concentrations were 4 or 5 (32,000 - 560,000 ppm) and 10 or 20 fish were exposed to each test concentration and control at a loading rate of 10 individuals per aquarium. In all tests, survival of organisms was monitored at intervals of 24 hr or less. Criteria for survival were movement (larval shrimp), pronounced response to prodding (other life stages of shrimp and blennies), and appendage movement (barnacles). Physical and chemical variables monitored during the tests were temperature, salinity, DO, and pH.

Test Series No. 2 (static laboratory tests; oxygen demand of formation water evaluated) was conducted with the barnacle (winter, Sample C of formation water, 18±2°C testing temperature) and crested blenny (spring, Sample D of formation water, 24±1°C testing temperature). Basic testing protocols were identical to those employed in Test Series No. 1 except that test and control media were not aerated. In both tests, 10 animals were exposed to each of six test concentrations (1,000 - 320,000 ppm formation water) and each control at a loading rate of five animals per aquarium.

Test Series No. 3 (flow-through laboratory test; oxygen demand of formation water not evaluated) was performed with subadult white shrimp in the fall. Sample B of formation water was tested at a temperature of 22±1°C. General testing protocols were the same as those used in Test Series No. 1 except that media were delivered to aquaria by peristaltic pumps at the rate of 42 ml/min (two volume changes per 24 hr). In the test, 40 animals were exposed to each of four test concentrations (34,000 - 300,000 ppm formation water) and each control at a loading rate of 20 animals per aquarium.

Test Series No. 4 (flow-through platform tests; oxygen demand of formation water not evaluated) was conducted with subadult brown shrimp and the barnacle during the spring. Sample H of formation water was tested at a temperature of 25-29°C. Tests (96-hr) were performed on a catwalk adjacent to the outfall of the platform. Receiving water was used as diluent water and as a single control, and normal randomization procedures were implemented to assign animals to test and control containers and position containers on the catwalk. Shrimp and barnacles were tested together in 38-l aquaria containing 30 l of media. Twenty shrimp and 20 barnacles were exposed to each of five test concentrations (43,000 - 390,000 ppm formation water) and the control at a loading rate of 10 individuals of each species per aquarium. Media were delivered to aquaria by peristaltic metering pumps at the rate of 500 ml/min (24 volume changes per 24 hr). Aeration was employed to maintain DO of media above 4 mg/l. Survival of organisms as well as usual physical and chemical variables was monitored at 24-hr intervals.

Data generated in acute toxicity tests with formation water were analyzed by several statistical techniques. LC50's and associated 95% confidence intervals were calculated by moving average, probit, or binomial methods (Stephan, 1978). If the moving average method was employed, the greatest data span consistent with the interpolative process was used in the calculation after redundant non-partial kills were eliminated from the data base. Yates' (1933) procedure was used to estimate missing data in Test Series No. 1 (LC50's for subadult white shrimp tested during the spring and the crested blenny tested in the winter). A nonparametric two-way analysis of variance - Friedman's (1937) rank test - was then employed to evaluate the effects of test organisms and experimental conditions on estimates of toxicity (LC50 values). An analysis based on the normal distribution was not considered for this evaluation since LC50 values could not be tested for homogeneity of variances or additivity of main effects, two important criteria that would need to be satisfied if statistical procedures based on the normal distribution were implemented. Another nonparametric technique - Spearman's coefficient of rank correlation (Steel and Torrie, 1960) - was used to describe correlation between toxicity of formation water tested in Test Series No. 1 and testing temperature.

Environmental concentrations of formation water discharged from platform 296-B were estimated by use of a hydrodynamic model developed by Smedes et al. (1980) for vertically distributed pollutants in the Buccaneer Field. The model assumes that pollutants are mixed in a volume of receiving water equal to approximately 13% of the volume occupied by the platform by the time that pollutants are transported to the "downcurrent" side of the platform. The model also employs conservative formulas to estimate concentrations of pollutants at intermediate distances (about 1-4 km) from the platform. Environmental concentrations of formation water were estimated as functions of time after discharge and distance from platform 296-B assuming that formation water is discharged from the platform at a rate of 2700 g/sec, which approximates 1459 bbl/day, the maximum reported volume of formation water discharged during 1975-1977 (Shell Oil Co., 1978). Environmental concentrations were estimated for a current speed of 5 cm/sec and hydrodynamic and meteorological conditions that prevailed during February 21, 1979, a day when initial dispersion of pollutants was determined to be relatively limited (Smedes et al., 1980).

Potential aquatic hazard associated with discharged formation water was evaluated by Lagrangian and Eulerian techniques. In the Lagrangian assessment, which is applicable to planktonic organisms, a time series of limiting permissible concentrations (LPC's) for a pollutant is derived by multiplying the lower limits of the 95% confidence intervals of exposure-time-dependent LC50's for a sensitive organism by an arbitrary but, presumably, conservative application (safety) factor (U.S. EPA and COE, 1977). LPC's are then graphically compared to estimated environmental concentrations of the pollutant plotted as a function of time after discharge. In the Eulerian evaluation, which is suitable for use with benthos, organisms attached to oil field structures, and fishes, the worst-case (lowest) LPC derived in the Lagrangian assessment is compared to estimated environmental concentrations of the pollutant at various distances from its point of discharge.

TABLE 1. PHYSICAL AND CHEMICAL CHARACTERISTICS OF FORMATION WATER EMPLOYED IN TOXICITY TESTS.

Physical or chemical constituent[a]	Concentration of constituent in formation water[b]								Range for all samples
	Sample A	Sample B	Sample C	Sample D	Sample E	Sample F	Sample G	Sample H	
Salinity ($^{o}/oo$)	42	44	40	42	42	40	41	40-42	40-44
SPM (mg/l)	49.8	39.3	46.2	39.0	30.5	27.0	16.5	20.0-50.0	16.5-50.0
Strontium (mg/l)	46.5	77.8	64.4	80.3	-	55.2	68.9	37.1-48.4	37.1-80.3
Iron (mg/l)	3.92	3.51	3.94	5.76	-	4.73	3.28	3.08-7.80	3.08-7.80
Barium (mg/l)	0.39	0.93	0.31	-	-	0.73	0.54	0.24-0.93	0.24-0.93
Alkanes (mg/l)	0.30	0.02	1.6	0.16	2.1	2.1	0.99	0.53-1.9	0.02-2.1
Pristane (μg/l)	24	3	213	7	145	167	74	43-105	3-213
Phytane (μg/l)	6	0	36	1	35	41	19	10-18	0-41
Acrolein	-	-	-	-	-	-	-	0	0

[a]Salinity, suspended particulate matter (SPM), and acrolein in Sample H were analyzed by Energy Resources Co. Inc. Other analyses were conducted by Southwest Research Institute (metals) and the University of Houston (alkanes, pristane, and phytane).

[b]Physical and chemical characteristics of Samples A-G are based on a single set of analyses. Characteristics of Sample H are predicated on five sets of analyses.

RESULTS

Formation water tested during the investigation was characterized by a golden color and an oily odor. The water had a salinity of 40-44°/oo and concentrations of suspended particulate matter ranging from 16.5 to 50.0 mg/l (Table 1). Most abundant metals in the water were strontium (37.1-80.3 mg/l), iron (3.08-7.80 mg/l), and barium (0.24-0.93 mg/l). Alkanes (total) were present at concentrations of 0.02-2.1 mg/l, while levels of pristane and phytane ranged from 3 to 213 μg/l and 0 to 41 μg/l, respectively. Acrolein was not detected in Sample H of formation water (the sample used in acute flow-through platform tests) even though chemical analyses were performed on the platform with fresh samples of material.

Receiving water employed during the study was characterized by a salinity of 30-34°/oo and a suspended particulate content of 0.43-2.80 mg/l. DO and pH were 4.7-9.3 mg/l and 7.5-8.2, respectively. Ocean water had a salinity identical to that of receiving water and a suspended particulate content of 0.33-1.65 mg/l. DO was 5.2-9.0 mg/l, and pH was 7.2-8.2.

Acute Toxicity of Discharged Formation Water

During each of the 31 acute toxicity tests performed during the investigation, survival of control organisms (animals exposed to receiving water or ocean water) was greater than 90%, thereby permitting further analysis of test results (U.S. EPA and COE, 1977). Organisms exposed to receiving water and ocean water experienced similar survival. In 26 tests in which paired controls were run, mean survival in receiving water was 95% while average survival in ocean water was 96%.

Estimates of acute toxicity of formation water were largely dependent on whether oxygen demand of the water was evaluated (Table 2). In Test Series No. 1 (oxygen demand not evaluated), 96-hr LC50's for the barnacle and crested blenny tested, respectively, in the winter and spring were approximately one order of magnitude greater than 96-hr LC50's for the same organisms tested under parallel experimental conditions in Test Series No. 2 (except that oxygen demand was evaluated).

When oxygen demand of formation water was disregarded, estimates of toxicity were not dramatically influenced by the method of exposing organisms to the water, i.e., static as compared to flow-through method. Subadult white shrimp evaluated in Test Series No. 1 (static laboratory tests) during the fall were characterized by a 96-hr LC50 of 61,000 ppm formation water, while the same organism tested under similar experimental

TABLE 2. MEDIAN LETHAL CONCENTRATIONS (LC50's) AND ASSOCIATED 95% CONFIDENCE INTERVALS FOR ORGANISMS ACUTELY EXPOSED TO FORMATION WATER UNDER VARIOUS EXPERIMENTAL CONDITIONS.

Organism	Season of test	Formation water used	Testing temp.	LC50[a,b]	95% Confidence interval[a,b]
Test Series No. 1[c]					
Brown shrimp					
Larva	Spring 1979	D	28	10,000	7,000-15,000
		E	28	12,000	9,000-18,000
		F	28	8,000	6,000-12,000
		G	28	8,000	5,000-11,000
Subadult	Summer 1978	A	25±1	94,000	63,000-172,000
	Fall 1978	B	22±1	60,000	0-100,000
	Winter 1979	C	18±2	183,000	130,000-279,000
	Spring 1979	D	24±1	61,000	47,000-76,000
Adult	Summer 1978	A	25±1	94,000	63,000-172,000
	Fall 1978	B	22±1	78,000	38,000-183,000
	Winter 1979	C	18±2	178,000	132,000-240,000
	Spring 1979	D	24±1	90,000	61,000-156,000
White shrimp					
Subadult	Summer 1978	A	25±1	56,000	51,000-62,000
	Fall 1978	B	22±1	61,000	48,000-76,000
	Winter 1979	C	18±1	133,000	67,000-366,000
Adult	Summer 1978	A	25±1	81,000	48,000-153,000
	Fall 1978	B	22±1	62,000	27,000-110,000
	Winter 1979	C	18±1	92,000	58,000-150,000
	Spring 1979	D	24±1	37,000	24,000-52,000
Barnacle	Summer 1978	A	25±1	33,000	25,000-38,000
	Fall 1978	B	22±1	84,000	68,000-104,000
	Winter 1979	C	18±2	154,000	111,000-222,000
	Spring 1979	D	24±1	60,000	49,000-71,000
Crested blenny	Summer 1978	A	25±1	158,000	100,000-320,000
	Fall 1978	B	22±1	408,000	320,000-560,000
	Spring, 1979	D	24±1	178,000	135,000-235,000
Test Series No. 2[d]					
Barnacle	Winter 1979	C	18±2	8,000	5,000-13,000
Cr. blenny	Spring 1979	D	24±1	7,000	5,000-12,000
Test Series No. 3[e]					
White shrimp					
Subadult	Fall 1978	B	22±1	62,000	48,000-76,000
Test Series No. 4[f]					
Brown shrimp					
Subadult	Spring 1979	H	25-29	44,000	25,000-60,000
Barnacle	Spring 1979	H	25-29	51,000	34,000-68,000

conditions in Test Series No. 3 (flow-through laboratory test) exhibited a 96-hr LC50 of 62,000 ppm (95% confidence interval for both LC50's was 48,000-76,000 ppm). Subadult brown shrimp evaluated in Test Series No. 1 (static laboratory exposure) during the spring exhibited a 96-hr LC50 of 61,000 ppm formation water (47,000-76,000 ppm) as compared to a 96-hr LC50 of 44,000 ppm (25,000-60,000 ppm) for the same organism evaluated in Test Series No. 4 (flow-through platform exposure) in the spring. Barnacles evaluated in Test Series No. 1 during the spring also displayed a higher 96-hr LC50 (60,000 ppm; 49,000-71,000 ppm) than animals tested in Test Series No. 4 during the same period (51,000 ppm; 34,000-68,000 ppm). However, these latter differences, even if significant, cannot be attributed exclusively to method of toxicant delivery since differences in other factors (e.g., sample of formation water used in test and testing temperature) may have been confounding influences.

Footnotes to Table 2.

[a]All LC50's and associated 95% confidence intervals are 96-hr values except in the case of larval brown shrimp, for which 48-hr values are reported. Units are ppm formation water.

[b]In most cases, LC50's and related confidence intervals were calculated by the moving average method. However, the binomial method was employed in Test Series No. 1 for subadult brown shrimp tested in the fall as well as for crested blennies tested in the summer and fall. The probit method was used for Test Series No. 4.

[c]Static laboratory tests; oxygen demand of formation water not evaluated. Except in the case of tests with larval brown shrimp, test and control media were aerated to maintain dissolved oxygen concentration (DO) above 4 mg/l. Aeration was not required to maintain a DO above 4 mg/l in tests with larval shrimp.

[d]Static laboratory tests; oxygen demand of formation water evaluated. Test and control media were not aerated. Although DO of control media remained above 4 mg/l during the tests, DO of test media decreased to 0.5-3.2 mg/l (barnacle) and 1.2-4.0 mg/l (crested blenny) by the end of the 96-hr testing period.

[e]Flow-through laboratory tests; oxygen demand of formation water not evaluated. Test and control media were aerated to maintain DO above 4 mg/l.

[f]Flow-through platform tests; oxygen demand of formation water not evaluated. Test and control media were aerated to maintain DO above 4 mg/l.

Organisms routinely exposed to formation water in Test Series No. 1 (static laboratory tests with oxygen demand of formation water disregarded) exhibited significantly different sensitivities as measured in terms of mean 96-hr LC50's (Table 3). The crested blenny was the least sensitive organism, with a mean LC50 of 269,000 ppm formation water. Progressively greater sensitivities characterized brown shrimp (mean LC50 = 100,000-110,000 ppm), the barnacle (mean LC50 = 83,000 ppm), and white shrimp (mean LC50 = 68,000-70,000 ppm). Subadult and adult life stages of both species of shrimp displayed relatively similar mean LC50's. Formation water usually affected all organisms except the blenny in a cumulative manner, i.e., the incipient LC50 (concentration of formation water that 50% of test organisms can survive indefinitely) was not reached by the end of the 96-hr testing period.

Experimental conditions established in Test Series No. 1 were also a source of statistically significant differences in mean 96-hr LC50's observed for routinely tested organisms (Table 3). Since several experimental variables were collectively partitioned into "block" effects in the statistical design of the test series, specific causes of these differences cannot be determined analytically. However, a perfect and significant (α = 0.01) negative linear correlation exists between 96-hr LC50 and testing temperature in the cases of the crested blenny, barnacle, and subadult white shrimp (Figure 2). Additional meaningful correlations could exist between 96-hr LC50 and attributes of other variables confounded in the statistical design, e.g., condition index of organisms, which may be seasonally related, or chemical differences in samples of formation water. In all probability, the correlation between 96-hr LC50 and testing temperature is best described by a complex structural model (Sokal and Rohlf, 1969), which would acknowledge that toxicity is influenced by several experimental factors. A nonsignificant negative linear correlation was observed between 96-hr LC50 and testing temperature in the case of most shrimp (subadult and adult brown shrimp and adult white shrimp). LC50's for these organisms exhibited similar temperature-related patterns with lowest LC50's associated with intermediate testing temperatures (Figure 2).

Larval brown shrimp were the most sensitive of all organisms exposed to formation water in Test Series No. 1 (Table 2). Mean 48-hr LC50 for larval shrimp tested with four samples of formation water was 9500 ppm formation water, a value approximately sevenfold lower than the mean 96-hr LC50's (68,000-70,000 ppm) obtained with subadult and adult white shrimp, the most sensitive of the organisms routinely evaluated in Test Series No. 1. As with most other organisms, formation water had a cumulative effect on larval shrimp.

TABLE 3. NONPARAMETRIC TWO-WAY ANALYSIS OF VARIANCE OF 96-HR MEDIAN LETHAL CONCENTRATIONS (LC50's) FOR ORGANISMS ROUTINELY EXPOSED TO FORMATION WATER IN TEST SERIES NO. 1.

1. 96-hr LC50's for Organisms (ppm formation water)

Organism (A=6): Experimental conditions (B=4)	Brown shrimp Subadult	Brown shrimp Adult	White shrimp Subadult	White shrimp Adult	Barnacle	Crested blenny	Mean
Summer 1978, Sample A of formation water, 25±1°C testing temp.	94,000	94,000	56,000	81,000	33,000	158,000	86,000
Fall 1978, Sample B of formation water, 22±1°C testing temp.	60,000	78,000	61,000	62,000	84,000	408,000	126,000
Winter 1979, Sample C of formation water, 18±2°C testing temp.	183,000	178,000	133,000	92,000	154,000	331,000[a]	178,000
Spring 1979, Sample D of formation water, 24±1°C testing temp.	61,000	90,000	29,000[a]	37,000	60,000	178,000	76,000
Mean	100,000	110,000	70,000	68,000	83,000	269,000	117,000

2. Analysis of Effects Attributable to Organisms (Friedman's rank test)

$$\chi^2_{(cal.)} = \frac{12}{AB(A+1)} \sum^{A}(\sum^{B}_{Rij})^2 - 3B(A+1)$$

$$= \frac{12}{6(4)(6+1)} (14.5^2 + 17.5^2 + 8^2 + 8^2 + 12^2 + 24^2) - 3(4)(6+1)$$

$$= \frac{12}{168} (1364.5) - 84.0 = 13.5^*,$$ as compared to $\chi^2_{(tab.)} = 11.1$ for $\alpha = 0.05$ and 5 df

3. Analysis of Effects Attributable to Experimental Conditions (Friedman's rank test)

$$\chi^2_{(cal.)} = \frac{12}{AB(B+1)} \sum^{B}(\sum^{A}_{Rij})^2 - 3A(B+1)$$

$$= \frac{12}{6(4)(4+1)} (13^2 + 14^2 + 23^2 + 10^2) - 3(6)(4+1)$$

$$= \frac{12}{120} (994.0) - 90.0 = 9.4^*,$$ as compared to $\chi^2_{(tab.)} = 7.8$ for $\alpha = 0.05$ and 3 df

[a]No toxicity tests were conducted with subadult white shrimp during the spring or with crested blennies during the winter. In order to facilitate statistical analyses, missing data were estimated according to methods described by Yates.

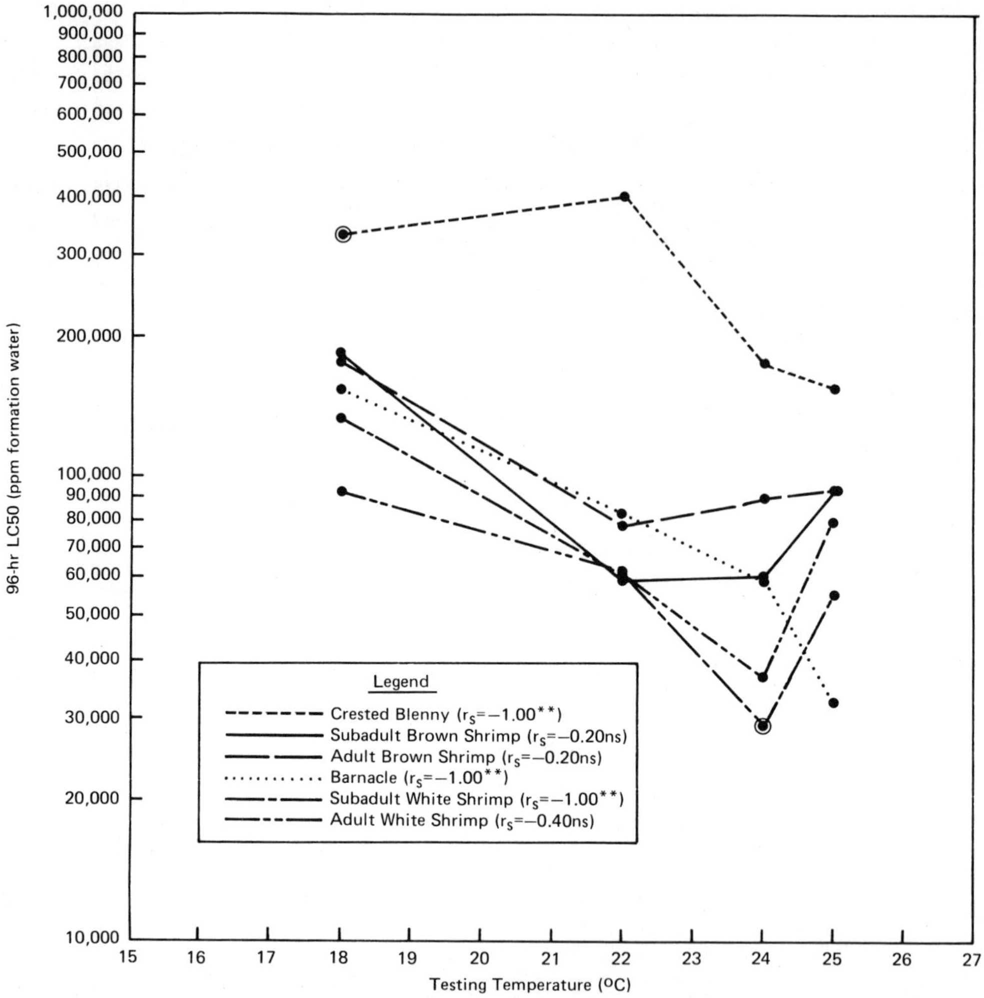

Figure 2. Nonparametric correlations between 96-hr median lethal concentrations (LC50's) and testing temperatures for organisms routinely exposed to formation water in Test Series No. 1. Correlation coefficients (r_s) were derived by Spearman's ranking technique. Estimates of missing data, indicated by circles, were not employed in the derivations.

Potential Aquatic Hazard Associated with Discharged Formation Water

Both Lagrangian and Eulerian assessments of potential aquatic hazard associated with discharged formation water are based on the worst-case (lowest) set of LC50 values obtained with larval brown shrimp (Table 4), the most sensitive organism evaluated in Test

TABLE 4. EXPOSURE-TIME-DEPENDENT MEDIAN LETHAL CONCENTRATIONS (LC50's) AND ASSOCIATED 95% CONFIDENCE INTERVALS FOR LARVAL BROWN SHRIMP EXPOSED TO SAMPLE G OF FORMATION WATER IN TEST SERIES NO. 1.

Exposure period (hr)	LC50[a]	95% Confidence interval[a]
3	160,000	100,000-320,000
12	117,000	96,000-145,000
24	37,000	27,000-52,000
40	13,000	10,000-18,000
48	8,000	5,000-11,000

[a]With the exception of the 3-hr period of exposure, LC50's and related confidence intervals were calculated by the moving average method. The binomial method was employed for the 3-hr computations. Units are ppm formation water.

Series No. 1. Coincidentally, this set of values is similar to data generated by organisms exposed to formation water in Test Series No. 2 (tests in which oxygen demand of formation water was measured).

Lagrangian Assessment. In the Lagrangian assessment of potential aquatic hazard (Figure 3), the LPC of formation water decreases from approximately 1000 ppm after the end of initial mixing (the 4-hr period immediately following discharge of a waste; U.S. EPA, 1977) to 50 ppm at 48 hr after discharge. During this time, discharged formation water is estimated to be diluted from a concentration of about 0.1 ppm to 0.001 ppm. The LPC of formation water is approximately four orders of magnitude greater than the estimated environmental concentration of formation water. In addition, lines representing the LPC and the dilution curve show no indication of converging after 48 hr. Consequently, it is concluded that discharge of formation water does not represent a potential hazard to plankton drifting through the wastewater plume.

The above-stated conclusion is based on the assumptions that the sensitivity of the one planktonic and six nonplanktonic organisms that were tested with formation water is representative of the sensitivity of all planktonic organisms likely to be exposed to formation water and that the use of an LPC predicted on the most sensitive of the tested organisms will protect the most sensitive of the indigenous organisms. Both assumptions espouse the thesis that an organism's sensitivity to a chemical substance

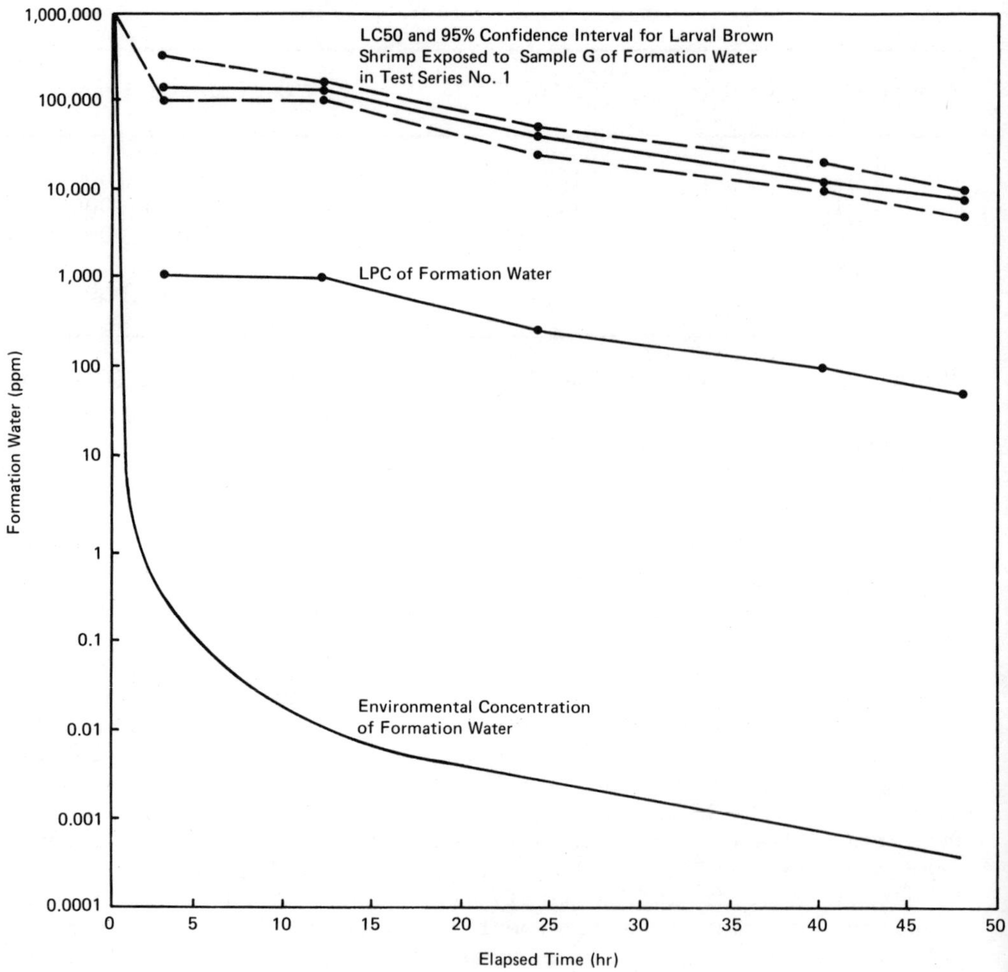

Figure 3. Temporal comparisons of bioassay-based limiting permissible concentrations (LPC's) and environmental concentrations of formation water. Median lethal concentrations (LC50's) and associated 95% confidence intervals are the lowest values generated in four bioassays with larval brown shrimp, the most sensitive of seven organisms evaluated in Test Series No. 1. LPC's were derived by multiplying lower limits of the 95% confidence intervals of the LC50's by a 1% application (safety) factor. Environmental concentrations of formation water were estimated by the conservative use of the dispersion model developed for vertically distributed pollutants in the Buccaneer Field.

and its mode of locomotion are not correlated and recognize the impracticability of testing more than a limited number of organisms in any aquatic hazard evaluation. Both assumptions are explicit in the ocean dumping regulations (U.S. EPA, 1977) and associated guidelines (U.S. EPA and COE, 1977).

Eulerian Assessment. The Eulerian assessment of potential aquatic hazard associated with discharged formation water employs 50 ppm as the single "exposure-time-independent" LPC. The dispersion model for vertically distributed pollutants in the Buccaneer Field (Smedes et al., 1980) cannot distinguish environmental concentrations of formation water in the immediate vicinity of the point of discharge. However, the environmental concentration of formation water is estimated to be approximately 4 ppm at 600 m from platform 296-B. Under the hydrological and meteorological conditions that existed on February 16 and 25, 1979, formation water is diluted to about 4 ppm by the time that it is transported 140 m from the platform. Therefore, it appears likely that discharged formation water is not a potential hazard to organisms with temporally unvarying (benthos and structure-associated organisms) and temporally unpredictable (most fishes) exposure patterns at the perimeter of the "release zone" (100 m from the wastewater outfall; U.S. EPA, 1973). Indeed, potential hazard may exist only within a few meters of the outfall.

DISCUSSION

The environmental assessment of the Buccaneer Field was supported by the EPA and the National Oceanic and Atmospheric Administration's (NOAA) Energy Resources Project on Fate and Effects of Contaminants in Marine Ecosystems and Organisms. The principal objectives of the fate and effects project are to identify and conduct studies that facilitate assessment of potential environmental impacts of petroleum-related discharges to coastal marine ecosystems (Wolfe, 1978). In addition, these studies are intended to have a broad applicability to different geographic areas.

The objectives of NOAA's fate and effects project had an important influence on the experimental design of the toxicity and aquatic hazard evaluations conducted during the investigation. Specifically, the evaluations were performed with effluent that is discharged to the marine ecosystem - formation water containing, in all probability, ammonium bisulfite, the biocide scavenger. No evaluations were made of the toxicity or aquatic hazard associated with the acrolein biocide which, based on available evidence (Magna Corp., 1978a,b; Middleditch and West, 1980; this investigation), is not discharged to the environment in important amounts and, like ammonium bisulfite, is not normally employed in oil-platform operations (Shell Oil Co., 1978).

A second important feature of the experimental design of the evaluations is the use of toxicity tests in which oxygen demand of the formation water (and ammonium bisulfite) was disregarded. The ultimate purpose of all toxicity tests is to determine whether a chemical substance has the potential for causing environmental damage. The reliability of such an evaluation is dependent on the degree to which the test duplicates, within the constraints of standard testing protocols, important environmental conditions that prevail when the substance is introduced into the environment. Thus, it is common practice to perform toxicity tests with representative indigenous organisms, ambient temperatures, and receiving water for diluent and control purposes. In addition, when the substance is a chemically complex effluent containing both conservative and nonconservative constituents, it must be decided whether to conduct tests that evaluate only the conservative constituents or additionally assess the nonconservative components, to which organisms may be exposed for only brief periods of time. The former decision entails a risk of underestimating potential environmental danger while the latter decision may result in an overestimation of potential environmental risk. The most sophisticated guidelines promulgated by the U.S. EPA and COE (1977) for resolving this issue pertain to oxygen demand, the nonconservative "constituent" of greatest concern in formation water. These guidelines provide for minimal levels of DO (4 mg/l) to be maintained, by aeration or other techniques, during testing of an effluent unless discharge of an effluent causes lower levels of DO to occur in the environment for a substantial period of time. Although these guidelines were initially developed to evaluate ocean dumping of dredged material, they have scientific applicability to any material that is disposed of in the ocean (or other aquatic environments) by barge or outfall. The guidelines have been or are presently being employed in Region II, EPA, to assess potential environmental damage associated with ocean disposal of such diverse materials as acid-iron waste, sewage sludge, and drilling fluids (Anderson, 1979). In all these cases, potential damage related to oxygen demand of the effluents is assessed through stoichiometric calculations which, in the case of materials allowed to be ocean-dumped, must indicate that DO, after allowance for initial mixing, will not be depressed by more than 25% below normally anticipated ambient conditions in the disposal area at the time of dumping (U.S. EPA, 1977).

Formation water discharged from platform 296-B did not noticeably depress DO in the vicinity of the platform (Danek and Tomlinson, 1980). Therefore, in conformance with the above-described guidelines, DO of test media in most acute toxicity tests and the long-term test was maintained above 4 mg/l (usually by aeration), thereby disregarding oxygen demand of the formation water. The use of aeration probably resulted in losses of volatile hydrocarbons initially present in the discharged formation water. However, as in the case of depressed DO, these relatively toxic constituents are unlikely to remain in the environment for any length of time (Middleditch and West, 1980). In addition, Smith and MacIntyre (1971) determined that losses of C_{10}-C_{15} hydrocarbons in No. 2 fuel oil were significantly greater in the environment than in a laboratory bubbler apparatus. If volatile organics in formation water behave similarly, the toxicity tests conducted during the investigation may have overestimated environmental effects attributable to the volatiles.

A third critical element of the experimental design of the toxicity and aquatic hazard evaluations is the major reliance that was placed upon static laboratory tests. Flow-through laboratory and platform tests were employed only to "calibrate" results obtained in static testing. The reliance on static laboratory tests was dictated by both standard testing protocols for ocean-discharged materials and practical considerations. The guidelines promulgated by the U.S. EPA and COE (1977) for evaluating material deposited in the ocean identify only static tests for assessing aqueous discharges. The use of receiving water and ocean water for diluent and control purposes restricted the number of flow-through tests that could be performed in the laboratory since the large volumes of waters required in flow-through testing could not be collected or properly stored. Only a limited number of flow-through tests could be conducted on platform 296-B because of frequently severe offshore weather conditions and lack of protected facilities on the platform.

The utility of static as compared to flow-through toxicity tests has become a controversial issue. Richards (1979) considered static tests to be little more than range-finding tests for more definitive flow-through tests. Other authorities (U.S. EPA and COE, 1977) have concluded that static tests can reliably be used to evaluate potential hazard associated with discharged materials. Studies conducted by Rice et al. (1978) shed additional light on the controversy. These investigators evaluated acute sensitivity of 30 species of Alaskan marine organisms to crude oil and No. 2 fuel oil. The tests, which were conducted under static conditions, identified sensitive and tolerant species. The sensitive species were primarily subtidal organisms (fishes and crustaceans) while the tolerant species were mostly intertidal invertebrates (snails, mussels, etc.). To determine the validity of the static tests for determining relative sensitivity of species to the oils, two sensitive organisms (pink salmon fry and a *Eualis* shrimp) and two tolerant organisms (black sea cucumber and a *Hemigrapsus* shore crab) were acutely exposed to toluene and naphthalene under static and flow-through conditions. The two sensitive organisms exposed to the petroleum fractions under flow-through conditions exhibited 96-hr LC50's that were approximately 70% of values obtained with static exposure. The two tolerant organisms displayed flow-through values that were about 35% of static values. Rice et al. concluded that evaluation of comparative sensitivities of diverse species (sensitive versus tolerant species) requires flow-through exposure and that static tests are not "that bad" for some species (sensitive species) considering the cost-benefit ratio. However, it must be emphasized that these conclusions pertain to two petroleum fractions that, because of their volatile (nonconservative) nature, would be expected to generate disparate LC50 values in static and flow-through testing. The conclusions cannot be extrapolated indiscriminately to complex materials containing toxicologically important conservative constituents.

The static laboratory tests that were performed during the investigation were conducted with a material containing many conservative constituents (mineral salts, sulfur, metals, nonvolatile hydrocarbons). Flow-through tests employed for calibration purposes gave no indication that flow-through exposure has a dramatic effect on test results. Therefore, we believe that reliance on static testing was an optimal testing strategy and that the benefits to be gained by more extensive flow-through testing would not have been commensurate with the increased cost or decreased data base that would have been associated with such testing.

The toxicity tests performed during the investigation were sometimes conducted under less than ideal circumstances. Subadult and adult shrimp were larger than 5-10 cm, the maximum size recommended for static and flow-through testing (U.S. EPA, 1978). Difficulties in collecting large numbers of crested blennies occasionally required that tests be conducted with fish for which the ratio of maximum length to minimum length was between two and three. Test organisms and test materials, e.g., formation water, receiving sediment, were usually collected during the same cruise; therefore, in order to conduct tests with fresh materials, holding and acclimation time for organisms was minimal. The limited availability of receiving water and ocean water sometimes required that tests be performed at high loading rates and, in the case of the flow-through laboratory test, a less than optimal flow rate. Although minimal holding and acclimation times, high loading rates, and the low flow rate in the laboratory test did not result in unacceptably high (>10%) control mortality in acute tests, these conditions may have caused sublethal stress in test animals, thereby resulting in overestimates of toxicity of formation water.

Results of the acute toxicity tests with formation water (Test Series No. 1) suggest that responses of subadult and adult shrimp to formation water may be governed by different temperature-related behavioral and/or physiological mechanisms than responses of the barnacle and crested blenny. A more definitive conclusion cannot be reached because of the relatively few temperatures at which tests were conducted and the confounding of effects associated with temperature and other experimental conditions. The relatively high sensitivity of larval brown shrimp to formation water is not unexpected since larval organisms are generally considered to be less tolerant than older life stages to oils and may be less tolerant than older animals to osmotic stress. The cumulative impact of formation water on all organisms except the crested blenny indicates a need for chronic testing and, in the interim, justifies the use of a conservative application factor in evaluations of potential hazard. The testing of additional organisms (e.g., larval white shrimp which, based on tests with older life stages of shrimp, may be more sensitive than larval brown shrimp to formation water) and verification of the hydrodynamic model for the Buccaneer Field would constitute additional refinements in the hazard evaluations. However, such refinements would appear unlikely to alter the basic conclusion reached in the evaluations, namely, that discharged formation water presents a minimal potential for environmental hazard. The validity of this conclusion is sustained by the ample margin of safety that exists between LPC's derived in the evaluations and estimated environmental concentrations of formation water.

This investigation and the previous evaluation (Zein-Eldin and Keney, 1978) of acute toxicity of formation water discharged from platform 296-B invite a number of comparisons. The previous evaluation and most of our investigation addressed toxicity of formation water that is unassociated with oxygen demand of the water. The previous study generated two sets of 96-hr LC50's for juvenile white shrimp. The first set of LC50's, 1750-6500 ppm formation water, was obtained with formation water that Shell had treated with two relatively conservative biocides (K-31™ and KC-14™) that were not scavenged. Consequently, these values probably represent combined toxicity of formation water and biocides. The second set of LC50's, >100,000 ppm formation water, was obtained with formation water that had not been treated with biocides and, therefore, reflects the "basic" toxicity of the water. This investigation generated one set of LC50's which, because Shell was treating formation water with a nonconservative biocide (acrolein) that was being scavenged effectively, represents toxicity of formation water and, probably, the scavenger, ammonium bisulfite. This set of LC50's includes six 96-hr values for subadult and adult white shrimp - 37,000-81,000 ppm formation water - that are roughly comparable to Zein-Eldin and Keney's data (two other LC50's for subadult and adult white shrimp were obtained at a substantially lower testing temperature than employed by Zein-Eldin and Keney). These comparisons suggest that Shell's current practice of treating formation water (acrolein and the scavenger, ammonium bisulfite) is more environmentally protective than its previous practice (K-31™ or KC-14™ without scavenger) and less environmentally protective than discharging untreated formation water.

Several generalizations can be derived from the investigation. Formation water is not characterized by high acute toxicity. For example, even when toxicity associated with oxygen demand of water is considered, 96-hr LC50's (7000-8000 ppm) are greater than 96-hr LC50's for components of drilling fluids (265-3500 ppm; Chesser and McKenzie, 1975), most of which are considered to be relatively nontoxic (Bureau of Land Management, 1976). However, the issue of major concern in assessing potential environmental impacts of discharged formation water is not relative or even absolute toxicity of the water but, rather, whether all conditions associated with its discharge, i.e., toxicity of water, volume discharged, and fate after discharge, indicate the potential for aquatic hazard. Although the basic components of this type of evaluation can readily be extrapolated to other discharges and environments, the conclusions are obviously case-specific. Lastly, it is important to recognize that a bioassay-based aquatic hazard evaluation can only assess potential hazard associated with discharge of an effluent to an aquatic system. If the evaluation indicates potential hazard to the system, the probability of hazard can best be determined by field studies in the vicinity of the discharge. Field studies are particularly important in hazard assessments of ocean outfalls, which may serve to attract and concentrate aquatic organisms.

ACKNOWLEDGEMENTS

This work is a result of research sponsored by the Environmental Protection Agency and Department of Commerce, National Oceanic and Atmospheric Administration, National Marine Fisheries Service, Southeast Fisheries Center, Galveston Laboratory under Contract No. 03-78-D08-0042.

REFERENCES

Anderson, P., U.S. Environmental Protection Agency (Region II), personal communication, 1979.

Bureau of Land Management, 1976, "Proposed 1977 Outer Continental Shelf Oil and Gas Lease Sale Offshore the North Atlantic States, OCS Sale No. 42, Volume 2, Draft Environmental Statement." U.S. Department of the Interior, Washington, D.C., 1192 pp.

Bureau of Land Management, 1979, "Proposed 1979 Outer Continental Shelf Oil and Gas Lease Sale Offshore the North Atlantic States, OCS Sale No. 42, Final Supplement to Environmental Statement." U.S. Department of the Interior, Washington, D.C., 701 pp.

Butler, P. A., 1965, "Effects of herbicides on estuarine fauna," Proc. Southern Weed Conf., 18:576.

Chesser, B. G., and McKenzie, W. H., 1975, "Use of a bioassay test in evaluating the toxicity of drilling fluid additives on Galveston Bay shrimp." Presented at the Conference on Environmental Aspects of Chemical Use in Well-Drilling Operations, Houston, Texas, May 21-23, 1975.

Danek, L. J., and Tomlinson, M. S., 1980, "Currents and hydrography of the Buccaneer Field and adjacent waters," in: Jackson, W. B., and Wilkens, E. P. (eds.), "Environmental Assessment of Buccaneer Gas and Oil Field in the Northwestern Gulf of Mexico, 1978-1979, Vol. VI." NOAA Technical Memorandum NMFS-SEFC-40, 33 pp. Available from: NTIS, Springfield, Virginia.

Friedman, M., 1937, "The use of ranks to avoid the assumption of normality implicit in the analysis of variance," J. Am. Stat. Assoc., 32:675.

Kutkuhn, J. H., 1962, "Gulf of Mexico commercial shrimp populations - trends and characteristics, 1956-59," in: "Fishery Bulletin 212." Fish and Wildlife Service, U.S. Department of the Interior, Washington, D.C., p. 343.

Magna Corporation, 1978a, communication from R. Clayton (Magna Corporation) to R. Grenn (Shell Oil Company), July 21.

Magna Corporation, 1978b, communication from R. Clayton (Magna Corporation) to R. Grenn (Shell Oil Company), September 15.

Magna Corporation, 1978c, "Magnacide B Microbiocide Description and Use Application Manual."

Middleditch, B. S., and West, D. L., 1980, "Hydrocarbons, biocides and sulfur," in: Jackson, W. B., and Wilkens, E. P. (eds.), "Environmental Assessment of Buccaneer Gas and Oil Field in the Northwestern Gulf of Mexico, 1978-1979, Vol. VII." NOAA Technical Memorandum NMFS-SEFC-41, 119 pp. Available from: NTIS, Springfield, Virginia.

Rice, S. D., Karinen, J. F., and Korn, S., 1978, "Acute and chronic toxicity, uptake and depuration, and sublethal response of Alaska marine organisms to petroleum hydrocarbons," in: "Marine Biological Effects of OCS Petroleum Development." NOAA, Boulder, Colorado, p. 11.

Richards, N. L., 1979, "Effects of chemicals used in oil and gas well-drilling operations in aquatic environments." Presented at the Fourth National Conference, Interagency Energy/Environment R&D Program, Washington, D.C., June 7-8.

Shell Oil Company, 1978, information provided to principal investigators concerning operations in the Buccaneer Field, Buccaneer Gas and Oil Field Study Workshop, Galveston, Texas, July 11-12.

Shell Oil Company, 1979, personal communication (from J. Burgbacher), January 26.

Smedes, G., Calman, J., and Beebe, J., 1980, "Hydrodynamic modeling," in: Jackson, W. B., and Wilkens, E. P. (eds.), "Environmental Assessment of Buccaneer Gas and Oil Field in the Northwestern Gulf of Mexico, 1978-1979." NOAA Technical Memorandum NMFS-SEFC-44, 57 pp. Available from: NTIS, Springfield, Virginia.

Smith, C. L., and MacIntyre, W. G., 1971, "Initial aging of fuel oil films on seawater," in: "Proceedings of the Joint Conference on Prevention and Control of Oil Spills." p. 457.

Sokal, R. R., and Rohlf, F. J., 1969, "Biometry." W. H. Freeman and Co., San Francisco, California.

Steel, R. G. D., and Torrie, J. H., 1960, "Principles and Procedures of Statistics." McGraw-Hill Co., New York.

Stephan, C. E., 1978, "Computer Program for Calculation of LC50 Values." Environmental Research Laboratory, U.S. Environmental Protection Agency, Duluth, Minnesota.

Tillery, J. B., 1980, "Trace metals," in: Jackson, W. B., and Wilkens, E. P. (eds.), "Environmental Assessment of Buccaneer Gas and Oil Field in the Northwestern Gulf of Mexico, 1978-1979, Vol. VIII." NOAA Technical Memorandum NMFS-SEFC-42, 93 pp. Available from: NTIS, Springfield, Virginia.

U.S. Environmental Protection Agency, 1973, "Ocean Dumping - Final Regulations and Criteria," Fed. Register, 38(190):28610.

U.S. Environmental Protection Agency, 1977, "Ocean Dumping - Final Revision of Regulations and Criteria," Fed. Register, 42(7):2462.

U.S. Environmental Protection Agency, 1978, "Bioassay Procedures for the Ocean Disposal Permit Program." Environmental Research Laboratory, Gulf Breeze, Florida, 121 pp.

U.S. Environmental Protection Agency and Corps of Engineers, 1977, "Ecological Evaluation of Proposed Discharge of Dredged Material into Ocean Waters; Implementation Manual for Section 103 of Public Law 92-532 (Marine Protection, Research, and Sanctuaries Act of 1972)." Environmental Effects Laboratory, U.S. Army Engineer Waterways Experiment Station, Vicksburg, Mississippi.

Wolfe, D. A., 1978, "An overview of NOAA's energy resources project on fate and effects of contaminants in marine ecosystems and organisms," in: "Marine Biological Effects of OCS Petroleum Development." NOAA, Boulder, Colorado, p. 223.

Workman, I. K., 1978, Galveston Laboratory, National Marine Fisheries Service, personal communication.

Yates, F., 1933, "The analysis of replicated experiments when the field results are incomplete," Empire J. Exper. Agric., 1:129.

Young, R. A., 1977, "Procedures for obtaining filtered particulate samples at sea." Laboratory Memorandum, Marine Geology and Geophysics Laboratory, May 5.

Zein-Eldin, Z. P., and Keney, P. M., 1978, "Environmental Assessment of an Active Oil Field in the Northwestern Gulf of Mexico - Bioassays of Buccaneer Oil Field Effluents with Juvenile and Adult Penaeid Shrimp." Progress report prepared for the National Marine Fisheries Service, Galveston, Texas.

ENVIRONMENTAL SYNTHESIS USING AN ECOSYSTEMS MODEL

Kenneth W Fucik

Science Applications, Inc.
2760 29th Street
Boulder, Colorado 80301

Ivan T. Show

Science Applications, Inc.
476 Prospect Street
La Jolla, California 92038

INTRODUCTION

The first attempt at using a multidisciplinary approach to evaluate the effects of offshore oil and gas production on the marine environment was conducted from 1972-1974 off the Louisiana coast. The results of the Offshore Ecology Investigation (OEI), as it was called, have recently been published (Ward et al., 1979). This was followed by a BLM-sponsored study in 1977-1979 in the same general area though smaller in scope than the OEI studies. The results of the BLM study are presently being evaluated (Defenbaugh, personal communication).

The Buccaneer program differs from these other studies in that a modeling work unit has been incorporated into the overall study design to aid in the synthesis and integration of the disciplinary data. Although models have been developed previously for various kinds of marine systems, this represents one of the first attempts at applying a model for marine environmental assessment.

Due to recent developments in the field of systems ecology, the direction taken by this work unit has changed somewhat since initiation of the effort. When this project began in 1978, one was obliged by the state-of-the-art to develop a simulation model

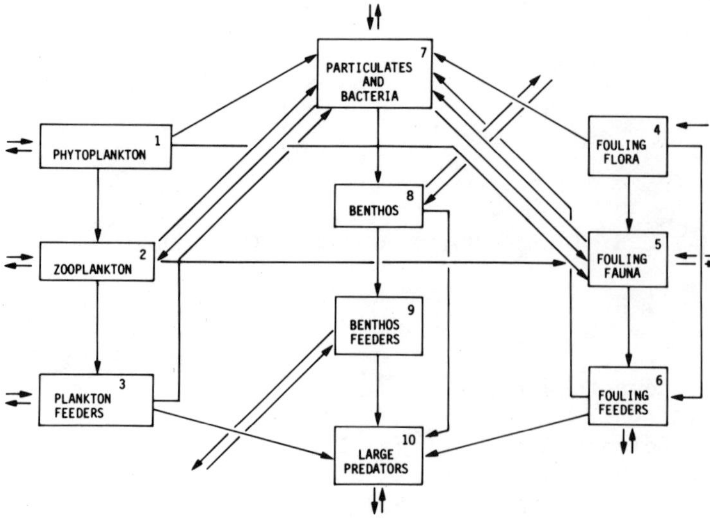

Figure 1. The conceptual model of the Buccaneer Field biological system.

that matched available sampling data as closely as possible and then to determine system behavior by varying parameters within the model and observing variations in the output. Stated another way, the direct use of most models has been to produce time series or steady state solutions for the state variables in the model. Model validity was then evaluated by comparison of model predictions with real world observations. The current trend, however, is to use an ecosystems model as the basis for analyzing and interpreting ecological structure and function (Patten, 1978; Patten and Finn, 1979; Patten et al., 1976). The basic approach uses flow analysis (Finn, 1976) to investigate the fate of material passing into and through the system. Barber (1978) has presented another approach to this same problem. Matis and Patten (1981) have developed a more sophisticated form of flow analysis in which the within-system environment of each system component is partitioned into components associated with particular inputs and outputs.

The model that is discussed in this paper uses this approach wherein model parameters are estimated from sampling data and then these estimates are used to analyze system structure and behavior by the techniques of Matis and Patten (1981).

TO \ FROM	0	1	2	3	4	5	6	7	8	9	10
0		A	A	M F		M	M	A	M F	M F	M F
1	A	P									
2	A C	K									
3	M		K								
4				P							
5	M C	K	K	K	K						
6	M				K	K					
7	A C	K	K	K	K	K	K				
8	M C							K			
9	M								K		
10	M						K		K	K	

Figure 2. Adjacency matrix for the Buccaneer Field system. A = advection; M = migration; K = material trophic transport; F = fishing; P = production. Compartment "0" refers to the environment external to the system.

CONSTRUCTION OF THE BUCCANEER FIELD MODEL

The overall Buccaneer Field model is comprised of a physical hydrodynamics component, a chemical component, and a biological system component. All of these components are coupled to produce the overall model. The objective of the hydrodynamics model is to describe the current and water movement dynamics in the Buccaneer Field and to relate these movements to chemical and biological distibutions. The chemical model describes the fate of oil that is introduced into the system as it is affected by various physical and biological factors. The biological systems model describes the trophic dynamic relationships that exist in the Buccaneer Field.

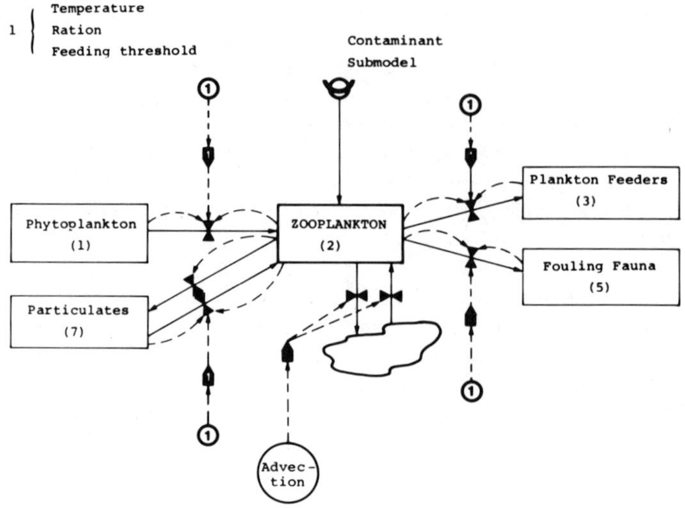

Figure 3. Forrester diagram for the herbivorous zooplankton model.

The Biological Model

The first step in the construction of the model was the development of a conceptual model of the Buccaneer Field. A conceptual model can be either a set of verbal statements or box and arrow diagrams that present our conception of the structure and function of the system. The conceptual design of the Buccaneer Field biological system consists of ten compartments (Figure 1). Our conceptual design differs from that of the LGL model of the Buccaneer Field (Gallaway and Margraf, 1979) in that we have placed the phytoplankton into a separate compartment rather than combining them with the holoplankton. Each of the compartments is influenced by interactions with components both within and outside the system. An adjacency matrix was constructed to define the interactions between the compartments (Figure 2). Each filled element of this matrix contains symbols that represent the nature of the interaction (e.g., primary production, advection, migration).

System interactions displayed in the adjacency matrix are controlled by variables such as light, temperature, or biotic density. The effects of these variables are conveniently displayed in Forrester diagrams. A complete set of these Forrester diagrams for each compartment in the model was constructed. However, only one of these is presented here as an example (Figure 3).

The Forrester diagram is the embodiment of the first stage of model development. The second stage was achieved in developing a quantitative static model. In doing so, the data base developed in other elements of the Buccaneer Field environmental program was used to construct an annual carbon flow budget based on the Forrester diagram. The actual method for translation to mathematical formulae and the manner in which other task data were used are not presented here but are given in detail elsewhere (Fucik and Show, 1980).

The static model describes system behavior over some limited period of time, in this case, one year. The quantitative dynamic model is capable of describing temporal changes in the behavior of the system by incorporating time-varying control functions instead of time averaged functions. A dynamic environs analysis was then applied to this form of the model.

The Hydrodynamics Model

The physical numerical model is based on the model developed by Leendertse et al. (1973) and Leendertse and Liu (1975). It is based on finite difference analogs of the vertically integrated form of the equations of motion, continuity, and conservation in an incompressible fluid. A detailed treatment of the theory and application of this model of the Buccaneer Field is given by Show (1979).

The Chemistry Model

The hydrocarbon modeling effort involved the development of algorithms for six distinct weathering processes (evaporation, dissolution, oil-water mixing, interactions with suspended particulate matter, photooxidation, and microbial degradation). This model is based on a model developed by Kolpack (1977).

The state variables in the model form a matrix, one dimension of which is comprised of chemical species (i.e., aliphatics, aromatics, branched/cyclic alkanes, alkyl-substituted aromatics) and the other by class based on molecular weight. The state variables are connected to one another and to chemical classifications outside of the matrix by what are termed transfer coefficients. These transfer coefficients are contributed to or controlled by the processes which produce time rates of change.

Diagrammatically, the system may be represented as in the following example.

Mean molecular weights

	101.4		130.1		187.5		264.5		452.1		681.0		772.0
Aliphatics	X_{11}	↔	X_{12}	↔	X_{13}	↔	X_{14}	↔	X_{15}	↔	X_{16}	↔	X_{17}
	↕		↕		↕		↕		↕		↕		↕
Aromatics	X_{21}	↔	X_{22}	↔	X_{23}	↔	X_{24}	↔	X_{25}	↔	X_{26}	↔	X_{27}
	↕		↕		↕		↕		↕		↕		↕
Branched/ cyclic alkanes	X_{31}	↔	X_{32}	↔	X_{33}	↔	X_{34}	↔	X_{35}	↔	X_{36}	↔	X_{37}
	↕		↕		↕		↕		↕		↕		↕
Alkyl-subst. aromatics	X_{41}	↔	X_{42}	↔	X_{43}	↔	X_{44}	↔	X_{45}	↔	X_{46}	↔	X_{47}

The X_{ij} represent the state variables (in volume or mass) and the arrows represent the transfers (in units of $time^{-1}$). In addition, each state variable has a potential transfer out of the state matrix.

Development of the hydrocarbon model is continuing under additional funding through the NOAA OSCEAP program.

Ecosystem Analysis

Although compartmental models have been widely used to simulate ecosystem structure and dynamics, the analysis of these models has advanced significantly only recently with the adoption of economic input-output analysis (Leontief, 1966) by system ecologists (Hannon, 1973; Patten, 1975; Finn, 1976; Patten et al., 1976; Barber, 1978). Patten (1978) defined the flow environment of compartments as input and output environs. Matis and Patten (1981) have developed an input-output environ analysis for linear compartment systems with stationary and static dynamics. Since this method partitions storages as well as flows and we are interested in both carbon and contaminant flows and storages, we have chosen to base our primary analysis of the Buccaneer Field ecosystem on these methods.

The dynamics of the system may be defined as

$$\frac{dX_i(t)}{dt} = \sum_{\substack{j=0\\ i\neq j}}^{n} f_{ij}(t) - \sum_{\substack{j=0\\ i\neq j}}^{n} f_{ji}(t)$$

where

$i = 1,2,\ldots,n$
$t \in (t_o, T) = \tau$
$X_i(t)$ = instantaneous storage of a conserved variable in compartment i at time t
f_{ij} = instantaneous non-negative flows from compartment j to i, $i \neq j$.

The subscript 0 indicates the external environment; let $Z_i = f_{io}$ and $Y_i = f_{oi}$, then the above equation may be rewritten as

$$\frac{dX_i(t)}{dt} = Z_i(t) + \Sigma f_{ij}(t) - \Sigma f_{ij}(t) - Y_i(t)$$

In the nonsteady state case, two formulations of the model are possible:

$$\frac{dX_i(t,Z_k,X_k(0))}{dt} = Z_i(t)\delta_{ik} + \sum_{j=1}^{n} f_{ij}(t,Z_k,X_k(t))$$

and

$$\frac{dX_i(t,Y_k,X_k(t))}{dt} = -\sum_{j=1}^{n} f_{ij}(t,Z_k,X_k(t)) - Y_i(t)\delta_{ki}$$

where

$i,k = 1,2,\ldots,n$
$\delta_{ik} = \delta_{ki} = 1$ if $i = k$, otherwise $\delta_{ik} = 0$.

From this partition, Matis and Patten (1981) define matrix manipulations that define, for each X_i, the amount of material in compartment i at time t_1 that has entered the system via compartment K since time t_o. The matrix manipulations also describe the amount of material X_i in compartment i at time t_o that exits the system via compartment k in the time interval $[t_o, t]$. These flows and storages are expressed in mass of material and as percentage of material.

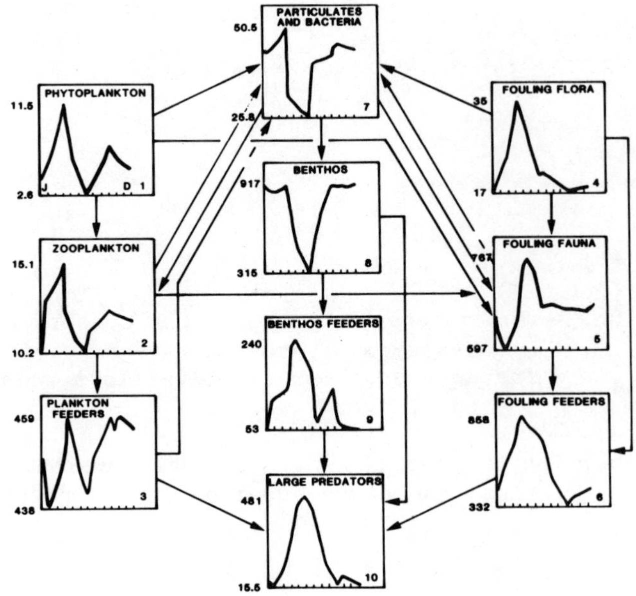

Figure 4. Seasonal biomass (g C/m^2) in the various compartments of the Buccaneer Field biological model integrated over 3-month periods.

RESULTS

The Biological Model

Seasonal Biomass Trends. Figure 4 shows the seasonal cycles for the various biological compartments as simulated by the biological model. Two basic patterns emerged that characterized the pelagic and fouling communities.

The pelagic community was characterized by a spring maximum, a summer minimum, and a secondary maximum in the fall. The fouling community, on the other hand, had only a spring maximum. The large predators, which are a part of the pelagic system, showed a pattern similar to that of the fouling community. This is likely due to emigration out of the system as well as increased fishing pressure.

The seasonal pattern in the particulate and benthic compartments was somewhat atypical of that in the pelagic compartments in that standing crops reached a minimum during the summer but remained relatively uniform during the rest of the year. The minimum in the benthic compartment probably results from increased predation from the benthic feeders. The pattern in the particulate compartment is probably due to decreased runoff from the bays in the winter and spring and stabilization of the water column due to formation of the thermocline thereby reducing resuspension of sediments in the summer (see Brooks et al., 1980).

The fouling fauna biomass was characterized by a rapid increase to a peak in the spring followed by a decline to a fairly stable population. Since the vast majority of this community is accounted for by barnacles, it is likely that this pattern reflects barnacle spawning and settling activity. If this is indeed the case, the leveling of the population may represent a barnacle biomass that is limited by the amount of available space for settlement and growth on the platform pilings. With a limited amount of substrate available for settling and growth, the barnacles compete for space as they grow. Increases in biomass during this period are probably balanced by the loss in numbers (see Howard et al., 1980).

Community Function. In attempting to understand how a system functions, it is important to understand the role of various organisms in cycling materials through the system. The biological model was designed to describe the flow and storage of carbon between the various compartments of the Buccaneer Field.

Figure 5 summarizes the flow of carbon in the Buccaneer Field system, integrated over three month periods. Because advection in this system is orders of magnitude greater than flows between compartments, flows in the system are given as net flows. The flow analysis indicated that carbon enters the system through five compartments (phytoplankton, zooplankton, plankton feeders, fouling flora, fouling fauna) and exits the system through the other five compartments. Except for the fouling fauna, those compartments which are net importers are either primary producers or plankton communities whose presence in the system is influenced by advection. The net exporters are primarily those species whose distributions are less directly affected by currents. The one exception was the particulates compartment. Increased particulates from the platforms may be a factor in this compartment being a net exporter.

Figure 6 depicts seasonal variations in the net inputs and outputs to the Buccaneer Field system. Flows into the system generally tend to be less variable than exports from the system. Whether this is a real feature of the system or an artifact due to a limited amount of data for these compartments is not known. External inputs and standing crops in the producer compartments vary seasonally yet these compartments show no net inputs or output during the year. This suggests that there is a rapid rate

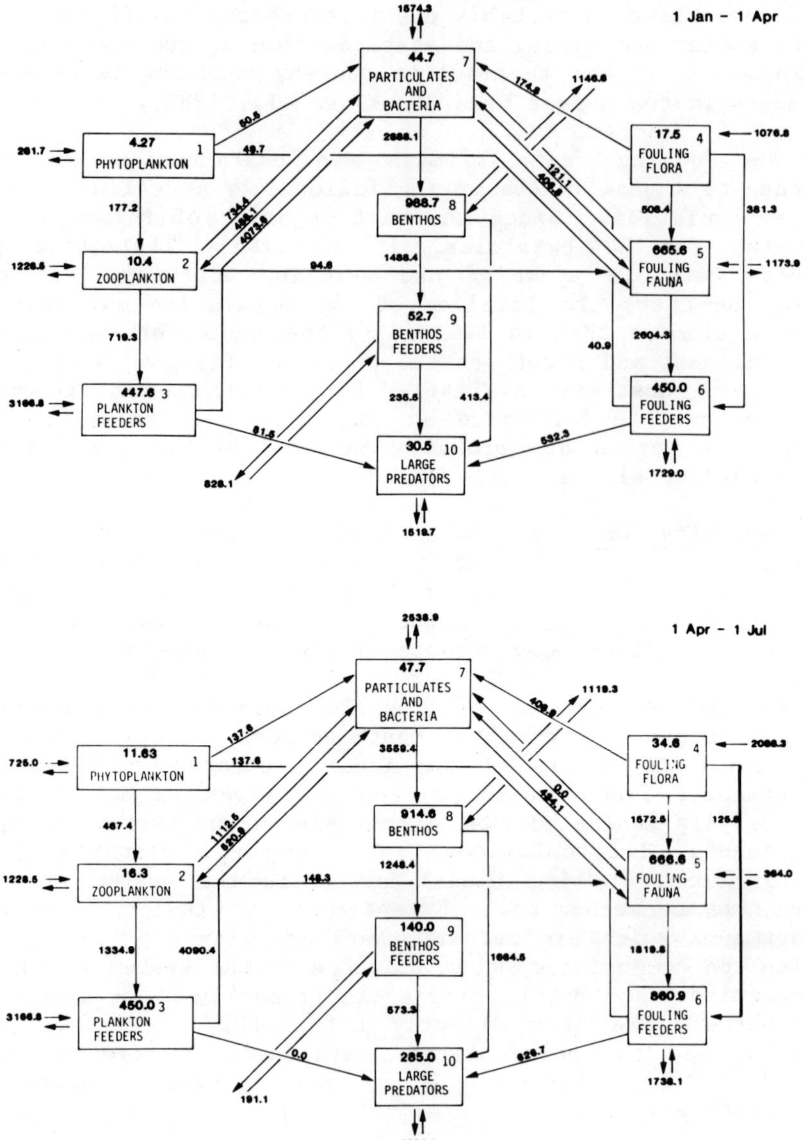

Figure 5. Quantitative carbon flows between compartments. Units are g C/m^2 integrated over 3-month periods.

of production that is balanced by grazing on these compartments. This also suggests a threshold level for grazing to occur.

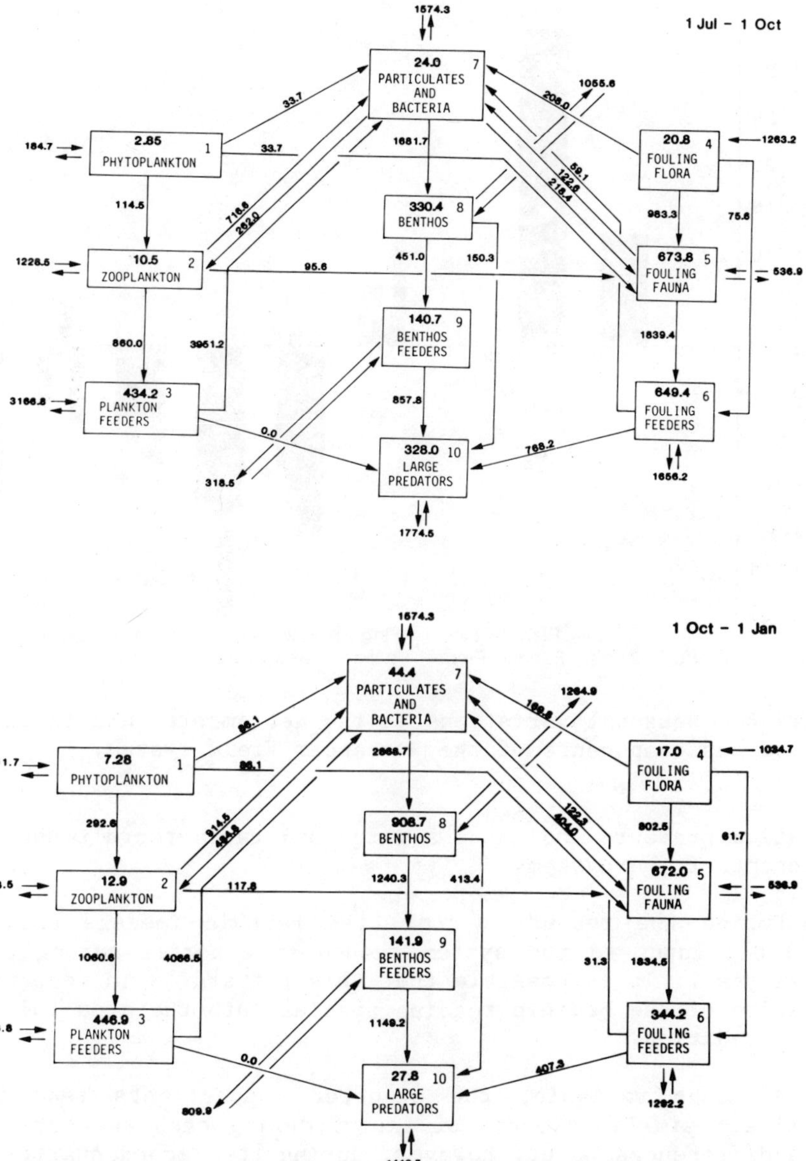

Figure 5 (continued).

It was surprising that the fouling fauna compartment tended to produce a net input to the system yet the benthic compartment was a net exporter. However, it is likely that the net exports in the benthic compartment can be explained by fishing of these species. The net inputs by the fouling fauna in spite of heavy

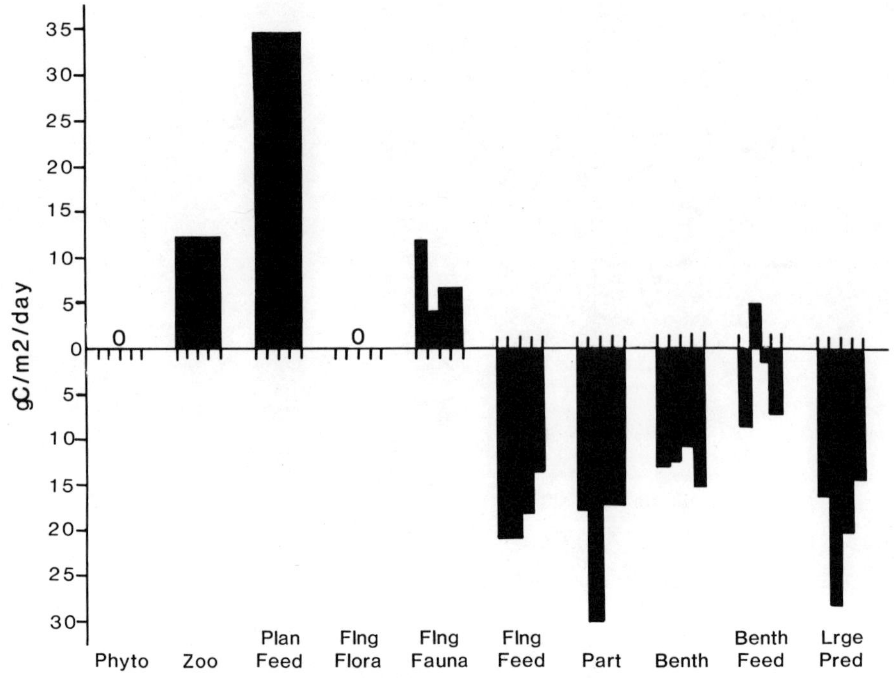

Figure 6. Seasonal variations in the net imports and exports from the various components of the Buccaneer Field system.

predation pressure indicates a large and rapid recruitment in this component of the system.

During the second quarter, the benthic feeders compartment was a net input to the system though an exporter during the rest of the year. It is possible that this pattern is in response to a migration of the benthic feeding species into the area followed by heavy predation.

Although variable, the exporter compartments tend to show relatively similar amounts of export during each quarter. Rather large differences occur, however, during the second quarter in the particulates and large predators compartments. For the particulate compartment, this may represent an excess production of particulates by the platform communities beyond what is required to support the biomass in the system. In the large predators compartment, much of this biomass is lost to the system through migration of these fish from the Buccaneer Field.

Carbon flows within the system are generally seasonally variant with the largest flows occurring during the spring.

TABLE 1. PERCENT OF MATERIALS STORED IN COMPARTMENT X_i THAT ORIGINATED IN THE PHYTOPLANKTON OR FOULING FLORA COMPARTMENTS.

	1 January - 1 April		1 April - 1 July		1 July - 1 October		1 October - 1 January	
	Phyto-plankton	Fouling flora	Phyto-plankton	Fouling flora	Phyto-plankton	Fouling flora	Phyto-plankton	Fouling flora
Phytoplankton	100	0	100	0	100	0	100	0
Zooplankton	77	10	89	7	75	11	86	7
Plankton feeders	73	12	87	7	72	11	83	6
Fouling flora	0	100	0	100	0	100	0	100
Fouling fauna	39	54	27	71	19	77	36	60
Fouling feeders	33	59	25	73	18	78	35	62
Particulates	70	14	81	13	68	17	80	11
Benthos	70	14	81	13	68	17	80	11
Benthic feeders	70	14	81	13	68	17	80	11
Large predators	55	33	69	26	46	43	71	21

Various patterns emerged in the flows between compartments which should be mentioned. The largest flows generally occurred in the benthic components of the system with the fouling community being intermediate and the smallest flows in the pelagic system. For the most part, the size of the flows increased with increasing trophic level in the pelagic and fouling components. The opposite trend was observed in the benthic components. Trends similar to those for the flows were observed in biomass levels in the pelagic, benthic, and fouling communities. Flows between the pelagic and fouling components were less than flows within these two components of the system.

During the second quarter (April-July) no flow of materials was evident from the fouling fauna to the particulate compartment. Similarly, during the last three quarters (April-December), no flows were observed from the plankton feeders to the large predators. Through the whole year the major flow from the plankton feeders was to the particulates compartments.

The distinctiveness of the pelagic community and fouling community within the system becomes fairly evident when storages within the system are considered. Table 1 shows where material that is stored in the system originates. It can be seen that about 70% of the material that is stored in the zooplankton, plankton feeders, particulates, benthos, and benthic feeders compartments originates in the phytoplankton compartment during the first and third quarters. The percentage is slightly less during the third quarter than during the first. This percentage increases to 80% or more during the second and fourth quarters. By comparison, of the materials stored in the fouling fauna and fouling feeders, an average of approximately 35, 26, 18, and 35%, respectively, originate in the phytoplankton during quarters 1 through 4. This proportionately larger dependence of the fouling species on the phytoplankton as compared to the fouling flora is probably due to the filter feeding behavior of the barnacles which comprise the major portion of the fouling fauna. Nevertheless, most of the stored materials in these two compartments arise from the fouling flora, with approximately 55, 72, 77, and 60% being measured in successive quarters.

The large predator compartment showed less of a dependence on the phytoplankton than did the other pelagic and benthic components although the major portions of the stored materials did originate in the phytoplankton. Ratios of material that originated in the phytoplankton versus fouling flora compartments during successive quarters was 55:33%; 69:26%; 46:43%; and 71:21%.

Of the stored material which exited the system, at least 95% or more of the materials in all compartments exited through the large predators. This was true for all quarters.

TABLE 2. RATES IN g/m^2-sec FOR VARIOUS PROCESSES ACTING ON OIL IN THE BUCCANEER FIELD AS CALCULATED BY THE MODEL.

Evaporation	2.8×10^{-5}	
Dissolution	1.2×10^{-4}	
Emulsification[a]	1.7×10^{-9}	1.4×10^{-7}
Sedimentation	2.8×10^{-11}	
Photooxidation	3.9×10^{-8}	
Microbial degradation	1.0×10^{-3}	

[a]Initial rate and rate after 24 hr; rates for all other processes remained constant.

The final analysis of the environs analysis considered that material which flowed directly through the system without being stored. Major flow patterns were identical to those described above for storages. This analysis confirms the major pathways of cycling within the system.

The Chemical Model

When oil enters the marine environment, it can undergo various physical, chemical, and biological modifications. The processes which were considered in the Buccaneer Field chemical model included evaporation of volatile components, sedimentation, dissolution, emulsification, photooxidation, and microbial degradation.

Rates were calculated for the various degradatory processes acting on the oil that enters the system (Table 2). Eight orders of magnitude difference existed between the various processes with microbial degradation being the most significant and sedimentation being the least significant. These rates tended to remain constant throughout the simulation except for the rate of emulsification which increased by two orders of magnitude over the 24 hr simulation period.

Over the 24 hr period approximately 10% of the total hydrocarbons are lost due to weathering. During this period, 16% of the aliphatics are lost, over 99% of the aromatics, 58% of the branched/cyclic alkanes, and 4% of the alkyl substituted aromatics (Figure 7).

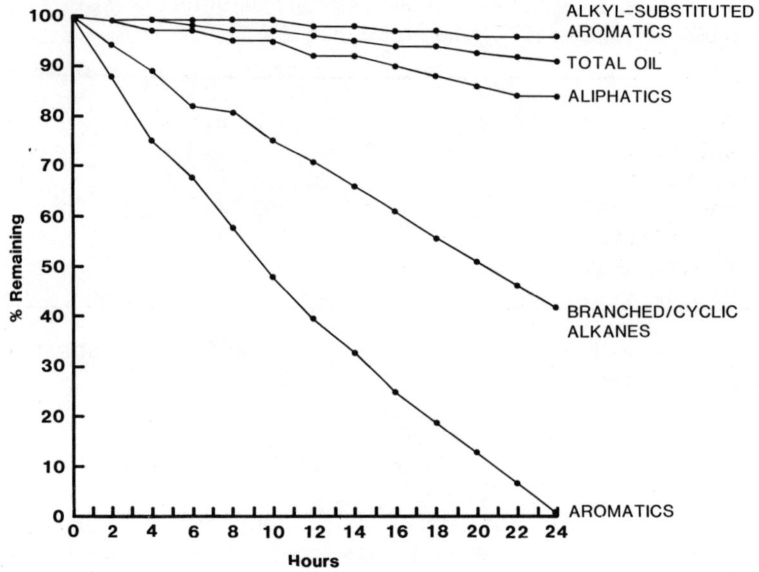

Figure 7. Rate of loss of various oil fractions as predicted by the chemical model over a 24-hour period.

DISCUSSION

Community Structure and Function

One of the most significant aspects of the biological model was the fact that the model showed two somewhat unique communities in the Buccaneer Field: one whose major inputs originated in the phytoplankton and one whose major inputs originated from the fouling flora. For species such as the sheepshead which feed primarily on the fouling community, the food source provided by the platforms would be the primary reason behind their presence in the Buccaneer Field. Similarly, the large predators compartment received a proportionately higher input from the fouling community than the other members of the pelagic community. The ratio between inputs from the pelagic and fouling components varied quarterly and ranged from highly divergent to very nearly equal. This likely reflects the position of these predators at the top of the food chain and a less selective feeding behavior. This allows these organisms to feed opportunistically on whatever species happens to be the most abundant at the time. The seasonal nature of the flows from the fouling flora to the other components of the fouling community also demonstrates the ability of the fouling fauna to utilize nutrient sources in the water column when macroalgal production is down.

Gallaway and Margraf (1979) constructed an earlier model of the Buccaneer Field biological system which suggested that the system was particulate based. Their model suggested that the planktonic, benthic, and floral components of the system accounted for less than 12% of the total throughflow in the system. Our model, however, suggests that the system is heavily dependent on the primary producers and the particulates are a net exporter in the system.

The importance of the phytoplankton and concomitant reduced importance of the particulates in the system is supported by the findings of Brooks et al. (1980) that very little of the particulate organic matter is noncellular material. In most cases, cellular biomass determined by ATP concentrations was greater than particulate matter determined by POC. Since ATP measuring techniques measure only living cells, it is unlikely that this particulate material was dead or decaying materials off the platform. The high levels of non-cellular organics that were observed in the spring were attributed by these same authors to fresh water inflow from Galveston Bay. This probably means that very little of the particulate material (15% as calculated in the model) that is produced by the fouling community gets into the water column.

This is not to say that the particulates are not an important part of the system. The fact that, on the average, only 15% of the particulate matter from the fouling community goes into the particulate compartment suggests a large amount of internal cycling of these particulates within the fouling community.

Numerous microhabitats are created by the irregular surfaces of the barnacle-covered platform legs. Much of the dead or decaying organic matter originating from the platform communities conceivably gets trapped in these cracks and crevices. This would account for the fact that much of this material never gets into the water column. This organic matter could, however, provide an abundant food source for the cryptic fauna.

The relatively large proportion of cycled material from the phytoplankton to the fouling community is probably explained by the feeding habits of the barnacles. These organisms are filter feeders and form the outside perimeter of the fouling community. If, as suggested, most of the organic materials produced by the fouling community do not get into the water column, then the barnacles must depend on the phytoplankton and zooplankton for the major portion of their nourishment. Hydrocarbon profiles of the barnacle tissues were typical of animals feeding on phytoplankton and detritus (Middleditch and West, 1980).

Fate of Hydrocarbons

Various physical, chemical, and biological transformations take place when oil enters the marine environment. Both the present model and that of Smedes et al. (1980) suggest that hydrocarbons are rapidly dispersed and carried out of the Buccaneer Field. However, concentration gradients for organic matter (Brooks et al., 1980), hydrocarbons, and sulfur (Middleditch and West, 1980) have been observed in the sediments in the immediate vicinity of the platforms. The fact that these substances are found in the immediate vicinity of the platform indicates a fairly rapid transport of these materials to the bottom. Sedimentation does occur around the platforms (Brooks et al., 1980) so it is likely that some of the hydrocarbons are reaching the bottom through adsorption to sediments. The fact that the majority of the inorganic suspended materials are in the 2-5 μM size range leaves some doubt, however, whether this could be a significant transport mechanism in view of the turbulence around the platforms. One possible mechanism of transport, however, may be through adsorption or incorporation into fecal matter. Parker et al. (1971) calculated that the transport of oil in zooplankton fecal material could result in the deposition of significant amounts of oil on the bottom. The model suggests a large transport of particulate matter (particularly from plankton feeders) into the benthic compartment. Gallaway and Martin (1980) found that the primary food source of the spadefish (the dominant plankton feeder in the Buccaneer Field) was a planktonic pteropod, *Carolina longirostris*. Harper (1977) found large numbers of these pteropod shells in his benthic samples. This implies a large amount of grazing by the spadefish and transport of large quantities of fecal material to the bottom.

Sulfur is the major element in the brine discharged from the platforms (Middleditch and West, 1980). It is possible that the hydrocarbons could be transported to the bottom through adsorption to these particles. Sizemore and Olsen (1980) found that sulfate reducing bacteria were especially abundant in the sediments at the platform sites. Hunt et al. (1973) concluded that degradation of petroleum would be extremely slow under anaerobic conditions. If these anaerobic conditions do exist in the sediments under the platforms, then this might account for the accumulation of hydrocarbons in this area.

The model of Smedes et al. (1980) suggested that hydrocarbons could be transported as far as the coast southwest of Galveston in a matter of a few days. Middleditch et al. (1979a) found petroleum alkane concentrations as high as 43 ppb within 3 km of the Buccaneer Field. Our chemistry model agrees with these findings as we observed only a 10% loss in total hydrocarbon levels over a 24 hr period. These levels are also consistent with

TABLE 3. AVERAGE STEADY STATE HYDROCARBON LEVELS IN EACH COMPARTMENT OF THE BUCCANEER FIELD.

Compartment	Total HC's (ppm)
1	16.4
2	12.5
3	12.0
4	16.9
5	6.6
6	5.6
7	2.4
8	2.4
9	2.4
10	5.6

levels measured during simulated weathering experiments by Sizemore et al. (1979), and Sizemore and Olsen (1980).

Hydrocarbons can be accumulated by marine organisms by transfer of the pollutants across membranes or through the diet. If food chain transfer of hydrocarbons occurs in the Buccaneer Field, then the flow of hydrocarbons through the system would follow the same major pathways as that described for carbon flows (i.e., the biological model). Data reported by Middleditch and West (1980) on hydrocarbons in fish and fouling organisms seems to suggest that there is food chain transfer of hydrocarbons in the Buccaneer Field. Steady state average values for hydrocarbons in each compartment were calculated using the biological and chemical models (Table 3). These results are probably somewhat limited since hydrocarbon concentration data for system importers was available only for zooplankton and the fouling fauna. Based on the model results, higher hydrocarbon concentrations are expected in the particulates and benthic components of the Buccaneer Field. However, the average steady state values that were calculated tended to be somewhat higher than the actual measured values for those compartments where comparisons were possible; also the actual hydrocarbon levels in many of the compartments were generally about equal (Middleditch et al., 1979a,b,c; Middleditch and West, 1980).

The discrepancy between the model and the measured values is not unexpected since we have no information concerning the fate of the hydrocarbons that are accumulated by the organisms. The fact that, for the most part, there are no large accumulations of hydrocarbons in the organisms would tend to suggest that most of the organisms are effectively depurating or metabolizing most of the accumulated hydrocarbons.

The model implies that the major flow of material in the Buccaneer Field system arises from the phytoplankton. The large amount of advection in the Buccaneer Field, however, probably means that a particular plankton's stay in the vicinity of the Buccaneer Field platforms is short-lived. This may mean a reduced chance that a contaminated organism will be preyed upon by an organism in or around the platforms. If this is true, then this would imply that the major pathway for hydrocarbon accumulations would be from the fouling flora through the fouling community. This, then would account for the low hydrocarbon levels in those species which do not feed on the fouling community.

Middleditch and West (1980) found that the fouling mat beneath the platform discharge had very high levels of petroleum alkanes (122 ppm). Hydrocarbon profiles were typical of unweathered hydrocarbons. Crested blennies, which feed on the mat, also showed unweathered petroleum profiles and had a mean alkane concentration considerably higher than that of other species of fish. In contrast to these other embers of the fouling community, however, the barnacles had very low petroleum alkane concentrations (2.48 ppm) and these were typical of weathered oil. This suggests different modes of petroleum accumulation in these species. Since the barnacles obtain their food from the water column, a resuspension and ingestion of particles containing weathered oil could account for the differences observed in the hydrocarbon profiles in the barnacles and the other fouling mat species. This is not unlike a situation observed by Fucik et al. (1977) in Trinity Bay, Texas. In this case, bivalves placed beneath an oil separator platform accumulated significant amounts of total naphthalenes over approximately three month periods. The weathered profile of the hydrocarbons in the bivalve tissues led the authors to speculate that the sediments were the major source of the accumulated hydrocarbons. Hydrocarbons were present in the water column, though at low levels (<11 ppb).

The bioassays conducted by Rose and Ward (this Volume) and Zein-Eldin and Keney (1979) indicate that very high concentrations of the brine discharged from the platforms were required to produce acute toxicities (LC50's ranged from 9500 ppm for larval brown shrimp to 100,000 ppm for adult brown shrimp). Rose and Ward concluded that the formation waters represent no acute hazard to plankton in the discharge plume; potential hazards to benthic structure-associated organisms may occur only within a few meters of the outfall. Gallaway et al. (1979) noted that evidence of adverse effects due to the produced water were evident in the immediate vicinity of the discharge. However, there are no comparative data from the bioassay studies or the field studies on the levels of petroleum aromatic hydrocarbons in the tissues of exposed organisms. The lack of data for the aromatics makes it difficult to evaluate the chronic, sublethal effects of hydrocarbons in the Buccaneer Field. However, based on the observations of other investigations of control, structures, and platforms with discharges, overall impacts appear to be minimal.

ACKNOWLEDGEMENTS

This work is a result of research sponsored by the Environmental Protection Agency and Department of Commerce, National Oceanic and Atmospheric Administration, National Marine Fisheries Service, Southeast Fisheries Center, Galveston Laboratory under Contract No. 03-78-D08-0036.

REFERENCES

Barber, M. C., 1978, "A Markovian model for ecosystem flow analysis," Ecol. Modelling, 5:193.

Brooks, J. M., Estes, E. L., and Huang, W. H., 1980, "Investigations of surficial sediments and suspended particulates at Buccaneer field," in: Jackson, W. B., and Wilkens, E. P. (eds.), "Environmental Assessment of Buccaneer Gas and Oil Field in the Northwestern Gulf of Mexico, 1978-1979, Vol. II." NOAA Technical Memorandum NMFS-SEFC-36, 261 pp. Available from: NTIS, Springfield, Virginia.

Finn, J. T., 1976, "Measures of ecosystem structure and function derived from analysis of flows," J. Theoret. Biol., 56:363.

Fucik, K. W., and Show, I. T., 1980, An ecosystem model for the Buccaneer Oil Field," in: Jackson, W. B., and Wilkens, E. P. (eds.), "Environmental Assessment of Buccaneer Gas and Oil Field in the Northwestern Gulf of Mexico, 1978-1979, Vol. IX." NOAA Technical Memorandum NMFS-SEFC-43, 105 pp. Available from: NTIS, Springfield, Virginia.

Fucik, K. W., Armstrong, H. W., and Neff, J. M., 1977, "Uptake of naphthalenes by the clam, *Rangia cuneata*, in the vicinity of an oil separator platform in Trinity Bay, Texas," in: "Proceedings of the 1977 Oil Spill Conference." Publication No. 4284, American Petroleum Institute, Washington, D.C., p. 637.

Gallaway, B. J., and Margraf, F. J., 1979, "Simulation modeling of biological communities associated with a production platform in the Buccaneer oil and gas field," in: Jackson, W. B. (ed.), "Environmental Assessment of an Active Oil Field in the Northwestern Gulf of Mexico, 1977-1978, Vol. II: Data Management and Biological Investigations." NOAA, pp. 2.5.1-1 to 2.5.1-67. Available from: NTIS, Springfield, Virginia, Accession No. PB80165970.

Gallaway, B. J., and Martin, L. R., 1980, "Effect of gas and oil field structures and effluents on pelagic and reef fishes, and demersal fishes and macrocrustaceans," in: Jackson, W. B., and Wilkens, E. P. (eds.), "Environmental Assessment of Buccaneer Gas and Oil Field in the Northwestern Gulf of Mexico, 1978-1979, Vol. III." NOAA Technical Memorandum NMFS-SEFC-37, 49 pp. Available from: NTIS, Springfield, Virginia.

Gallaway, B. J., Johnson, M. F., Howard, R. L., Martin, L. R., and Boland, G. S., 1979, "A study of the effects of Buccaneer Oil Field structures and associated effluents on biofouling communities and the Atlantic spadefish (*Chaetodipterus faber*)," in: Jackson, W. B. (ed.), "Environmental Assessment of an Active Oil Field in the Northwestern Gulf of Mexico, 1977-1978. Volume II: Data Management and Biological Investigations." NOAA, pp. 2.3.8-1 to 2.3.8-126. Available from: NTIS, Springfield, Virginia, Accession No. PB80165970.

Hannon, B., 1973, "The Structure of Ecosystems," J. Theoret. Biol., 41:535.

Harper, D. E., Jr., 1977, "Distribution and abundance of macrobenthic and meiobenthic organisms," in: Jackson, W. B. (ed.), "Environmental Assessment of an Active Oil Field in the Northwestern Gulf of Mexico, 1976-1977." NOAA, pp. 175-273. Available from: NTIS, Springfield, Virginia, Accession No. PB283890.

Howard, R. L., Boland, G. S., Gallaway, B. J., and Dennis, G. D., 1980, "Effects of gas and oil field structures and effluents on fouling community production and function," in: Jackson, W. B., and Wilkens, E. P. (eds.), "Environmental Assessment of Buccaneer Gas and Oil Field in the Northwestern Gulf of Mexico, 1978-1979, Vol. V." NOAA Technical Memorandum NMFS-SEFC-39, 60 pp. Available from: NTIS, Springfield, Virginia.

Hunt, P. G., Koutz, F. R., Marrmann, R. P., and Martin, T. G., 1973, "Microbial Degradation of Petroleum in Continental Shelf Sediments." U.S. Department of Commerce, 16 pp. Available from: NTIS, Springfield, Virginia, Accession No. AD772698.

Kolpack, R., 1977, "Fate of Oil in a Water Environment. Phase II. A Dynamic Model of the Mass Balance for Released Oil." Draft report to the American Petroleum Institute, Washington, D.C.

Leendertse, J. J., and Liu, S. K., 1975, "A Three-Dimensional Model for Estuaries and Coastal Seas. Vol. II. Aspects of Computation." Rand Corp. Report, Santa Monica, California.

Leendertse, J. J., Alexander, R. C., and Liu, S. K., 1973, "A Three-Dimensional Model for Estuaries and Coastal Seas. Vol. I. Principles of Computation." Rand Corp. Report, Santa Monica, California.

Leontieff, W. W., 1966, "Input-Output Economics." Oxford, London.

Matis, J. H., and Patten, B. C., 1981, "Environs analysis of linear compartmental systems: the static, time invariant case," Intl. Statistical Inst. Bull., in press.

Middleditch, B. S., Basile, B., and Missler, S. R., 1979a, "Determine levels, pathways, and bioaccumulation of selected discharge constituents (non-metals) in the marine ecosystem in the oil field/hydrocarbon modeling," in: Jackson, W. B. (ed.), "Environmental Assessment of an Active Oil Field in the Northwestern Gulf of Mexico, 1977-1978, Vol. III, Chemical and Physical Investigations." NOAA, pp. 2.4.1-1 to 2.4.1-302. Available from: NTIS, Springfield, Virginia, Accession No. PB80107899.

Middleditch, B. S., Chang, E. S., and Basile, B., 1979b, "Alkanes in barnacles (*Balanus tintinnabulum*) from the Buccaneer Oilfield," Bull. Environ. Contam. Toxicol., 23:6

Middleditch, B. S., Chang, E. S., Basile, B., and Missler, S. R., 1979c, "Alkanes in fish from the Buccaneer Oilfield," Bull. Environ. Contam. Toxicol., 22:249

Middleditch, B. S., and West, D. L., 1980, "Hydrocarbons, biocides and sulfur," in: Jackson, W. B., and Wilkens, E. P. (eds.), "Environmental Assessment of Buccaneer Gas and Oil Field in the Northwestern Gulf of Mexico, 1978-1979, Vol. 8." NOAA Technical Memorandum NMFS-SEFC-41, 119 pp. Available from: NTIS, Springfield, Virginia.

Parker, C. A., Freegarde, M., and Hatchard, C. G., 1971, "The effect of some chemical and biological factors on the degradation of crude oil at sea," in: Hepple, P. (ed.), "Water Pollution by Oil." Institute of Petroleum, London, p. 237.

Patten, B. C., 1975, "Ecosystem as a co-evolutionary unit: A theme for teaching systems ecology," in: Innis, G. S. (ed.), "New Directions in the Analysis of Ecological Systems, Part 1." Society for Computer Simulation, La Jolla, California.

Patten, B. C., 1978, "Systems approach to the concept of environment," Ohio J. Sci., 78:206.

Patten, B. C., and Finn, J. T., 1979, "System approach to continental shelf ecosystems," in: Halfon, E. (ed.), "Theoretical Systems Ecology." Academic Press, N.Y., p. 183.

Patten, B. D., Bossenman, R. W., Finn, J. T., and Cale, W. G., 1976, "Propagation of cause in ecosystems," in: Patten, B. C. (ed.), "Systems Analysis and Simulation in Ecology." Vol. IV, Academic Press, N.Y.

Show, I. T., 1979, "An application of compartmental models to meso-scale marine ecosystems," in: Matis, J. H., Patten, B. C., and White, G. C. (eds.), "Compartmental Analysis of Ecosystem Models." International Cooperative Publishing House, Fairland, Maryland.

Sizemore, R. K., and Olsen, K., 1980, "Bacterial Communities," in: Jackson, W. B., and Wilkens, E. P. (eds.), "Environmental Assessment of Buccaneer Gas and Oil Field in the Northwestern Gulf of Mexico, 1978-1979, Vol. IV." NOAA Technical Memorandum NMFS-SEFC-38, 32 pp. Available from: NTIS, Springfield, Virginia.

Sizemore, R. K., Hollaway, S. L., and Faw, G. M., 1979, "Determine and compare groups of bacteria from an active oil field and unaltered area," in: Jackson, W. B. (ed.), "Environmental Assessment of an Active Oil Field in the Northwestern Gulf of Mexico, 1977-1978, Vol. II: Data Management and Biological Investigations." NOAA, pp. 2.3.7-1 to 2.3.7-41. Available from: NTIS, Springfield, Virginia, Accession No. PB80165970.

Smedes, G., Calman, J., and Beebe, J., 1980, "Hydrodynamic modeling," in: Jackson, W. B., and Wilkens, E. P. (eds.), "Environmental Assessment of Buccaneer Gas and Oil Field in the Northwestern Gulf of Mexico, 1978-1979, Vol. X." NOAA Technical Memorandum NMFS-SEFC-44, 57 pp. Available from: NTIS, Springfield, Virginia.

Ward, C. H., Bender, M. E., and Reish, D. J., 1979, "The Offshore Ecology Investigation. Effects of Oil Drilling and Production in a Coastal Environment." Rice University Series, 65 (4 and 5), Houston, Texas.

Zein-Eldin, Z. P., and Keney, P. M., 1979, "Bioassay of Buccaneer oil field effluents with penaeid shrimp," in: Jackson, W. B. (ed.), "Environmental Assessment of an Active Oil Field in the Northwestern Gulf of Mexico, 1977-1978, Vol. II: Data Management and Biological Investigations." NOAA, pp. 2.3.4-1 to 2.3.4-25. Available from: NTIS, Springfield, Virginia, Accession No. PB80165970.

CURRENTS AND HYDROGRAPHY OF THE BUCCANEER FIELD AND ADJACENT WATERS

Larry J. Danek[a] and Michael S. Tomlinson[a]

Hazleton Environmental Sciences
1500 Frontage Road
Northbrook, Illinois 60062

INTRODUCTION

The Currents and Hydrographic Work Unit conducted between July 1978 and May 1979 was one of 11 separate tasks sponsored by the National Marine Fisheries Service to investigate potential chronic effects of oil production in the marine environment.

The goal of the Currents and Hydrography Work Unit was to provide a description of the existing hydrographic environment in and around the Buccaneer Gas and Oil Field. To achieve this goal, seasonal field studies were conducted which included measurement of currents, wind, waves, total suspended solids, and hydrographic parameters including temperature, salinity, conductivity, dissolved oxygen, pH, and transmissivity. The monitoring was conducted in an effort to describe and predict the water movement and energy regimes related to suspension, deposition, resuspension, and transport of surficial sediments and other particulate matter.

METHODS AND MATERIALS

Four oceanographic cruises were conducted in the Buccaneer Field, Gulf of Mexico, from July 1978 to May 1979. During the

[a]Present address: Environmental Science and Engineering, Inc., P.O. Box ESE, Gainesville, Florida 32602.

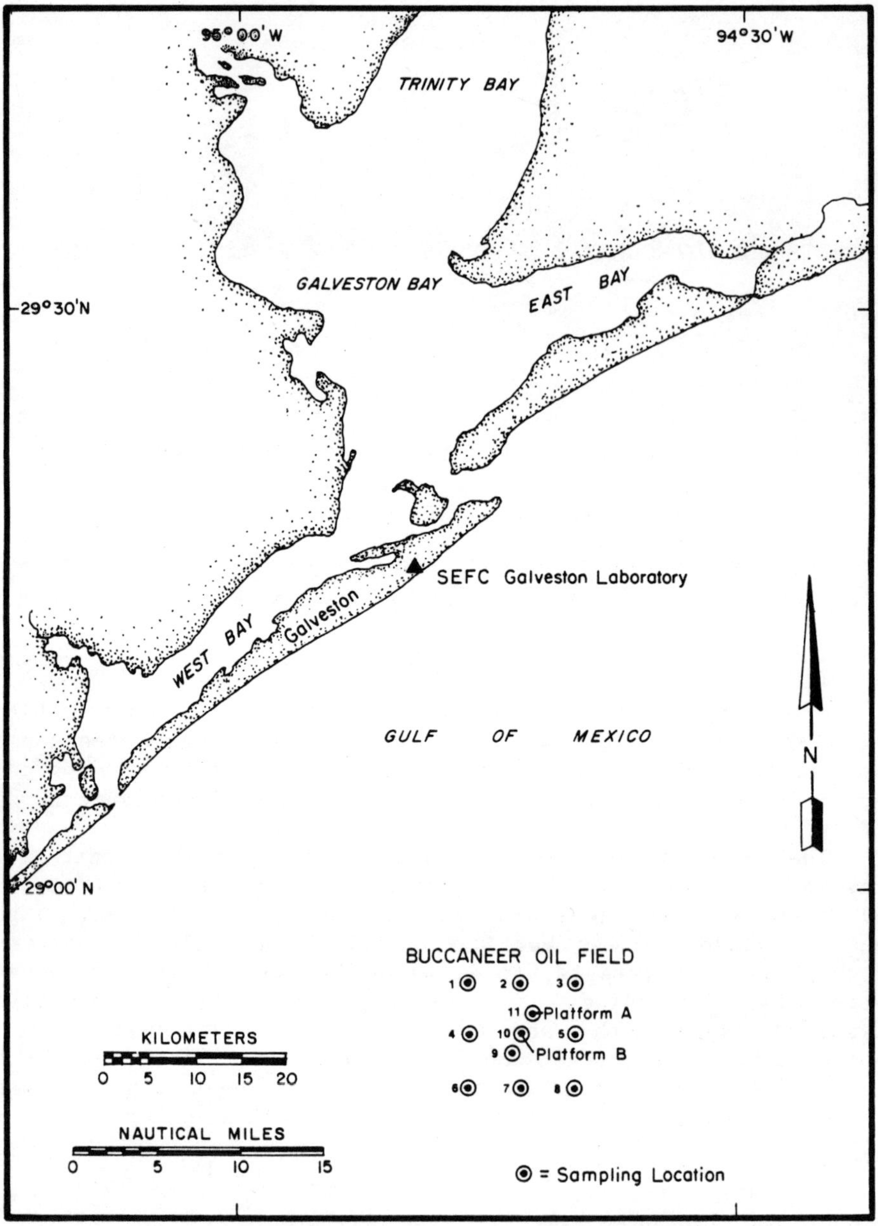

Figure 1. Site location and hydrographic sampling grid for the Currents and Hydrography Study conducted by Hazleton Environmental Sciences in the Buccaneer Gas and Oil Field, Gulf of Mexico.

TABLE 1. THE ACCURACY, RANGE, AND TIME CONSTANTS OF THE PROBES ON THE INTEROCEAN MODEL 513D CSTD.

Parameter	Range	Accuracy	Time constant
Conductivity	0-65 mmho/cm	±0.05	10 msec
Salinity	0-40°/oo	±0.05	1.4 sec
Temperature	-5-45°C	±0.05	60 msec
Depth	0-200 m	±1	50 msec
Dissolved oxygen	0-40 ppm	±0.2	10 sec
pH	2-14	±0.1	200 msec
Transmissivity	0-100%	±1	400 msec

first cruise from 25 to 27 July 1978 and the third cruise on 14 February 1979, six types of measurements were made from the vessel GUS III or from platform 296-B belonging to Shell Oil Company. The measurements included (1) hydrographic sampling consisting of in situ measurements of conductivity, salinity, temperature, dissolved oxygen, pH, and transmissivity, (2) electromagnetic current meter measurements, (3) total suspended solids, (4) continuous wind measurements, (5) continuous current meter measurements, and (6) wave measurements. The wind, waves, and continuous current measurements were made for a period of at least 30 days during the first and third field studies. Only hydrographic sampling, total suspended solids, and electromagnetic current meter measurements were made during the second cruise (25 to 26 October 1978) and the fourth cruise (8 to 9 May 1979). The methods used in collecting and processing the data were as follows.

Hydrographic Measurements

Conductivity, salinity, temperature, depth, dissolved oxygen (DO), pH, and transmissivity were measured at 11 sampling stations within a 6 nautical mile (11 km) square grid centered on the Buccaneer Field (Figure 1). An InterOcean Model 513D CSTD Probe was used to measure each parameter at approximately one second intervals as the probe was lowered to within 1 m of the bottom. The measurements were relayed to an InterOcean Model 690M Digital Data Recorder which recorded the signals on magnetic tape. The CSTD probe was also linked with an InterOcean Model 514D CSTD Readout Unit. This unit provided digital and analog readout of individual parameters so that the measured signals could be monitored and selected parameters could be recorded manually if desired.

The CSTD data were recorded on magnetic tape as the probe descended. When the probe was within 1 m of the bottom the cassette recorder was turned off. Values of each of the physical parameters were then recorded manually in a log book as a backup. The CSTD was then raised consecutively to depths of 10 and 1.5 m, and the physical parameters at each level were again recorded in the log book. The accuracy, range, and time constants of the probes used on the CSTD are given in Table 1.

To ensure the accuracy of the collected data all of the probes were calibrated by the manufacturer. In addition, the pH, DO, conductivity, and transmissivity probes were calibrated in the field against known standards. The results of the field calibrations were then used during data processing to adjust the recorded data.

Tables and vertical and horizontal contour plots of salinity, temperature, DO, pH, and transmissivity were computer generated from the data collected at the 11 stations to aid in the interpretation of the results. Some of these are presented in this report to help illustrate the hydrographic features of the area.

Total Suspended Solids

Water samples for total suspended solids were collected with a Jabsco Model No. 36600-0030 water pump rated at 8 gallons per minute. A graduated weighted plastic intake line was lowered to several depths from which the samples were pumped. Sufficient time was allowed to flush the intake tube before the water sample was collected in a rinsed glass bottle. The sample was then labeled and refrigerated until it was analyzed.

The technique for analyzing the water samples was that described in "Standard Methods for the Examination of Water and Wastewater," Fourteenth Edition (Taras, 1975). Briefly, the method is as follows: a Reeve-Angel Glass Fiber Filter (Grade No. 934-AH) is placed in a Gooch crucible, the apparatus is rinsed with distilled water, dried at 105°C, and the tare weight of the apparatus is obtained. A 500 ml aliquot of well mixed sample is filtered under vacuum, then the apparatus is again dried at 105°C and reweighed. This weight minus the tare weight times 2 yields the total suspended solids value in mg/l. These data were then tabulated for presentation and comparison.

Current Meter Measurements

Vertical profiles of the currents were taken during the ebb and flood stages of the tide at 6 or more stations near platform

296-B with a Marsh-McBirney Model 727 Electromagnetic Current Meter. Currents were normally measured at depths of 2, 5, 10, 15, 16, 17, 18, and 19 m. The time, orientation of the probe, and x and y velocity components were recorded on field data logs. These data were then keypunched and loaded onto a DEC-10 computer system which then produced vertical profile plots of the currents to aid in interpretation of the data.

The current meter has a stated threshold velocity of 0.5 cm/sec, which is also the resolution of the recorded velocity components. Absolute accuracy of the measurements is specified as being ±2% of the instrument readout. Maximum long-term drift, an inherent instrument error, is approximately 1.9 cm/sec. Consequently, under worst case conditions of large zero-drift, the measurements could be in error by as much as 2 cm/sec. Additional error can be induced into the measurements if there is substantial vertical movement of the instrument such as that induced by wave action on the boat.

Continuous current measurements were made at one station approximately 45 m south of the quarters section of platform 296-B with a multidepth array using three ENDECO Type 105 Current Meters. The ENDECO Type 105 Current Meter is an axial flow, ducted impeller instrument. Analog values of impeller rotation, magnetic bearing of the instrument, and a 24-hour mark comprise the data that are recorded on the 16-mm film. Each meter was calibrated by ENDECO just prior to the study. Direction calibrations were made by comparing indicated directions with real magnetic headings every 15 degrees of swing. Speed calibrations were conducted using a synchronous motor attached to the tail-shaft. The threshold speed for the impeller is 2.5 cm/sec with a resolution of 2.5 cm/sec and an accuracy of ±3% of full scale. Current direction accuracy is ±5° at threshold speed, ±3.6° above threshold speed, and is resolvable to ±3.6°.

One meter for each depth was placed at 4.5, 10.5, and 18 m below the surface. The bottom current meter was approximately 2 m above the sea floor. The mooring was anchored with a 200 kg weight and the mooring line was supported by a subsurface float. A 60 m ground line ran from the 200 kg weight to the leg of platform 296-B to facilitate retrieval of the array.

Following retrieval of the instruments, the data films were sent to ENDECO for processing. The processed data consisting of speed and direction values for each 30-minute interval during the recording period were entered onto the computer system. Tables of current speed and direction, joint frequency distribution tables, progressive vector plots, and current speed and direction plots were generated from the data.

Meteorological Measurements

Meteorological data were collected continuously for approximately 30 days during the first and third field studies with a Meteorology Research Incorporated Model 1071 Mechanical Weather Station. This device was mounted 30 m above the sea surface on platform 296-B. Analog values of wind run and wind direction were continuously recorded. The threshold of the instrument was 0.34 m/sec for both the vane and cup anemometer. The speed was recorded with an accuracy of ±2% of the measured value and the direction with an accuracy of ±3.6° and a resolution of 15°.

The data were returned to the laboratory and converted to hourly average values from the analog record. The results were then entered into the computer data base. Tables of wind speed and direction, joint frequency distribution tables, and progressive vector plots were then generated from the data.

Periodic meteorological observations were made from aboard the vessel GUS III during the study to supplement the continuous observations recorded on platform 296-B. Wind speed was measured using a hand held Dwyer Wind Gauge and the wind direction was determined from a magnetic compass and corrected to degrees from true north. Wet bulb and dry bulb temperatures were measured with a Bendix psychrometer. Cloud cover, wave height, and wave direction were estimated from visual observations.

Wave Measurements

Wave measurements were made with a Bass Engineering Model WG/100M self-contained wave measuring and recording system that was installed in approximately 5 m of water on well jacket 288-18. The instrument sensed pressure fluctuations with a Bourdon tube pressure transducer whose signal was transformed with an optical lever system to produce a variable voltage output. The operation of the optical lever system is described in detail by Bass and Byrnes (1974). This system determined water surface variations with a precision of ±6 mm and a resolution of ±3 mm. The timing was controlled by a crystal clock that had an accuracy of ±0.01%. The wave field was sampled every 4 hr for a 9.6 min interval during which time water height measurements were taken every 1.0 sec. The data were recorded on a magnetic cassette that was later decoded and entered onto the computer system.

The data were first automatically scanned to remove all bad characters from the data and also to check for proper timing sequence that precedes each data set. The resulting data sets were then detrended and the mean was subtracted, which left only the pressure fluctuations about a zero mean. The residual pressure readings were subsequently plotted and examined visually to remove any outlying points. All erroneous points were replaced by a linear interpolation of the two adjacent points. The clean data sets obtained after the editing process were used for subsequent analyses.

The consecutive zero-up-crossing method was used to analyze the wave data. This method defined the point where the water-level signal changed from negative to positive as the beginning of a wave and the next zero-up-crossing as the end of the wave. Each wave height was determined by calculating the difference between the maximum and minimum water level values between consecutive zero-up-crossings, and the wave period was simply the time interval between up-crossings. The entire data set for each 9.6 min recording interval was analyzed in this manner and, typically, 100 waves were tabulated. These waves were then sorted according to wave height and the highest one-third were averaged to determine the significant wave height for that recording interval. The periods of the highest one-third waves were then averaged to determine the significant wave period.

Since the wave gauge was located 5 m below the surface and since the pressure signals from water-level fluctuations decrease with depth in the water column, the measured wave height values were corrected to estimate the actual wave heights at the water surface. The corrections for this attenuation of the pressure fluctuations with depth were made as described by Kim and Simons (1974). The wave height at the surface, H, is related to the measured wave depth, H_z, by the equation:

$$H = \frac{\cosh (KD)}{\cosh K(D-Z)} H_z \qquad (1)$$

where: D = total water depth
K = wave number
Z = depth of the sensor

The wave number was determined implicitly from the dispersion equation:

$$\omega^2 = gK \tanh (KD) \qquad (2)$$

where: ω = wave frequency = 2π/period
g = acceleration due to gravity

The procedure followed was to first determine H_z from the zero-up-crossing method and then compute H from Equation 1. Equation 2 was used to calculate K, which used the wave period determined from the zero-up-crossing routine. An estimation of the maximum value of the elliptical velocity of the wave field at the bottom was then made by using a variation of Equation 1 to compute the

TABLE 2. DATA COLLECTION SUMMARY INDICATING DATA TYPE, CRUISE, AND NUMBER OF STATIONS FOR EACH DATA TYPE.

	July 1978	October 1978	February 1979	May 1979
Hydrography				
Conductivity	11	11	11	11
Salinity	11	11	11	11
Temperature	11	11	11	11
Dissolved oxygen	11	11	11	11
pH	11	11	0	11
Transmissivity	11	11	11	11
Total suspended solids	2	2	2	2
Current measurements				
Continuous	1	-	1	-
Electromagnetic	7	6	7	6
Wave measurements	1	-	1	-
Meteorological measurements				
Continuous	1	-	1	-
Shipboard	11	11	11	11

theoretical amplitude of the motion at the bottom, H_b. The amplitudes of the elliptical velocities were then estimated according to:

$$\text{Orbital Velocity (OV)} = \frac{1}{2}\,\omega H_b = \pi H_b/\text{Period} \qquad (3)$$

A similar method for computing orbital velocities is described in detail by Kinsman (1965).

The final result of wave record analysis was a tabulation of significant wave height, wave period, and OV estimations near the bottom recorded every 4 hr while the instrument was in operation.

RESULTS AND DISCUSSION

The results of the data collected during the four cruises (25-27 July and 25-26 October 1978 and 14 February and 8-9 May 1979) are presented in the following section. The order of presentation is as follows: hydrography (including salinity, temperature, dissolved oxygen, pH, and transmissivity), currents, total suspended solids, and wind and waves. A summary of the parameters that were collected during each cruise and the number of stations sampled is presented in Table 2.

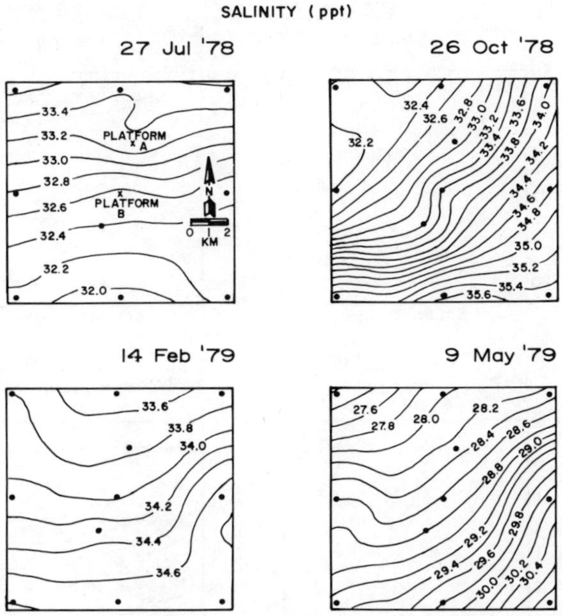

Figure 2. Horizontal contour plots of surface salinity in the Buccaneer Field for July and October 1978 and February and May 1979.

Hydrography

The salinity in the area surrounding the Buccaneer Field generally increased to the south (Figure 2). This trend has been reported by Capurro and Reid (1972) and is attributed to fresh-water runoff from the land surrounding the Gulf of Mexico (Capurro and Reid, 1972; Temple et al., 1977; Martin, 1977). Most of this freshwater flows west and may freshen the Gulf of Mexico as far south as the Mexican border. An exception to this general salinity distribution pattern was observed during July 1978 as the water was more saline to the north. This reversal in the salinity gradient was probably caused by local currents altering the salinity distribution as previous observations indicate that the currents in July are generally to the northeast rather than the predominant southwesterly direction (Armstrong and Hamilton, 1979).

Although the salinity trends for October 1978, February 1979, and May 1979 all revealed an increase in salinity to the south, the magnitude of the salinity gradient and the average salinity varied from season to season. During October, the surface salinity ranged from 32.2°/oo in the north to 35.6°/oo in the south, which is an increase of 3.4°/oo across 6 nautical miles

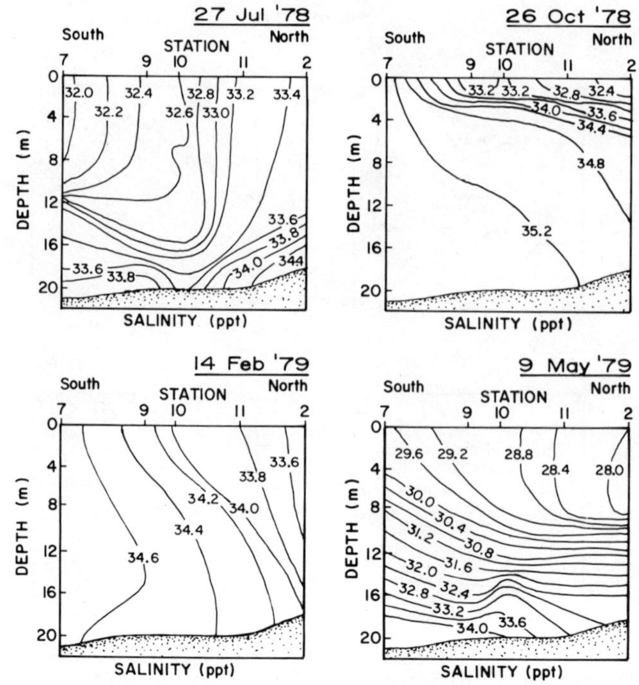

Figure 3. Vertical contour plots of salinity of a north-south cross section through the middle of the Buccaneer Field for July and October 1978 and February and May 1979.

(11 km). Not only was the salinity gradient reversed in July 1978, but the magnitude of the gradient was only one third as large as the salinity gradient observed in October. The July salinities ranged from 32.0^{o}/oo in the south to 33.4^{o}/oo in the north.

The surface salinity in February 1979, during the low runoff period, ranged from 33.5^{o}/oo in the north to 34.8^{o}/oo in the south. However, the May 1979 salinity, during the high runoff period, ranged from 27.2^{o}/oo in the north to 30.9^{o}/oo in the south. The average surface salinity in May of 29.0^{o}/oo was well below the February average salinity of 34.2^{o}/oo. In fact, the May 1979 average surface salinity was approximately 4-5°/oo lower than the average surface salinities determined from the July and October 1978 and February 1979 sampling periods. This 4-5°/oo decrease in surface salinity has been reported for the month of May by Temple et al. (1977) and was attributed to increased runoff just prior to, and during, the month of May. Temple also reported that these marked decreases in salinity were related to peak outflow from the Mississippi River, but with a time lag of 1-2 months. This is quite plausible as the coastal freshwater

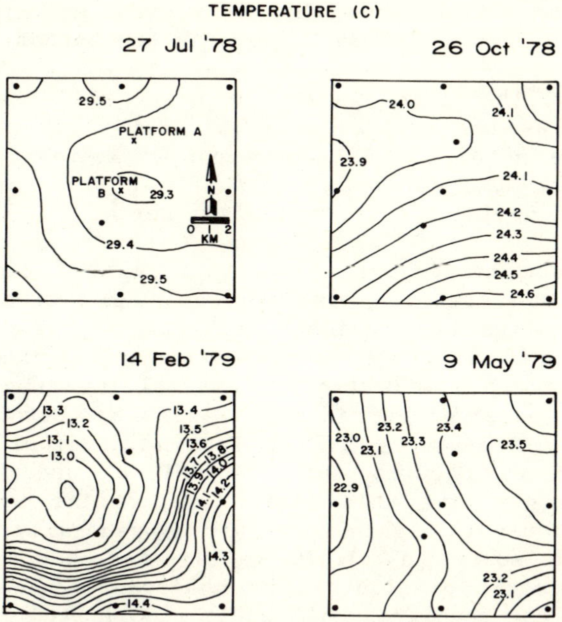

Figure 4. Horizontal contour plots of surface temperature in the Buccaneer Field for July and October 1978 and February and May 1979.

generally flows west along the Gulf Coast (Capurro and Reid, 1972). This report is substantiated by various studies which indicate that the currents generally flow in a westward direction (U.S. Department of Commerce, 1973; SEADOCK, 1975). These currents would have the tendency to force the freshwater to closely follow the shoreline, which would also explain the rather large salinity gradients observed across the 11 km study area.

Vertical cross sections through the middle of the study area (from Station 2 south to Station 7) shown in Figure 3 revealed varied salinity distributions. The July 1978 salinity contours were apparently disrupted by local currents which caused the reverse trend in the horizontal salinity gradient. The northeasterly currents also produced a confluence of water masses at the site as the fresh surface water from the south mixed with the more saline bottom water from the north.

During October 1978 the fresher water edge intruding from the north basically affected only the top 4 m of water. This was not the case during the winter (February 1979) as there was very little vertical stratification of salinity. The salinity, however, did increase to the south throughout the water column. Intense vertical stratification was evident at the depth interval from 8 m to the bottom during May 1979. The salinity ranged from 27.2^{o}/oo at the surface to 34.0^{o}/oo near the bottom.

Generally, the bottom waters did not vary in salinity seasonally as much as the surface waters. This observation was also made by Temple et al. (1977) who reported average bottom salinities of between 33.0^{o}/oo and 35.0^{o}/oo. This was very nearly the range observed during the four sampling periods.

The surface temperature distributions were far more complex than the observed salinity patterns as illustrated in Figure 4. The north-south trends so evident with salinity, and the result of freshwater runoff from the land, were nearly nonexistent at the surface. However, a weak trend of decreasing temperature toward the north was observed at mid-depth and near-bottom for July and October 1978 and February 1979. This latitudinal variation has been reported by Capurro and Reid (1972) and Temple et al. (1977). The only time this trend was distinct near the surface was during October 1978 when temperatures generally decreased from 24.7°C in the south to 23.8°C in the northwest. The largest surface temperature gradients were observed in February 1979 and were probably the result of cold water runoff from land during the winter months.

The surface temperatures also showed the effects of seasonal cooling as the average temperature of 24.2°C in October was approximately 5.5° cooler than the surface temperatures in July 1978. The average surface temperature in July of 29.5°C was 0.5° above the historical average temperature for July (U.S. Department of Commerce, 1973). The average surface temperatures in October 1978 (24.2°C) and February (13.7°C) and May 1979 (23.2°C) were below the historical surface temperatures of 27.0, 18.4, and 24.5°C, respectively (U.S. Department of Commerce, 1973). However, none of these temperatures were below the reported minimum temperature for their respective months.

It should be noted that the average temperatures calculated for the four cruises were each based on a single day's observations and that, due to the nonsynoptic nature of the survey (usually 10 hr), some diurnal warming and cooling in the surface water may have occurred.

The vertical cross sections running north to south through the Buccaneer Field revealed varied temperature distributions (Figure 5). During the July 1978 cruise the temperature was nearly constant throughout the water. A slight temperature inversion (approximately 0.3° warmer at the bottom) was evident during October 1978. Temperature inversions for the top 20 m of water are normal for the months of January, February, and November according to the U.S. Department of Commerce (1973). The apparent

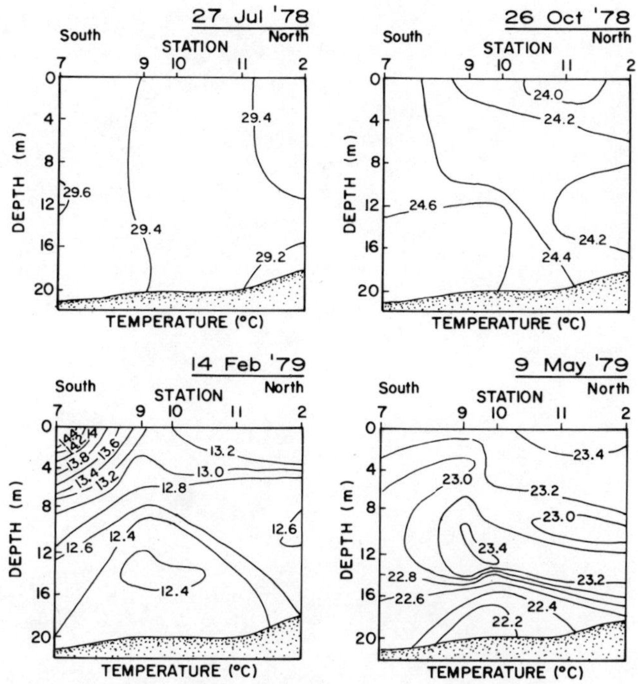

Figure 5. Vertical contour plots of temperature of a north-south cross section through the middle of the Buccaneer Field for July and October 1978 and February and May 1979.

instability (due to temperature) of the water column in October was offset by a very fresh layer of less dense surface water.

The temperature during February 1979 generally decreased with depth in the southern part of this study area. There was apparently an intrusion of warmer surface water entering the study area from the southeast. During May 1979 the vertical distribution of temperature was much more complex. A fresher, colder (yet less dense) layer of water intruded from the north over the top of warmer, more saline water. This resulted in the vertical temperature distribution near platform 288-A of warm surface waters (23.4°C) which decreased to 23.0°C at 9 m; the temperature then increased to near surface temperatures at 13 m before dropping to 22.4°C near the bottom.

The dissolved oxygen (DO) concentrations in the Buccaneer Field generally ranged from 4.0 to 6.0 mg/l. The exception to this occurred in February 1979 when water temperatures of approximately 13°C were at least 10° cooler than the temperatures measured during the other three cruises. These lower temperatures resulted in DO concentrations ranging from 8.1 to 9.1 mg/l.

Currents

Previous studies of currents in the Gulf of Mexico indicate that the nearshore currents along the coast of Texas are predominantly to the west and south with a reversal in May through August with flows to the northeast (U.S. Department of Energy, 1978; Armstrong and Hamilton, 1979). The general seasonal circulation patterns for the area of the Buccaneer Field are given in Figure 7 (from SEADOCK, 1975). These circulation patterns, however, should only be considered as general indications of the expected currents as the actual direction of flow is quite variable (U.S. Department of Commerce, 1973). The currents are influenced not only by the large scale circulation in the Gulf, but also by regular tidal changes, changes in the wind, river discharges, density gradients, and meanders from deep water currents, all of which create a variable current system. The vertical variation of the currents in the area also seems variable and unpredictable (U.S. Department of Energy, 1978) with frequent reversals in flow within various density layers throughout the water column. Drift buoy measurements in the area in 1976, however, indicated that there was very little vertical variation in the currents (Armstrong and Hamilton, 1979); simultaneous measurements with current meters on platform 288-A, though, indicated that there were periods of layering with contrasting directions of flow. Such discrepancies in results indicate that either the water currents in the area are highly variable or the current meter measurements may have been influenced by the proximity of the platform.

The average current speeds that have been reported for the area generally ranged from 10 to 30 cm/sec. The maximum speeds measured in the general area ranged from 53 cm/sec (U.S. Department of Energy, 1978) to 180 cm/sec (Armstrong and Hamilton, 1979) with speeds much farther offshore reaching in excess of 250 cm/sec (U.S. Department of Commerce, 1973). Spectral analysis of current meter data (Armstrong and Hamilton, 1979) indicated that the semidiurnal and diurnal tidal frequencies were the major periodic components of the velocity field. Other energy concentrations were observed at periods of 5 and 10.5 days.

The water current observations that were made during the 1978-1979 studies by Hazleton Environmental Sciences generally agreed with many of the previous observations; however, there were several discrepancies in the results. Progressive vector plots (PROVECS) of the current meter data from 26 July to 30 August 1978 (Figure 8) indicated that the net drift during this period was to the southwest at all depths. Most previous studies had indicated that the flow was to the northeast during this season. The PROVECS also indicated that the direction of flow was quite uniform with depth, which disagrees with previous observations that reported frequent periods of layered flows during the summer

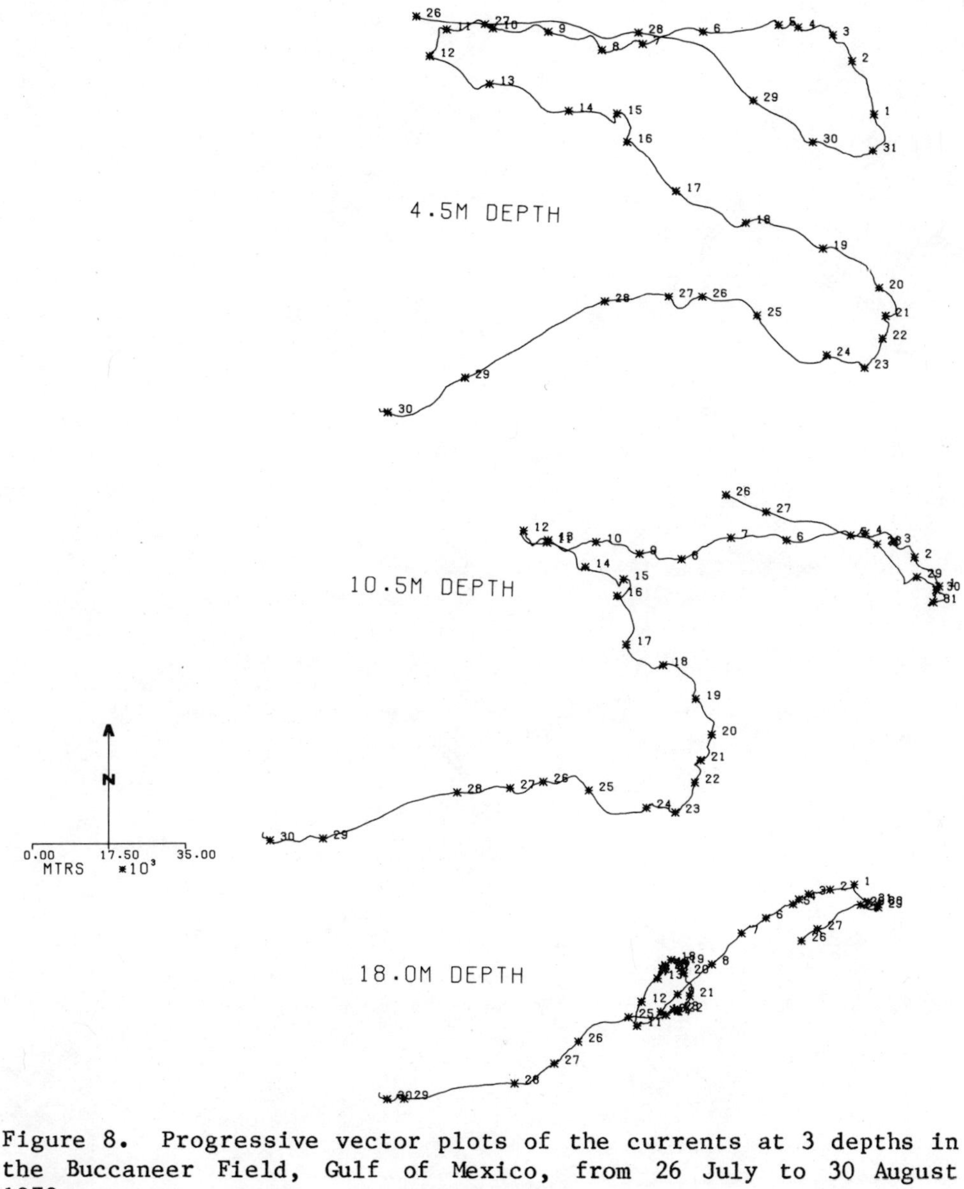

Figure 8. Progressive vector plots of the currents at 3 depths in the Buccaneer Field, Gulf of Mexico, from 26 July to 30 August 1978.

(Armstrong and Hamilton, 1979). The average current speeds varied from 18 cm/sec near the surface to 7.7 cm/sec near the bottom. The maximum recorded speed during the July-August recording period was 66.0 cm/sec which occurred on 28 August during tropical storm Debra. The current speed near the bottom exceeded 26 cm/sec more than 2% of the time; this speed is sufficient to resuspend and

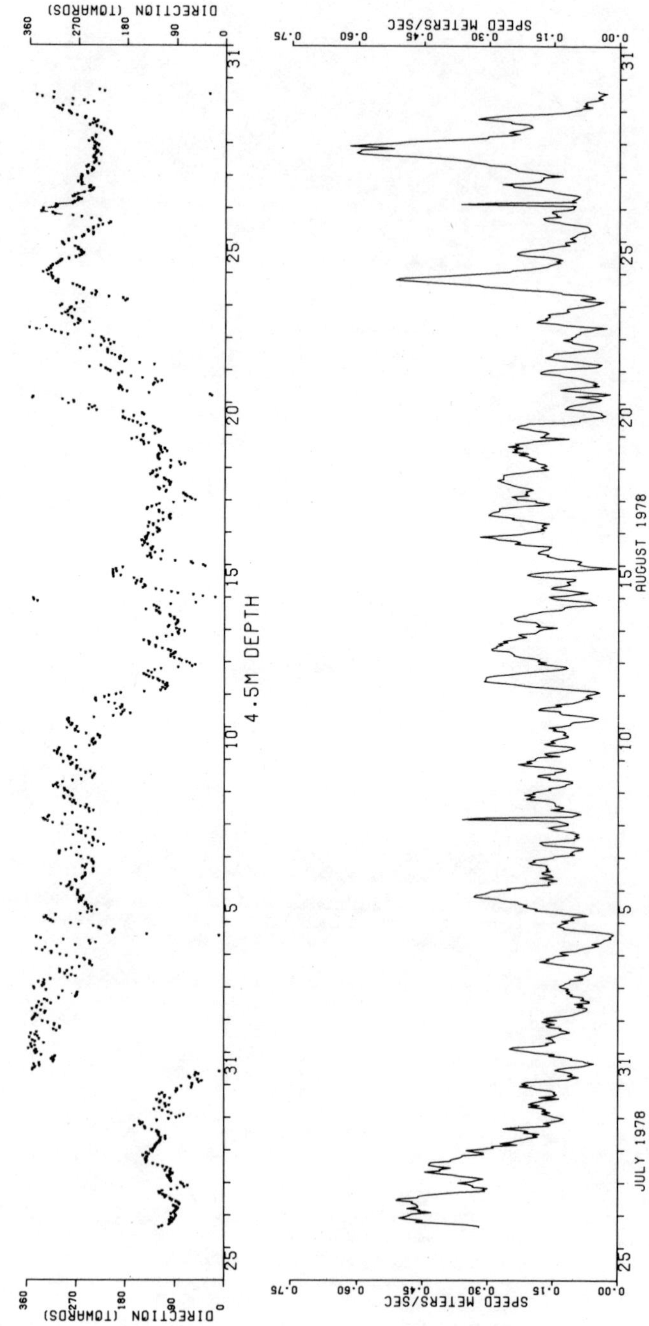
DIRECTION (TOWARDS)
SPEED METERS/SEC
4.5M DEPTH
JULY 1978
AUGUST 1978

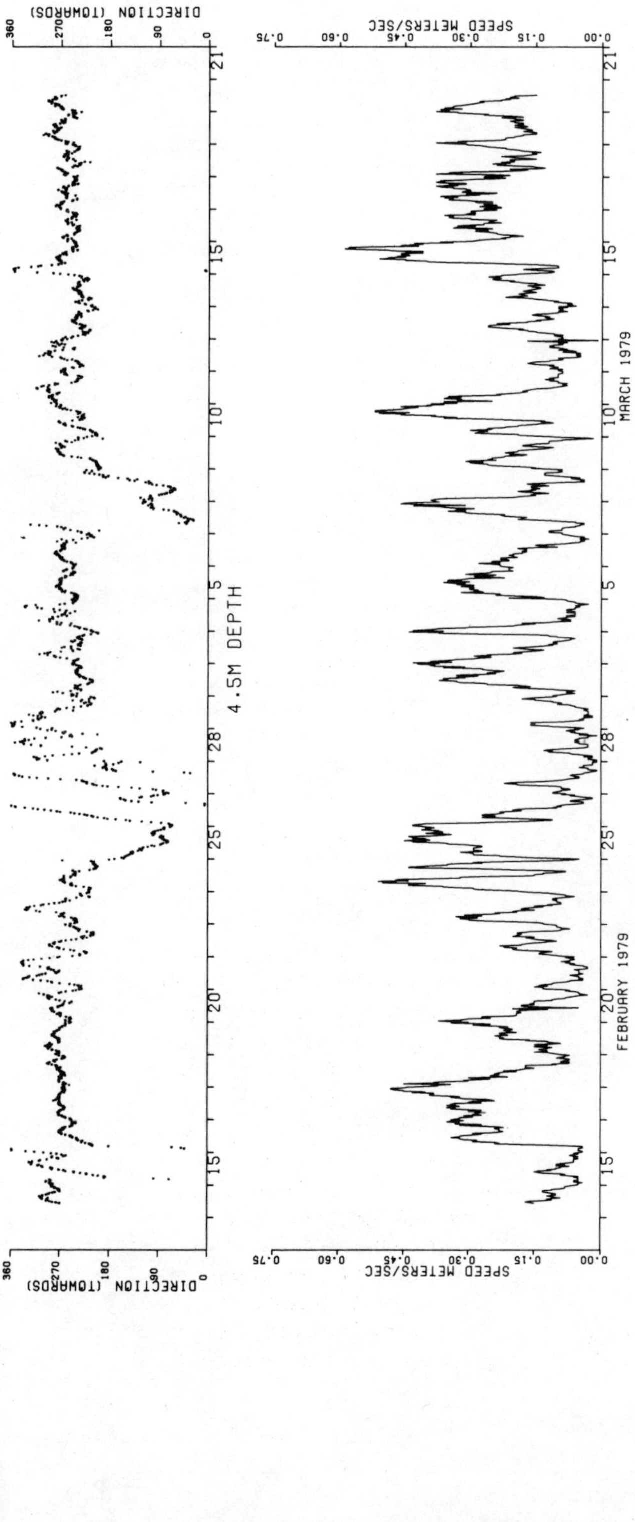

Figure 10. Current speed-direction plots for the 4.5 m depth in the Buccaneer Field, Gulf of Mexico, from 26 July to 30 August 1978 and from 14 February to 20 March 1979.

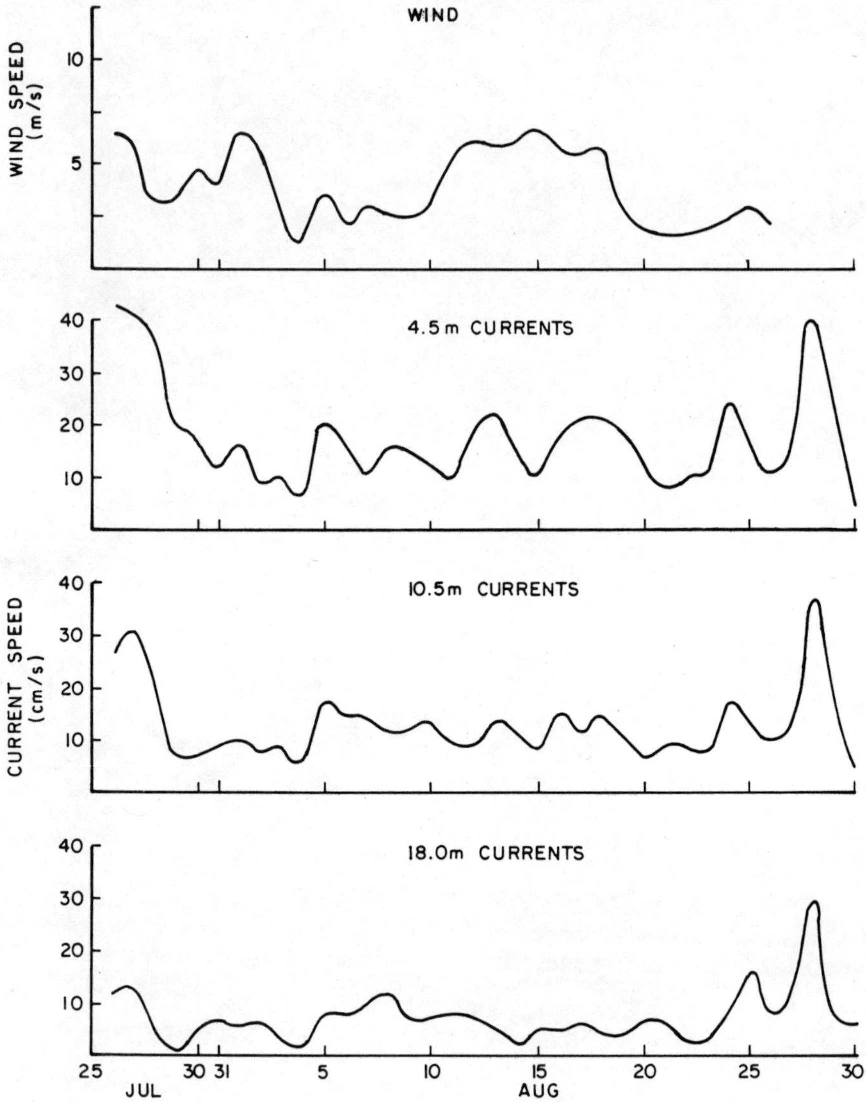

Figure 12. Daily mean wind and current speeds for the Buccaneer Field from 26 July to 30 August 1978.

the shift in wind, the water was released. The release of water coupled with the change in wind direction produced the easterly currents and the increase in current speed. Consequently, all the major increases and decreases in the observed current speeds could be related directly to changes in the wind. A similar response to changing wind direction was observed in November 1976 (Armstrong and Hamilton, 1979) as current speeds up to 180 cm/sec were reported.

A comparison of the wind and current speeds measured during the July-August 1978 recording period is presented in Figure 12. The correlation between the wind and currents is not as apparent in the results. A closer examination of the original wind record, however, suggested that there was considerable interference from the platform on the measured wind field which may account for the poor comparison. The instrument was moved for the February 1979 measurements to eliminate the interference. The results from the current meters again show very good correlation between depths except for a general decrease in speed with depth. The large increase in speed on 28 August was the result of tropical storm Debra which produced currents as high as 66 cm/sec.

The electromagnetic current meter measurements taken during the four field studies also indicated that the current direction was generally uniform with depth and that the current speed typically decreased with depth. There was a reversal in flow between upper and lower layers on 26 October 1978, but this was the only time that such a velocity shear was observed.

A comparison of the electromagnetic current meter measurements with the data from the continuous recording current meters indicated that there was very good agreement in the measured directions of flow. Most measurements at comparable depths agreed within 20°, and the maximum difference was only about 40°. Considering that the continuous recording meters measured an hourly average and the vertical profiles were instantaneous, the agreement was quite good. The measured current speeds did not agree as well as the continuous measurements. The continuous measurements, on the average, were about 31% lower. The higher values from the electromagnetic current meter measurements were probably caused by vertical and horizontal motions of the boat during measuring and also by interference from wave orbital velocities. Consequently, the vertical profile measurements appeared to be biased, somewhat, toward higher values.

Total Suspended Solids

The total suspended solids (TSS) measurements made during the hydrographic sampling indicated that the TSS values that were recorded were generally less than 5.0 mg/l with values frequently being less than 1.0 mg/l. The highest value that was measured was 12.0 mg/l and occurred in July 1978. The lowest values occurred in October 1979 with most of the measurements being around 1.0 mg/l. In general the measurements taken during the hydrography surveys were higher than those reported for the Surficial

during February as the values exceeded 20 cm/sec 9% of the time as compared to only 4% for July 1978. The maximum computed orbital velocity was nearly 40 cm/sec which is sufficient energy to readily resuspend unconsolidated sediments (Weller, 1960) in the range of 0.05 to 2 mm (coarse silt to coarse sand). The historical wave data indicate that wave heights can reach 7 m for this area. Assuming a wave period of 8 sec for these waves, an orbital velocity of well over 1 m/sec at the water sediment interface is possible. The available wave energy in conjunction with the water currents indicate that most unconsolidated sediments or particulate contamination from the Buccaneer Field can readily be flushed from the area especially under storm conditions.

CONCLUSIONS

The results of the currents and wave measurements indicated the bottom sediments in the vicinity of the Buccaneer Field are subject to considerable erosional stress. Theoretical calculations of wave orbital velocities that exist at the water-sediment interface indicate that during the two periods of wave observations the scouring velocity at the bottom exceeded 20 cm/sec about 7.5% of the time. The maximum calculated speed near the bottom was nearly 40 cm/sec. Historical wave information for the area indicates that waves greater than 7 m occur in the area which can readily generate orbital velocities at the bottom in excess of 1 m/sec. The currents, however, are the primary source of sediment resuspension and transport as the current speeds near the bottom exceeded 26 cm/sec about 3.5% of the time during the two periods of observation. Previous measurements in the area showed the currents reached as high as 180 cm/sec, and historical information indicates the currents can exceed 2 m/sec. These results reveal that most unconsolidated sediments and particulate matter resulting from oil production will be periodically flushed from the area. The results further indicate that the resuspended material will generally be transported to the southwest although there is considerable variability in the direction of flow. The average rate of transport is about 2 g/sec per meter of water surface. This transport rate, however, can increase dramatically during storm conditions.

The water currents in the area were influenced primarily by the wind and tides. Both diurnal and semi-diurnal tidal periods were apparent in the current meter records. The wind, however, was the dominant factor controlling the currents as nearly all major changes in current speeds could be related to changes in wind speed or direction. Even though increasing wind speed generally produced increasing current speeds, the currents did not always flow in the same direction as the wind, which indicates the shoreline and bathymetry were also an important influence on the currents.

The hydrographic surveys near the Buccaneer Field indicated horizontal salinity variations as great as 3.4°/oo across the study area, which is typical for this location. The fresher water was usually to the north and was the result of fresh water runoff from sources as far away as the Mississippi River that generally flow southwest along the coast. In July, however, the salinity gradient was reversed with fresher water to the south of the study area. This reversal was the result of a changing current system as the expected currents for July are to the northeast while during the other seasons the expected flow is to the southwest. The northeast flow in July apparently upset the general trend of southwesterly flow of freshwater along the coast and produced the reversal in salinity gradients. There was also considerable vertical variation in the salinity, especially in the spring, as fresher surface waters penetrated the study area from the north.

This one-year study of the currents and hydrography near the Buccaneer Gas and Oil Field indicated that the area can best be described as a mixing zone for coastal and offshore waters. The area is also a high energy regime with scouring from waves and currents with a potential for sedimentation during the more quiescent periods. The changing currents which are influenced by the tides, winds, and shoreline, continually change the location of the mixing zone. Consequently, the extent of the penetration of the fresh coastal water and, therefore, the distribution of salinity and temperature are quite variable. This variability in conjunction with the normal seasonal changes results in a highly variable hydrographic environment.

ACKNOWLEDGEMENTS

This work is a result of research sponsored by the Environmental Protection Agency and Department of Commerce, National Oceanic and Atmospheric Administration, National Marine Fisheries Service, Southeast Fisheries Center, Galveston Laboratory under Contract No. 03-78-D08-0038.

REFERENCES

Armstrong, R. S., and Hamilton, R. C., 1979, "Describe seasonal circulation patterns in the oil field," in: Jackson, W. B. (ed.), "Environmental Assessment of an Active Oil Field in the Northwestern Gulf of Mexico, 1977-1978, Vol. III: Chemical and Physical Investigations." NOAA, pp. 2.3.9-1 to 2.3.9-250. Available from: NTIS, Springfield, Virginia, Accession No. PB80107899.

Bass, J. C., and Byrnes, R. M., 1974, "Precision electro-optical wave and tidal gauge," A.S.C.E. Symp. on Ocean Wave Measurement and Analysis, New Orleans, 2:190.

Capurro, L. R. A., and Reid, J. L. (eds.), 1972, "Contributions on the Physical Oceanography of the Gulf of Mexico." Gulf Publishing Co., Houston, Texas, 288 pp.

Kim, Y. Y., and Simons, L. H., 1974, "Sea state measurements from pressure records." A.S.C.E. Symp. on Ocean Wave Measurement and Analysis, New Orleans, 1:40.

Kinsman, B., 1965, "Wind Waves: their Generation and Propagation on the Ocean Surface." Prentice-Hall, Englewood Cliffs, New Jersey, 676 pp.

Martin, J. A., 1977, "Hydrography of the Buccaneer Oil Field and adjacent waters," in Jackson, W. B. (ed.), "Environmental Assessment of an Active Oil Field in the Northwestern Gulf of Mexico, 1976-1977." NOAA, pp. 575-632. Available from: NTIS, Springfield, Virginia, Accession No. PB283890.

SEADOCK, Inc., 1975, "Seadock Environmental Report for Texas Offshore Crude Oil Unloading Facility." Houston, Texas.

Sverdrup, H. U., Johnson, M. W., and Fleming, R. H., 1946, "The Oceans: their Physics, Chemistry and General Biology." Prentice-Hall, Englewood Cliffs, New Jersey.

Taras, M. J. (ed.), 1975, "Standard Methods for the Examination of Water and Wastewater." 14th ed., Amer. Public Health Assoc., Washington, D.C.

Temple, R. F., Harrington, D. L., and Martin, J. A., 1977, "Monthly Temperature and Salinity Measurements of Continental Shelf Waters of the Northwestern Gulf of Mexico, 1963-1965." NOAA Technical Report NMFS-SSRF-707, 26 pp.

U.S. Department of Commerce, 1973, "Environmental Conditions within Specified Geographical Regions: Offshore East and West Coasts of the United States and in the Gulf of Mexico." U.S. Government Printing Office, 735 pp.

U.S. Department of Energy, 1978, "Final Environmental Impact Statement: Strategic Petroleum Reserve: Seaway Group Salt Domes, Brazoria County, Texas, DOE/EIS-0021, Vol. 3, App. G." Washington, D.C.

Weller, J. M., 1960, "Stratigraphic Principles and Practice." Harper and Row, New York, 725 pp.

HYDRODYNAMIC MODELING OF DISCHARGES

Gerald W. Smedes, Richard P. Herbst,[a] and Jack Calman[b]

Environmental Research and Technology, Inc.
1111 Third Avenue, Suite 2770,
Seattle, Washington 98101

INTRODUCTION

An operational hydrodynamic model is an essential component of the Buccaneer Gas and Oil Field Study, conducted under the auspices of the National Marine Fisheries Service (NMFS). Such a model is needed to quantify the distributions of various kinds of pollutants in space and time in the marine ecosystem, and to predict how these will ultimately impinge upon and be absorbed by the system. Because actual field measurements and monitoring programs are limited in the extent of data which can be obtained, model development allows for reduced data gathering since only a relative small number of points are needed to verify the model predictions. Once a model is shown to be reliable and valid, it can be used to describe the distribution and extent of various parameters associated with the developed well site.

There are many near-field, circulation, and dispersion models which could be adapted to the Buccaneer Field situation. However, initial efforts with some of these models indicated operational difficulties and high costs in computer time. Since these constraints were incompatible with the needs and resources of the NMFS study program for the Buccaneer Field, an alternate approach was pursued. The model we developed was designed to aid in the

[a]Present address: Regulatory Affairs Department, Exxon Minerals Co., P.O. Box 2180, Houston, Texas 77001.

[b]Present address: Applied Physics Laboratory, Johns Hopkins University, Johns Hopkins Road, Laurel, Maryland 20810.

assessment of the environmental impact of pollutants released during routine operations of a producing offshore platform. With appropriate input data, the model could also be applied to short-term, accidental discharges of materials.

Pollutants discharged from a Buccaneer Field platform will be dispersed in different ways, depending upon their solubility, density, and size of particles, among other properties. This model does not attempt to describe in detail the fate of specific pollutants nor their detailed vertical distribution in the water column. The model describes the dispersion of three classes of pollutants, namely: those which mix thoroughly and are distributed vertically in the water column; those which float on the surface of the water; and those which sink to the bottom. The model is designed to operate in a time-sequential mode, using subsurface current data at each time step.

METHODS

Pollutants Distributed in the Water Column

Initial mixing of a pollutant discharged from the platform is affected by the wake created by the platform's structural members. Because of the complicated structure of the platform and flow patterns around it, mixing in the wake can be described only approximately. Dye discharge experiments (Armstrong and Hamilton, 1979) have shown that by the time a dye discharged at the platform emerges on the downstream side, it is mixed in a volume of water approximately equal to 1/8 of the volume occupied by the platform. Thus, the initial concentration of pollutant in the neighborhood of the platform can be estimated as follows: Let Q (g/sec) be the rate of discharge of mass of pollutant, U (cm/sec) be the ambient current, L be the length of the platform, and z the water depth. The time, t, for the pollutant to be swept past the platform is:

$$t = \frac{L}{U} \tag{1}$$

During this time the mass, M, of pollutant discharged is:

$$M = Qt = \frac{QL}{U} \tag{2}$$

This mass of pollutant is mixed initially into a volume, V_o, of water approximately equal to:

$$V_o = \frac{1}{8} L^2 z \tag{3}$$

TABLE 1. NEAR-FIELD DISPERSION PARAMETERS.

	Summer	Fall	Winter	Spring[a]
$\dot{r}$ (m/hr)	54	21	30	42
h (m)	5	7	10	7.5

[a]Measurements were not made in spring. The values listed are averaged summer and winter values.

leading to an initial concentration, χ_o,

$$\chi_o = \frac{M}{V_o} = \frac{8Q}{ULz} \tag{4}$$

For example, if U = 5 cm/sec, L = 50 m, and z = 20 m, then

$$\chi_o = 1.6\text{x}10^7 Q \tag{5}$$

If the rate of discharge, Q, were 100 g/sec, the initial concentration would be 16 ppm.

Observations of the expansion of a patch of dye introduced into the water about 2 km from the platform have revealed seasonal variations in the dispersion pattern (Armstrong and Hamilton, 1979). Although the shape of the dye patch is elliptical, an "equivalent radius" may be defined so that the area, A, of the dye patch is given by $A = \pi r^2$. Some of the observations on the rate of change of the equivalent radius, $\dot{r}$, and of the depth of penetration of the dye, h, are listed in Table 1. If an injection of a dye (or discharge of a pollutant) results in an initial mixed volume V_o of concentration χ_o then, at a later time,

$$\frac{\chi}{\chi_o} = \frac{V_o}{V} \tag{6}$$

Knowing the rate of expansion and depth of penetration, we can estimate

$$V = \pi(\dot{r}t)^2 h \tag{7}$$

so that

$$\frac{\chi}{\chi_o} = \frac{V_o}{\pi(\dot{r}t)^2 h} \tag{8}$$

Combining (3) and (7) we have,

$$\frac{\chi}{\chi_o} = \frac{L^2 z}{8\pi h \dot{r}^2 t^2} \tag{9}$$

an expression valid in any consistent set of units. The dye dispersion observations were made over several hours, say t_o, so the formula should be valid for any time $t \leq 3$ hours.

Downstream from the immediate vicinity of the platform, mixing occurs in the turbulent wake of the platform. This mixing is described by a similarity theory. Taking L as the width of the platform, similarity theory predicts, at downstream distances x such that

$$\frac{x}{L} \geq 80 \tag{10}$$

the half-width ℓ of the wake increases as

$$\frac{\ell}{L} = \frac{1}{4}\left(\frac{2x}{L}\right)^{1/2} \tag{11}$$

(Tennekes and Lumley, 1972). The concentration, χ, of pollutant in the wake is determined from the relationship,

$$\frac{\chi}{\chi_o} = \left(\frac{2x}{L}\right)^{-1/2} \tag{12}$$

(Monin and Yaglom, 1974). With L = 50 m, then at a downstream distance of

$$x = \left(\frac{x}{L}\right) L = 80\ (50\ \text{m}) = 4\text{x}10^3\ \text{m} \tag{13}$$

from the platform, the maximum concentration occurring in the wake of width

$$2\ell = \frac{1}{2}L\left(\frac{2x}{L}\right)^{1/2} = 320\ \text{m} \tag{14}$$

is

$$\chi = 0.08\chi_o \tag{15}$$

With the example discharge rate of Q = 100 g/sec, the concentration at distance x = 4 km downstream in the wake is

$$\chi = 1.3\ \text{ppm} \tag{16}$$

This concentration is very approximate; by the time the pollutant has travelled so far downstream, other effects of wind and settling affect its distribution.

Before these far-field approximations can be used, however, they must be matched to the near-field approximations given above so that the computed concentrations are continuous. The near-field approximations may be written (see equation 8):

$$\hat{\chi}_N = A_s t^{-2} \tag{17}$$

where A_s is a constant which varies with the season and $\hat{x} = x/x_o$. This approximation is valid for several hours. The far-field approximation may be written (from equation 12) as

$$\hat{\chi}_F = C_s (\hat{x})^{-1/2} \tag{18}$$

where C_s is a constant which depends upon the season and $x = x/L$. This approximation is valid after about 4 km from the platform. These solutions must now be matched as follows. Suppose that at some distance $\hat{x}_M$ it is required that the two computed concentrations be the same, that is, $\hat{\chi}_N = \hat{\chi}_F$ at $\hat{x} = \hat{x}_M$

$$A_s t^{-2} = C_s \, \hat{x}_M^{-1/2} \tag{19}$$

or

$$C_s = A_s t^{-2} \, \hat{x}_M^{1/2} = \hat{\chi}_N \, \hat{x}_M^{1/2} \tag{20}$$

Back-substitution gives

$$\hat{\chi}_F = \hat{\chi}_N \left(\frac{\hat{x}_M}{\hat{x}}\right)^{1/2} \tag{21}$$

Combining all results

$$\hat{\chi} = \hat{\chi}_N \qquad \text{for } \hat{x} < \hat{x}_M \tag{22a}$$

$$\hat{\chi} = \hat{\chi}_N \left(\frac{x_M}{\hat{x}}\right)^{1/2} \qquad \text{for } \hat{x} > \hat{x}_M \tag{22b}$$

It remains to determine the matching distance, $\hat{x}_M$. The far-field solution is assumed for $x > 4$ km, and the near-field solution is assumed valid for several hours. For a typical current speed of 6 cm/sec, a particle will have travelled about 650 m, corresponding to a nondimensional distance $\hat{x} = x/L = 13$. From equation 22, it is clear that the far-field concentration estimate is always smaller than that which would be computed using the near-field formula. To be conservative, then, we choose $\hat{x}_M$ (which must be between 650 m and 4 km) to be equal to 650 m, so that concentrations at intermediate distances from the platform will be overpredicted.

During the time in which the pollutant plume is being diluted, it is swept downstream in a continuously changing current. Thus, at each step, the pollutant is advected in different directions over different incremental distances. The model keeps track of this trajectory, and thereby identifies different impact areas which might be affected.

To calculate subsurface pollutant trajectories, the model requires measured (or assumed) values of the speed and direction of subsurface currents (e.g., at 10 m depth) at each time step. If v is the speed and θ the bearing of the subsurface current, then the eastward (Δx_i) and northward (Δy_i) displacements during the ith time interval Δt are given by

$$\Delta x_i = v_i \, \Delta t \, \sin\theta_i \tag{23}$$

$$\Delta y_i = v_i \, \Delta t \, \cos\theta_i \tag{24}$$

The model expresses these displacements as incremental changes in the range ΔR and bearing Δϕ of the pollutant during the ith time step:

$$\Delta R_i = [(\Delta x_i)^2 + (\Delta y_i)^2]^{1/2} \tag{25}$$

$$\Delta\phi_i = \tan^{-1} (\Delta x_i / \Delta y_i) \tag{26}$$

The position of the pollutant relative to the source is given by the net range and bearing:

$$R = \Sigma_i \Delta\vec{R}_i = [(\Sigma_i \Delta x_i)^2 + (\Sigma_i \Delta y_i)^2]^{1/2} \tag{27}$$

$$\phi = \tan^{-1}\left(\frac{\Sigma_i \Delta x_i}{\Sigma_i \Delta y_i}\right) \tag{28}$$

The values for each incremental and net range and bearing are computed and printed for each ith time.

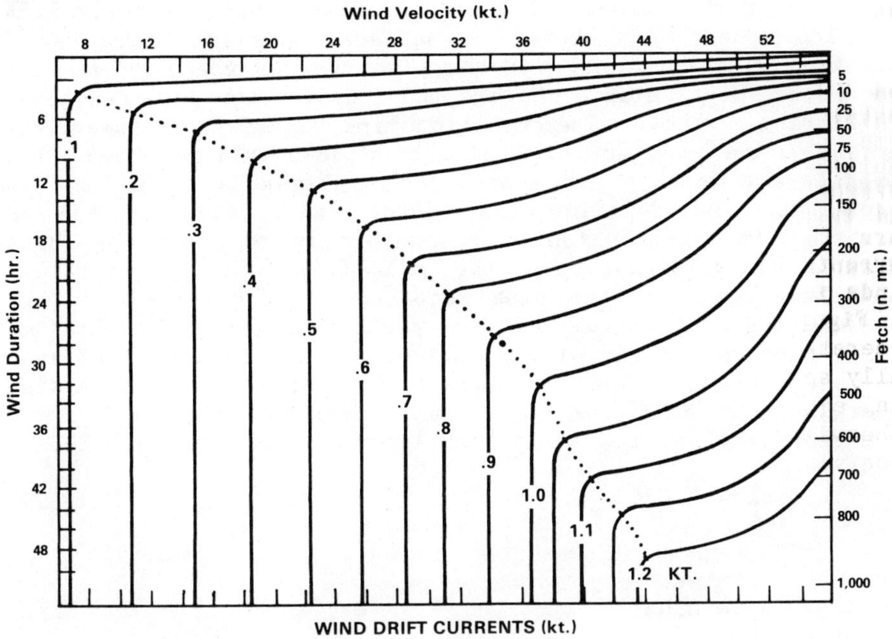

Figure 1. Relationship of wind velocity, duration, and fetch to wind drift currents (after James, 1968).

Floating Pollutants

Surface currents are often different from subsurface currents, so the trajectory of floating pollutants is calculated independently. The trajectory of the floating pollutants is tracked in the same manner as the trajectory of the vertically mixed pollutants. The essential difference involves relating the observed wind to the surface drift current. Floating pollutants are subject to the direct action of the wind. They are carried along by the wind-driven surface currents, which do not penetrate throughout the depth of the water column. The direction of movement is assumed to be directly with the wind. The model must describe the trajectory of a mass of floating pollutants as the wind changes speed and direction over time.

The procedure sequence used in the model to calculate floating pollutant trajectories is as follows: first, the size and number of time intervals for which the trajectory is to be computed are specified; second, the speed and direction of the wind at each time step is read in: the program then computes the corresponding drift current induced by the wind, displaces the mass of pollutant accordingly, and keeps track of the total movement of the pollutant mass.

Wind drift currents will attain a speed approximately 3.5% of the wind speed 10 m above the surface, provided there has been sustained wind movement. If the wind changes speed or direction, some time must elapse before the drift current reaches its equilibrium value. The relationships among wind speed, drift currents, and wind duration have been described by James (1968), and relevant results are reproduced in Figure 1. (It is assumed here that wind duration, not fetch, will dominate the drift current. This assumption is reasonable for most of the observed winds in the western part of the Gulf of Mexico.) The dotted line in Figure 1 denotes the time required for a given wind field to generate its equilibrium drift current, T_e. The program automatically accounts for the wind duration: if it has been long enough, the equilibrium value of the drift current v_D is assigned; otherwise the drift velocity is reduced by the ratio of the observed duration, T_w, to the equilibrium duration, T_e. Thus

$$v_D = 0.035\, v_w, \qquad T_w > T_e \tag{29}$$

$$v_D = 0.035\, v_w \frac{T_w}{T_e}, \qquad T_w < T_e \tag{30}$$

If the wind changes speed, the following method described by James is used. If the wind increases, the program computes the time it would take for the higher wind speed to generate the drift current existing at the time of the increase. This gives an "equivalent duration" period for the higher wind speed, and the computation proceeds as before. If the wind speed decreases, it is assumed that the drift current immediately drops to the equilibrium value corresponding to the lower wind speed. Changes in wind direction are handled by treating the two orthogonal components of wind speed separately in the manner described above.

The initial discharge is assumed to be confined to within one meter of the surface. As in the case of vertically mixed pollutants, it is assumed that the initial mixing occurs in a volume equal to 1/8 of that occupied by the platform (to 1 m depth).

Settling Particles

The purpose of modeling the sediment settling velocities is to estimate how far from the source particles of various sizes will be carried by the current before they are deposited on the sea floor. The model assumes that the small sediment particles are simply carried along by the horizontal current while they are settling towards the bottom. The horizontal distance, x_s, travelled by the sediments before settling to the bottom is

$$x_s = Ut_s \tag{31}$$

where t_s is the time required to settle to the bottom and U is the ambient current. The time, t_s, for settling is

$$t_s = \frac{D}{W_s} \tag{32}$$

where D is the water depth and W_s is the settling velocity.

The settling velocities for small particles follow Stoke's law,

$$W_s = \frac{gd^2}{18\nu}(\delta_s - \delta_w) \tag{33}$$

where g is the gravitational acceleration, ν is the kinematic viscosity of water, d is the diameter of the sediment particles, and δ_s and δ_w are the specific gravities of the sediment and seawater, respectively. Taking $g = 981$ cm/sec^2, $\nu = 1.31\text{x}10^2$ cm^2/sec (at 10°C), and representative values of $\delta_s = 2.65$ (sand, silt), and $\delta_w = 1.025$ (10°C, 33°/oo) we find

$$W_s \simeq 6.8\text{x}10^3\ d^2 \tag{34}$$

in cgs units. A particle of diameter 0.012 cm will have a settling velocity of 1 cm/sec. For larger particles, the approximations upon which equation 34 is based are not totally reliable.

In the model calculations, the values of ν, δ_s, and δ_w may be specified, and equation 33 is used; otherwise, the default value incorporated in equation 34 is used to compute settling velocities. Unless d is specified, the program computes w_s, t_s, and x_s for particle sizes from 0.001 to 0.015 cm in diameter.

Model Output

Output from the model includes calculations for the concentration and trajectory of vertically mixed pollutants at each time step, trajectories of the floating and settling pollutants, and a summary table. A computer graphics software package has also been used, to generate graphs of wake width, concentrations of vertically mixed pollutants as a function of time, and the range and trajectory of subsurface and floating pollutants.

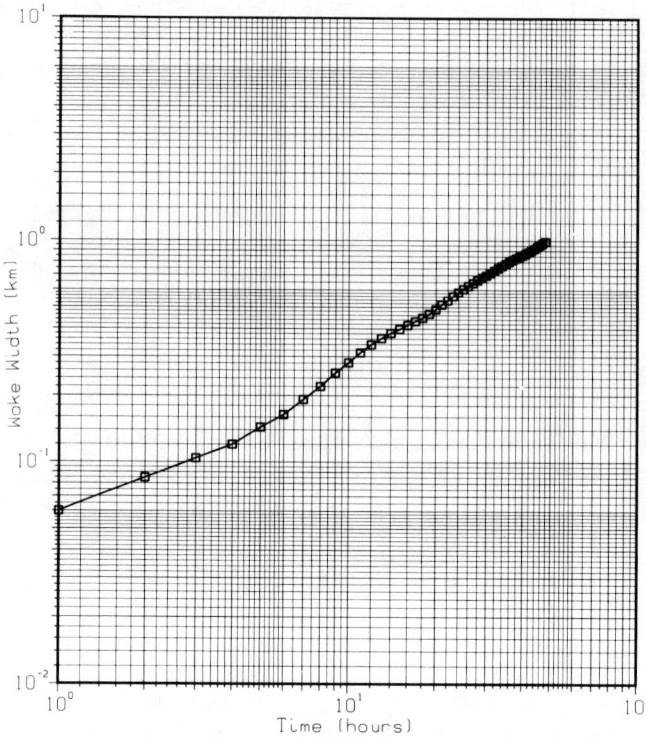

Figure 2. Growth of pollutant wake downstream from platform, February 16, 1979.

RESULTS AND DISCUSSION

The dispersion model for the Buccaneer Gas and Oil Field was run using data for six two-day periods - three each in winter (February) and summer (August). Figure 2 shows the growth of the pollutant wake as vertically distributed pollutants are transported downstream from the platform. This figure indicates that the wake can spread to approximately 1 km in width within 48 hours. No appreciable differences in the rate or extent of wake spreading are indicated between seasons.

Figure 3 shows the decrease in pollutant concentration (nondimensional) over time, for vertically distributed pollutants. Concentrations decrease by a factor of approximately 10^6 in 48 hours, as pollutants are dispersed downstream. Initial turbulent mixing reduces concentrations by 10^2 in the immediate vicinity of the platform, and the remainder of the dispersion occurs as currents carry the pollutants downstream. The overall patterns are similar for all cases examined. Graphs showing reduction of concentration over range (not shown) also indicate a

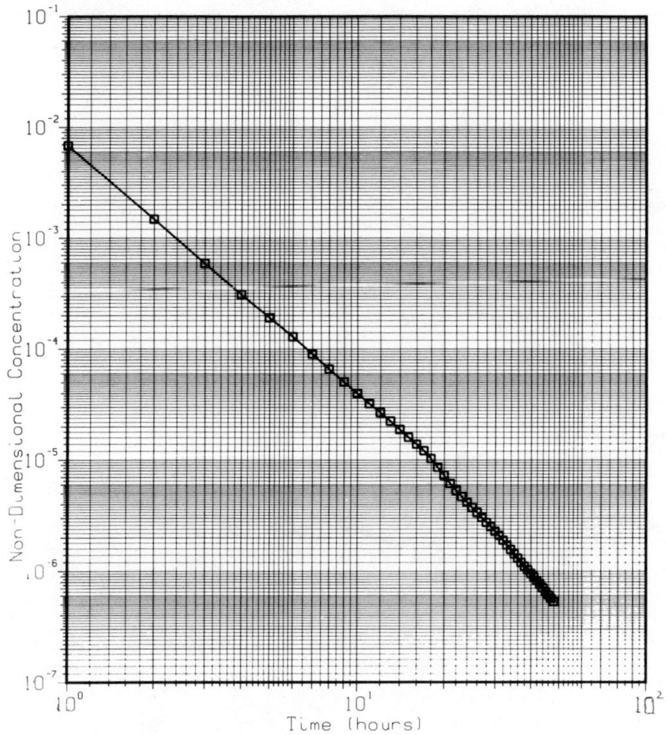

Figure 3. Concentration of vertically distributed pollutants over time, August 4, 1978.

regular rate of reduction, but the total distance transported varies more among the specific cases examined.

The trajectories of subsurface pollutants are summarized in Figures 4 and 5. These figures demonstrate an expected variability in discrete trajectories for each sample date. However, it is also clearly evident that vertically mixed pollutants are transported over greater distances in the winter season than in the summer. The winter trajectories generally paralleled the coastline, and no short-term impact on the Texas coast is indicated.

Floating pollutant trajectories for the sample dates are summarized in Figures 6 and 7. Seasonal differences in trajectories are even more marked than those for vertically distributed pollutants. The floating pollutants are carried much greater distances by wind drift currents in February than in August. These results also indicate that the coast southwest of Galveston could be affected by floating pollutants within about two days in winter.

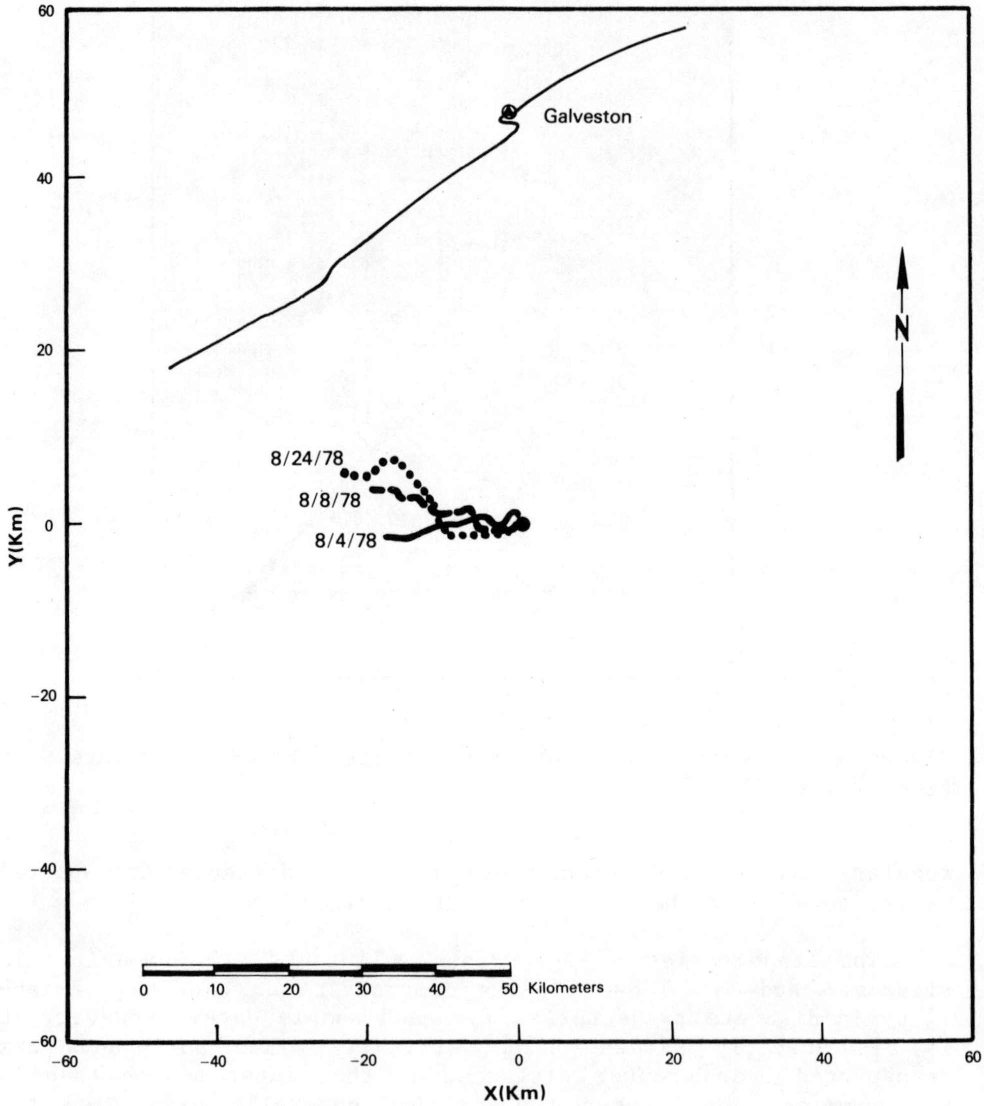

Figure 4. Subsurface pollutant trajectories, August 1978.

Actual field measurements and monitoring programs are limited in the extent of data which can be obtained. The development of models allows for reduced data gathering, since only a relatively small number of data points is needed to verify model predictions. Specifically, the development of applicable hydrodynamic models is necessary to meet all of the objectives and goals of the study program in assessing the environmental impacts of an active oil and gas field in the northwestern Gulf of Mexico. These

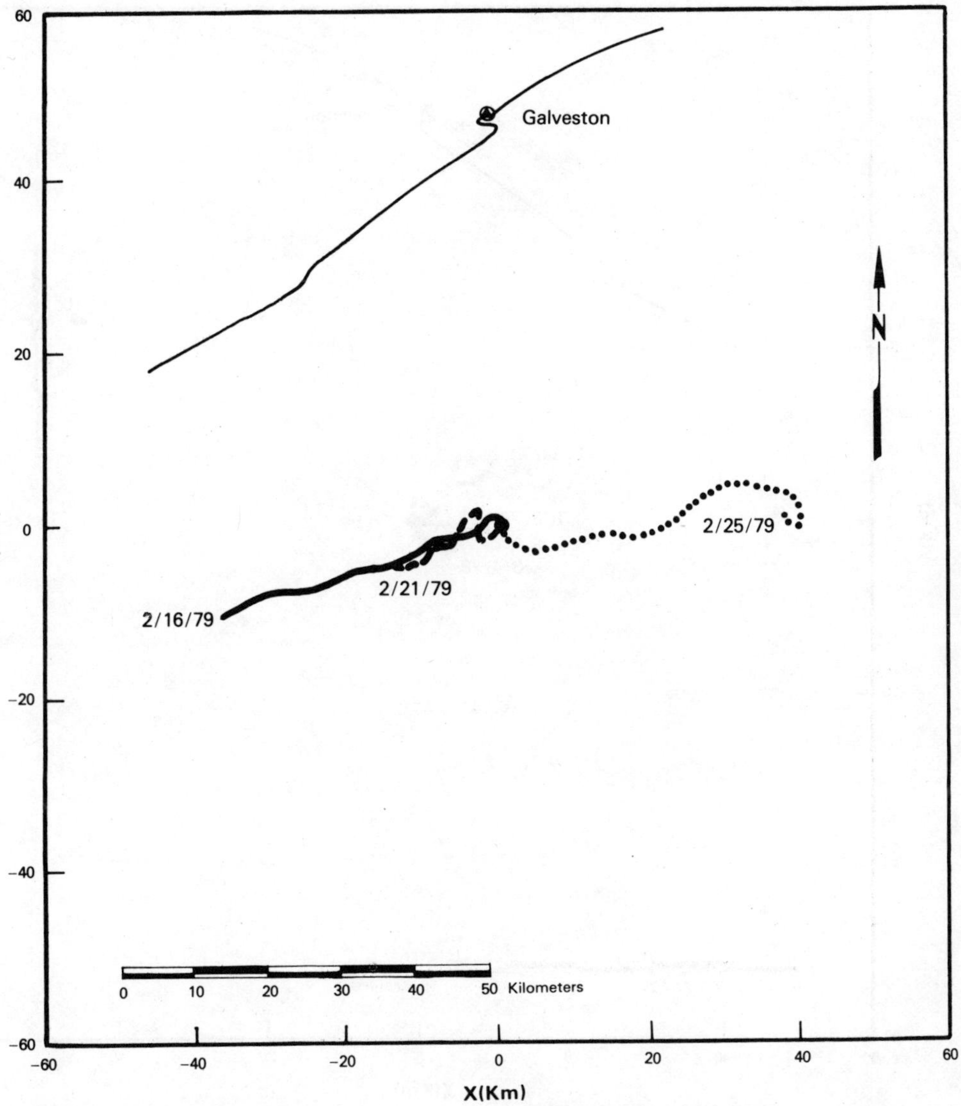

Figure 5. Subsurface pollutant trajectories, February 1979.

models are needed in order to project the extent of environmental alterations associated with the development of an active oil and gas field, to quantify their distribution in time and space in the marine ecosystem, and to be able to predict how these sources and their distribution will ultimately impinge upon and be absorbed by the system.

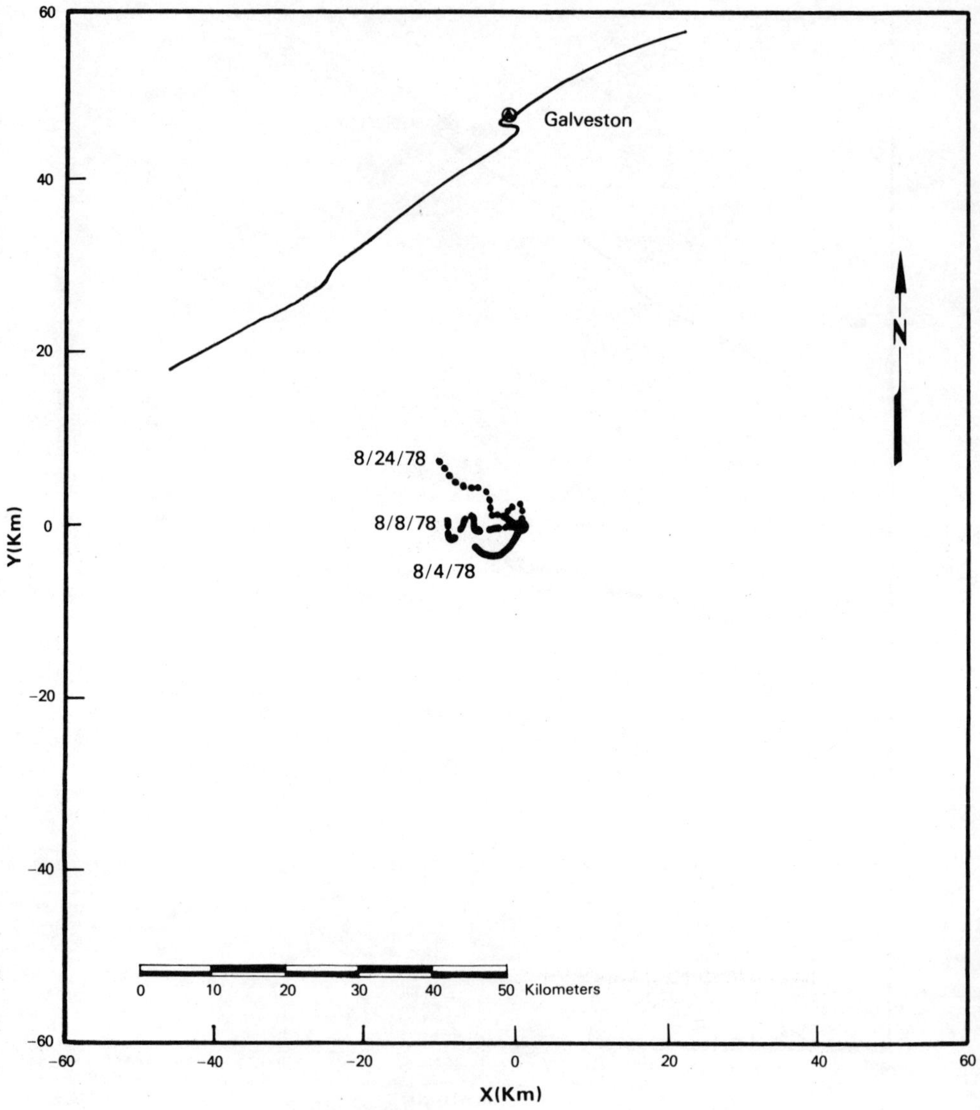

Figure 6. Floating pollutant trajectories, August 1978.

SUMMARY

The hydrodynamic model of the Buccaneer Gas and Oil Field provides a significant tool which can be used in conjunction with the results from other elements of the study program to evaluate the effects of operational contaminant discharges on the surrounding marine ecosystem.

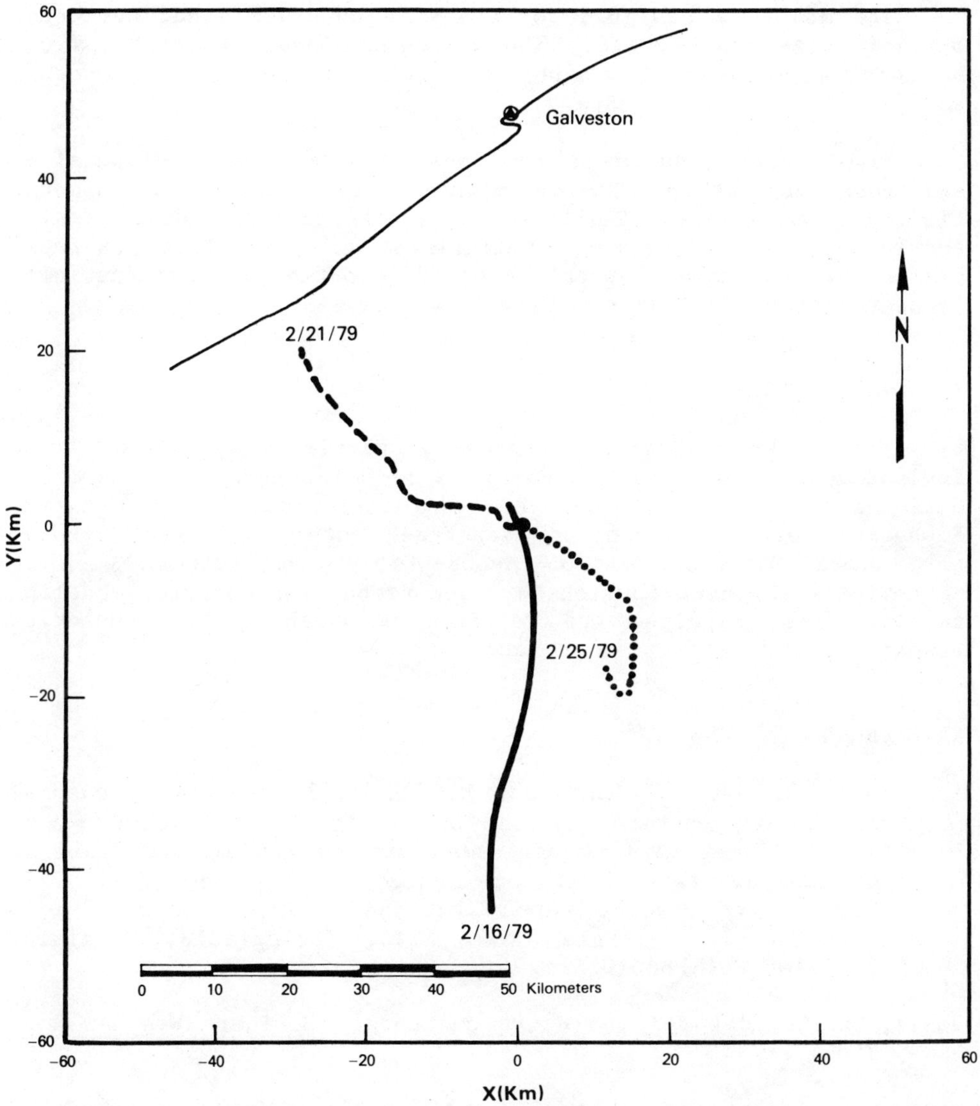

Figure 7. Floating pollutant trajectories, February 1979.

An understanding of the physical dispersion and transport processes can be used to indicate the marine resource areas most likely to be impacted by Buccaneer Field operations. Results of the modeling analyses can also be incorporated in a physical mass transport model to identify the pollutant input to various biological components of the ecosystem.

The model is designed for use with observed wind, currents, particle size inputs, etc. The source of these observations can be local measurements, weather service predictions, or any other sources of appropriate data.

Preliminary results of the model indicate rapid dispersion, and long-range transport of minute quantities of pollutants. Limited data on seasonal differences in the patterns of dispersion and transport indicate both similarities and differences in these processes. Pollutants appear to be transported over substantially greater distances in winter than in summer.

ACKNOWLEDGEMENTS

This work is a result of research partially sponsored by the Environmental Protection Agency and Department of Commerce, National Oceanic and Atmospheric Administration, National Marine Fisheries Service, Southeast Fisheries Center, Galveston Laboratory under Contract No. 03-78-D08-0035. The authors wish to acknowledge the contributions of Jack Beebe, for computer modeling and computer graphics, and Patricia McConnehey, for manuscript preparation.

REFERENCES

Armstrong, R. S., and Hamilton, R. C., 1979, "Describe seasonal circulation patterns in the oil field," in: Jackson, W. B. (ed.), "Environmental Assessment of an Active Oil Field in the Northwestern Gulf of Mexico, 1977-1978, Vol. III: Chemical and Physical Investigations." NOAA, pp. 2.3.9-1 to 2.3.9-250. Available from: NTIS, Springfield, Virginia, Accession No. PB80107899.

James, R. W., 1968, "Wind drift currents," Mariners Weather Log, 12:8.

Monin, A. S., and Yaglom, Y. A., 1974, "Statistical Fluid Mechanics: Mechanics of Turbulence." The MIT Press, Cambridge, Massachusetts.

Tennekes, H., and Lumley, J. H., 1972, "A First Course in Turbulence." The MIT Press, Cambridge, Massachusetts.

TRANSPORT AND DISPERSION OF POTENTIAL CONTAMINANTS

Reed S. Armstrong

Atlantic Environmental Group
National Marine Fisheries Service
National Oceanic and Atmospheric Administration
Narragansett, Rhode Island 02882

INTRODUCTION

During the first three years of the Buccaneer Gas and Oil Field Study (1976-1979) dye studies, moored and profiling current meters, anemometers, wave measurements, and drifting buoys were employed to determine the dynamic characteristics and energetics of the waters in and around this region of the continental shelf off Galveston, Texas. This variety of observations has afforded an opportunity to define the transport mechanics and to develop models for describing the spread and movement of potential contaminants that may have entered the water from petroleum operations in the Field.

Smedes has described (this Volume) modeling results for dispersion and transport for the case of instantaneous releases of potential pollutants. The focus of the work presented here is toward describing dilution and distribution patterns that could be expected to occur after periods of continuous discharge or from frequent, sporadic releases.

METHODS AND RESULTS

For dynamic considerations, pollutants that may enter the waters from petroleum operations would be of the form of (1) dissolved and suspended contaminants, (2) floating and surface film pollutants, (3) sinking, particulate matter, or (4) resuspended, sediment material. Based on observations from the Buccaneer Field Study, empirical models for continuous releases were developed for each of these modes for dispersing pollutants.

Dissolved and Suspended Materials

Dye studies, conducted in conjunction with drifting buoy deployments in the Buccaneer Gas and Oil Field Study indicated that dissolved and suspended substances are transported with the current, and spreading rates are greatest along the wind and current direction and least transverse to the current (Armstrong, 1979). Axial spreading rates of point source releases were linear with time and did not seem to vary with tidal current reversals. Averaged spreading rates amounted to ±70 m/hr along the current direction and ±25 m/hr transverse to the current, with current speeds of about 12 cm/sec. Dye study results indicated that discharge from the Buccaneer Gas and Oil Field quickly mixes to a depth of about 10 m and does not seem to go deeper. To estimate the transport and dilution of potential contaminants in dissolved and suspended form, three classes of conditions were modeled:

Discharge without currents. This would result in the least spread and dispersion, and highest concentrations in the local environment, and would develop in a case without wind or current. For such a situation suspended and dissolved substances would diffuse in a circularly spreading plume which undergoes no net transport. Although such an event is highly unlikely in the Buccaneer Field, conditions approaching this circumstance may develop for brief periods. Using the spreading rate transverse to the wind and current derived from the dye studies (25 m/hr) as a realistic approximation for radial dispersion in a static situation, dilution, concentration, and spread magnitudes can be estimated. For this estimation, it is assumed that all discharged material becomes homogeneously distributed through the upper 10 m of the water and never penetrates deeper. It is also assumed that as the discharge plume expands radially another 25 m each hour, 1/t (t = time in hours) of an hour's discharge disperses into this expanded ring, with 1/t of that discharge volume accumulating with later discharge in each inner ring. Example results of plume dimensions, dilutions, and resulting concentrations are listed in Table 1.

At the rates shown in Table 1, it would take about 10 weeks for the discharge plume to reach the coast (about 40 km away), where dilution would be about $10^{-12}/m^3$ and, using the example in Table 1 of continuous discharge of 1 kg/hr, contaminant concentration at the coast would be about 10^{-10} g/m^3. In this case, with continuous discharge, concentrations at fixed distances from the source would increase at a rate proportional to the natural log of length of time of discharge.

Discharge with steady current. This model represents the condition in which continuous discharges of dissolved and suspended contaminants are transported and dispersed by a non-

TABLE 1. DILUTION AND EXAMPLE CONCENTRATIONS OF DISSOLVED AND SUSPENDED MATERIALS FOR CONTINUOUS DISCHARGE WITHOUT CURRENTS.

Discharge duration (hr)	Discharge plume dimension (radial distance in meters from source)	Average dilution per m^3		Average concentration[a] of pollutant with discharge of 1 kg/hr (g/m^3)	
		outer edge of plume	within 25 m of source	outer edge of plume	within 25 m of source
1	25	-	$5.1x10^{-4}$	-	$5.1x10^{-2}$
3	75	$3.4x10^{-5}$	$9.3x10^{-4}$	$3.4x10^{-3}$	$9.3x10^{-2}$
10	250	$2.7x10^{-6}$	$1.5x10^{-3}$	$2.7x10^{-4}$	0.15
30	750	$2.9x10^{-7}$	$2.0x10^{-3}$	$2.9x10^{-5}$	0.20
100	2500	$2.6x10^{-8}$	$2.7x10^{-3}$	$2.6x10^{-6}$	0.27

[a]Assuming discharged material homogeneously mixed through upper 10 m of water column.

varying current. By tracking the growth and spread of a dye plume as it progresses along a current trajectory, we can determine dispersion for continuous discharge with a steady current. Averaging results from the three dye studies conducted in the Buccaneer Gas and Oil Field Study indicates that after an hour of continuous discharge, contaminants would be stretched along a strip 0.45 km long (0.45 km/hr average current); at the head, dispersion would spread the material 25 m to both sides of the current trajectory and 70 m downstream, in advance of the current vector, in a semi-elliptical shape. Since the dye studies indicated that these dispersion dimensions increase linearly with time, for at least a few hours, the plume resulting from the hour's discharge would extend in a triangular shape from the elliptical head back to the discharge point for the apex of the triangle. The area encompassed by this plume after an hour would be 1.4×10^4 m^2, giving a dilution of 7.1×10^{-5} m^{-3}. Assuming, for example, a release of 1 kg of contaminant in the hour, and assuming that this material is homogeneously spread through the upper 10 m of the water, and that currents and dispersion are invariant through the upper 10 m, then after an hour the average concentration within the plume would be 7.5×10^{-3} g/m^3. Since the current speed is greater than the rates of dispersion, accumulation near the source cannot occur. (The 0.45 km/hr current speed, from dye study conditions used above, is less than or equal to mean scalar current speeds at all depths from the current meter records of the Buccaneer Gas and Oil Field Study.)

If, after an hour of discharge, release of contaminants stopped, then the plume would be transported away from the source with the current and enlarged by dispersion. Each succeeding hour, assuming continued linear spreading, the plume would expand ±25 m laterally and ±70 m longitudinally to the current vector. The areal growth of the plume each hour would be defined by a series of concentric, egg-shaped rings as we follow the plume. If we assume that during each successive hour 1/t of the pollutant disperses into the enlarged ring area, then we can calculate progressive dilution at distances outward from the center of the plume. Computed dilutions at distances radially away from the center of the plume (along the current direction and transverse to the current) resulting from an hour of discharge, and for continuing discharges for other time increments, are shown in the nomograph of Figure 1. For the curves in Figure 1, it was assumed that the dilution at half the radial distance across a ring equalled the average dilution value for that ring.

Discharge with varying currents. Observations in the Buccaneer Field indicate that the currents are not steady but vary considerably in speed and direction. As evidenced in the preceding discussion, dispersion of dissolved and suspended materials is much more rapidly accomplished by spreading along a

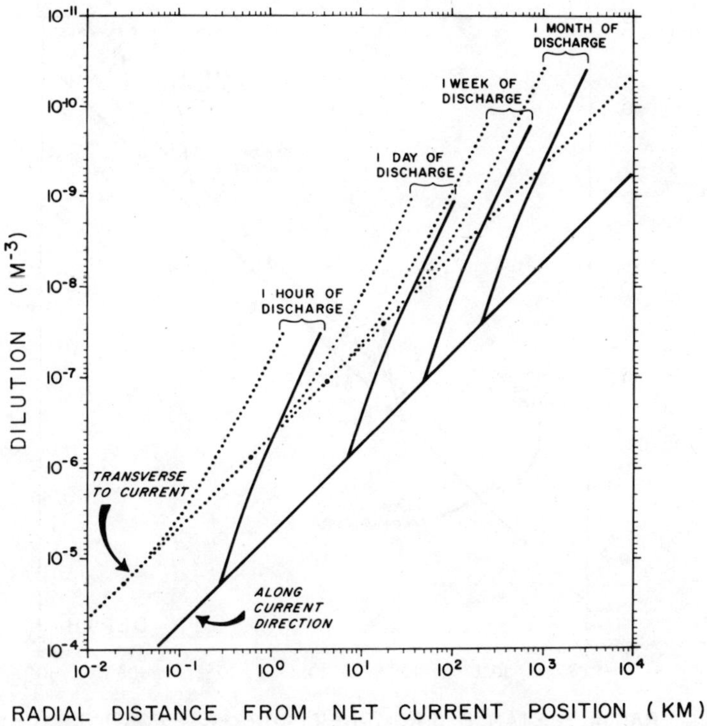

Figure 1. Dilution of dissolved and suspended materials at the Buccaneer Gas and Oil Field resulting from continuous discharge with steady current.

current vector than by mixing around the vector. Therefore, the distribution and dilution resulting from continuous discharge from a point source can be fairly well determined if we can define the areal distribution of the time-varying current vectors.

The distribution of current vectors at the Buccaneer Field can be estimated from the statistical compilations of the year's worth of hourly current meter data recorded in 1976 (Hamilton, 1979). Using the method of bivariate analysis with variance and covariance values of alongshore (NE-SW) and normal-to-shore (NW-SE) components of the currents, components of equal frequency ellipses were calculated using tables of F-test for Equality of Variances with derived eigenvalues (characteristic roots). This approach allows calculating axial magnitudes and axes orientation of a series of concentric ellipses, each of which would be expected to include a certain proportion of the current vectors (F-test confidence regions), assuming normal distributions in the data. The concept is that, after an appropriate period of time, the currents would have transported and dispersed a percentage of

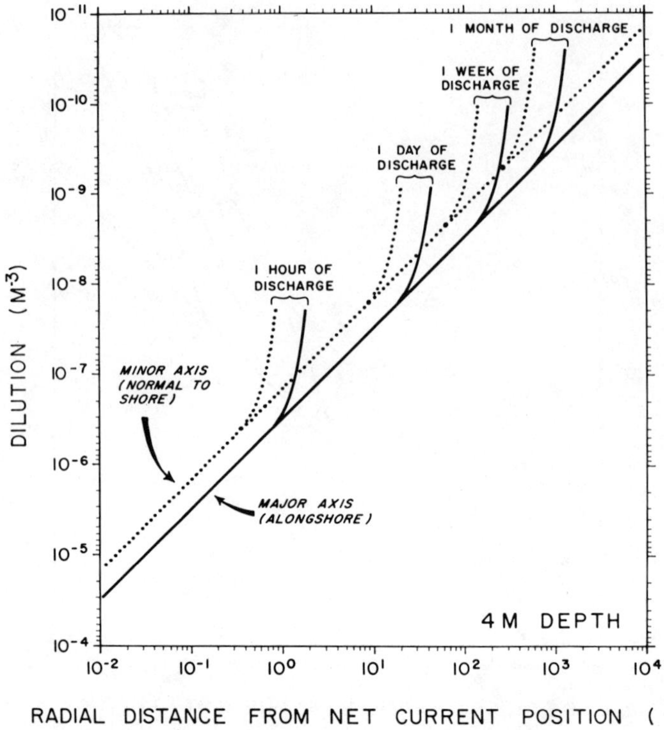

Figure 2. Dilution of dissolved and suspended materials at the Buccaneer Gas and Oil Field resulting from continuous discharge with annually varying currents.

continuously discharged material over the area encompassed by that confidence ellipse. Dilution values for a depth level can be calculated by dividing the area within a percentile ellipse into that percent of a unit area. Dilution computations used here were for elliptical rings (differences between pairs of concentric ellipses). The resulting value of dilution for a ring was taken to be the value at mid-distance, axially, across the ring. Axial distances for dilution values are measured away from the net displacement (mean current vector) position and, therefore, refer to a moving coordinate system.

Since the axial components of confidence ellipses are in units of speed, multiplication by time (ellipse axes and unit area for dilution calculations) gives dilution at distances away from the center of the discharge plume for various periods of continued discharge. Nomographs of dilution for variance values for the whole year's records for the current meters at 4, 10, and 18 m were developed. Figure 2 is shown for the currents at a depth of 4 m. Curves drawn in Figure 2 are for dilution along the

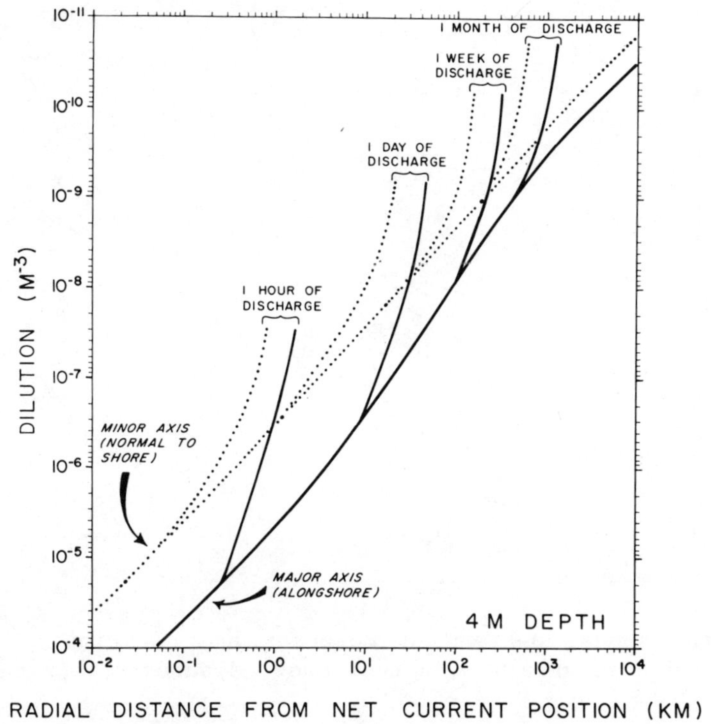

Figure 3. Dilution of dissolved and suspended materials at the Buccaneer Gas and Oil Field resulting from continuous discharge as currents shift from steady flow to annually varying flow.

principal axes of the confidence ellipses, with the major axis aligned approximately alongshore, and in the direction of mean transport and the minor axis directed generally normal-to-shore. Curves for 4 and 10 m depths were nearly identical, but dilution along the major axis near the bottom, at 18 m, was only about half as much as in the upper waters.

The current meter data used above only represent the Eulerian field of motion and conditions within the Buccaneer Field. To test the validity of using these results in the Lagrangian sense and outside the Field, confidence ellipses and dilutions were calculated from the records of daily movements of all of the drifting buoys deployed around the Field and tracked for distances as far away as about 150 km. Dilution curves derived from the buoy movements were very similar to those for the 4 and 10 m current meter records; therfore, it is considered that the results from the current meter data are representative for conditions of dispersion over much of the continental shelf off eastern Texas.

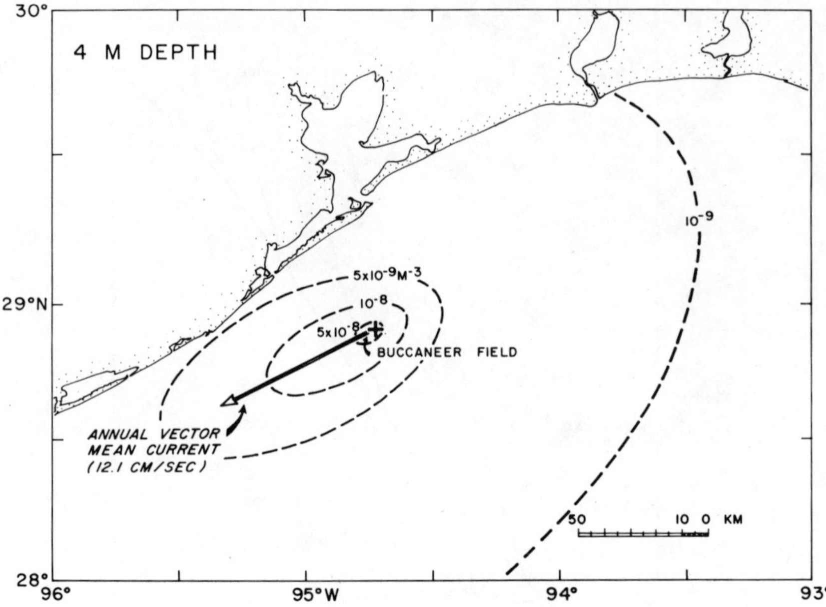

Figure 4. Annual pattern of movement and dilution of dissolved and suspended materials with continuous discharge (4 m depth).

However, dilution values from the 4 and 10 m current meters, as exemplified by Figure 2, indicate about a 20-fold greater dilution along the current direction and about 30-fold greater transverse to the mean current as compared to the results derived from the dye study measurements for the case of dispersion along a steady current trajectory (Figure 1). With continuous discharge it should be expected that, for the first 2-3 hours, dispersion and dilution of dissolved and suspended contaminants should proceed similar to that depicted for a steady current (Figure 1). At longer periods currents begin to vary and, after a year, features as in Figure 2 from the current meter data should be more appropriate. Therefore, a combination from these two models should give the most realistic depiction. As an intermediate condition between the steady current and the annual variance in the current meter records, dilution curves were developed by the method of bivariate analysis using averages of monthly values of variance in the current meter data. Annual values showed about twofold greater dilution than did the monthly mean values. From these three sets of curves (annual and monthly mean current meter variance and dye study interpretation) combination dilution nomographs were constructed (Figure 3, for upper waters) to depict dilution of continuous discharge as dynamic conditions in the waters change from a state characterized by steady flow, persisting for two hours, to one of varying currents.

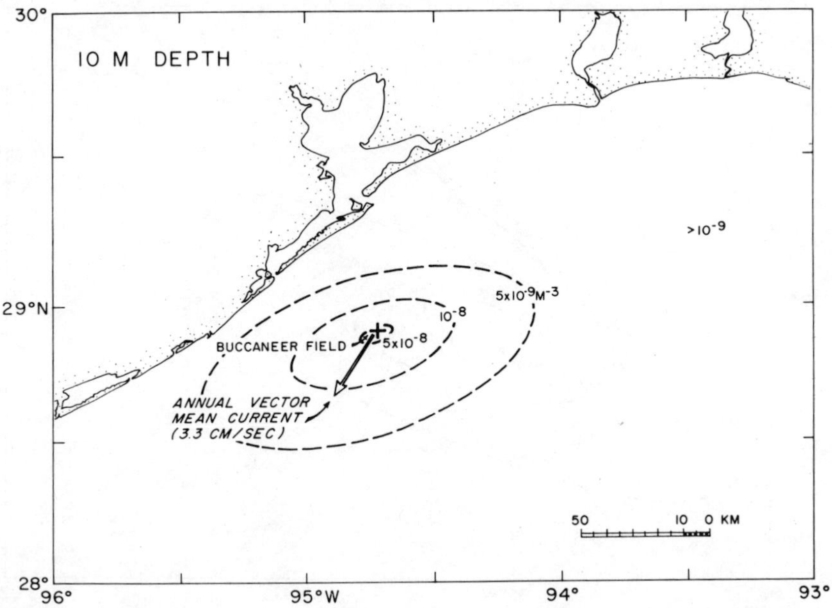

Figure 5. Annual pattern of movement and dilution of dissolved and suspended materials with continuous discharge (10 m depth).

If we know the rate of discharge of dissolved and suspended material and the resulting vertical distribution in the water column, concentrations resulting from continuous discharge can be estimated from the dilution nomographs. Concentration values from the nomographs are calculated by multiplying the hourly rate that the contaminant is introduced into a 1 m depth layer by the dilution value, from the nomograph, with concentration in units of discharged contaminants per cubic meter. Distances in the dilution charts are measured along and normal to the net transport (mean current vector). For example, with discharge of dissolved and suspended contaminant at a rate of 1 kg/hr and assuming this material becomes evenly distributed through the upper 10 m of the water column, then the contaminant would be introduced at the rate of 100 g/hr into each 1 m depth layer. From Figure 3, after 1 hr, concentration would be 1.9×10^{-3} g/m^3 (1.9×10^{-5} m^{-3} dilution) at distances of 270 m downstream and 25 m lateral to the net transport location after the hour and, following the curves for 1 hr of discharge, concentration of 10^{-5} g/m^3 (10^{-7} m^{-3} dilution) would occur at distances of 1.3 km along and 0.65 km transverse to the mean current, as measured away from the net transport location. After a week of continuous discharge at this discharge rate, assuming the mean current is parallel to the coast, concentrations of about 5×10^{-7} g/m^3 (5×10^{-9} m^{-3} dilution) should be expected along the coast 40 km lateral to the current.

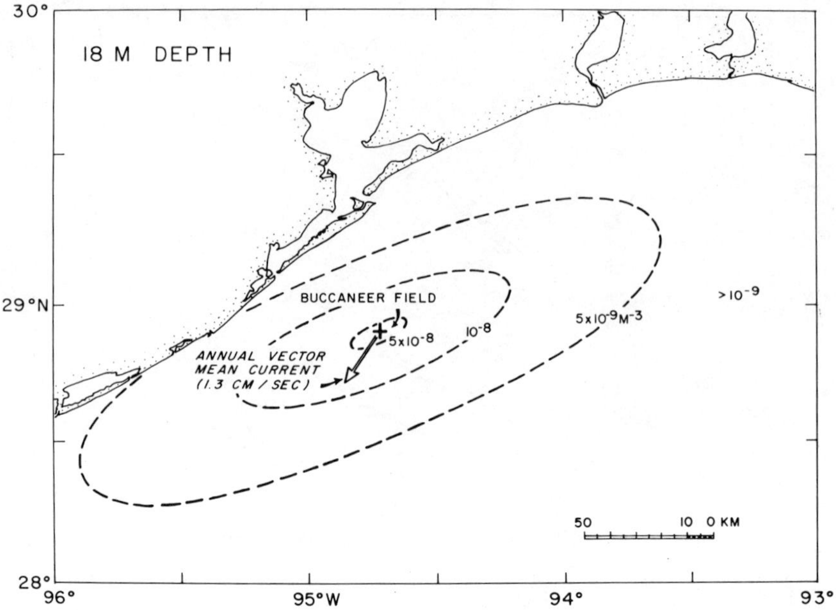

Figure 6. Annual pattern of movement and dilution of dissolved and suspended materials with continuous discharge (18 m depth).

From Figure 3 and its counterparts for 10 m and 18 m depths dilution maps can be constructed for a given mean current. Using the annual mean vector currents from the current meter records of 1976 [12.1 cm/sec at 4 m depth, 3.3 cm/sec at 10 m, and 1.3 cm/sec at 18 m, directed toward west-southwest at 4 m and toward southwest at 10 and 18 m, from Hamilton (1979)] distribution maps of dilution levels were constructed to represent the conditions resulting from continuous discharge of dissolved and suspended contaminants (Figures 4-6). Dilution levels over the areas shown in Figures 4-6 would develop within about one week. Since the patterns shown in these maps would develop within a week but were derived from annual means of currents, the distributions shown should not be expected to actually occur. However, the features indicated of rapid and wide dispersal without accumulation should be representative and realistic. Spreading is indicated to be mainly alongshore throughout the water column.

Floating and Surface Film Pollutants

As part of the modeling study, conducted by Environmental Research and Technology, Inc. in the Buccaneer Gas and Oil Field (Smedes et al., 1980), the transport and dispersion of instantaneous releases of surface-borne contaminants was examined. They

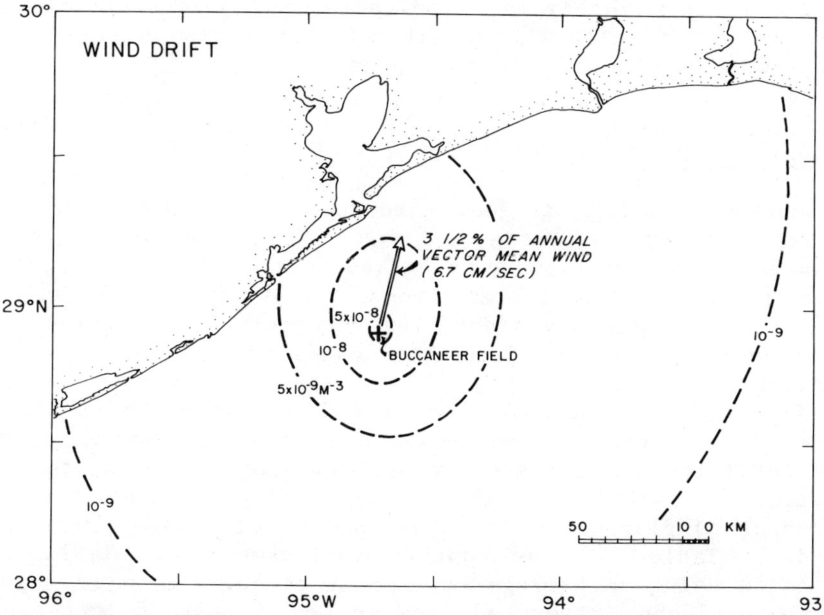

Figure 7. Annual pattern of movement and dilution of floating and surface film releases with continuous discharge (wind drift).

concluded that wind drift was the principal driving force with transport at 3.5% of the wind speed and in the same direction as the wind.

To consider the pattern and distribution of surface-borne materials resulting from continuous discharge, a dilution nomograph was constructed using the method of bivariate analysis with annual values of variance listed by Hamilton (1979) for hourly wind observations at the Buccaneer Field in 1976. The procedures used for developing the dilution nomograph were the same as employed with current meter data in the preceding subsection, except that axial components computed for confidence ellipses of winds were reduced to the 3.5% level, following the conclusions drawn by Smedes et al. (1980) for surface layer wind drift. The dilution curves and values were similar to those for the 4 m current meter record of Figure 2, except that values along the major and minor axes were almost equal (confidence ellipses almost circular). Using 3.5% of the annual vector mean wind (6.7 cm/sec) for a transport vector, a dilution chart was constructed (Figure 7) to display the average distribution of floating and surface film contaminants that might be expected from continuous release, based on the annual variability of wind drift.

The pattern in Figure 7 indicates that widespread dispersion is expected but, with the normal mean southerly winds, northerly transport toward land would be typical.

Sinking Particulate Contaminants

Another modeling project conducted in the Buccaneer Gas and Oil Field Study by Smedes et al. (1980) involved estimating distances from the source required for particulate matter to settle to the bottom after transport by the currents. From Stoke's law, they derived settling velocities for particulates by size fractions, ranging from fine sands to clay size particles. Expanding on these results, sinking times for particles of each size fraction to reach the bottom (20 m) can be calculated and, for a given current speed structure, the horizontal distances these particles would travel before settling to the bottom can be computed. Resulting settling velocities, sinking times, and horizontal displacements for fine sand to clay size particles are listed in Table 2. Horizontal displacements in Table 2 were calculated using an average current speed for the water column of 15 cm/sec (depth integrated, scalar mean speed of Buccaneer Gas and Oil Field current meter records of 1976). From Table 2, particulate contaminants coarser than fine sands probably sink to the bottom and accumulate almost directly below the discharge and particles finer than medium silts would behave essentially as suspended material.

The techniques used for dissolved and suspended materials involving bivariate analysis of current meter records can be employed to estimate accumulations of continuous or frequent, sporadic releases of sinking particulate contaminants that could be expected from dispersion and transport by the varying currents in the Buccaneer Gas and Oil Field. As an example of potential distributions, the cumulative thickness of the layer of settling particulates that would result from one year of continuous releases at the rate of 1 m^3/day of all size fractions was calculated. For these calculations it was assumed that the influx was comprised of equal volumes of each of the 7 size fractions listed in Table 2 so that 52 m^3 of each size fraction was released into the waters during the year. Distributional pattern of current variability was defined from depth integrated averages of confidence ellipse axes from the annual variance of the current meter data at 4, 10, and 18 m depths. Net transport was computed as the depth integrated average of the annual vector means of the current meter data. Values of sinking time to reach the bottom, in Table 2, were then used to calculate, for each size fraction, the thickness of particulate accumulation on the bottom after a year at distances from the source of input. Total accumulation of all sinking particulates (thickness of contaminated sediments) was

TABLE 2. SETTLING RATES AND TRANSPORT OF SINKING PARTICULATE CONTAMINANTS.

Particle size (microns)	Mean settling velocity (cm/sec)[a]	Sinking time to bottom (hr)	Horizontal displacement (km)[b]
>120 (coarse fractions)	>1	<0.5	<0.3
120 (fine sand)	1.0	0.56	0.3
60-100 (very fine sand)	0.462	1.20	0.65
20-60 (coarse silt)	0.136	4.08	2.1
6-20 (medium silt)	1.43×10^{-2}	38.9	21
2-6 (fine silt)	1.36×10^{-3}	408	210
<2 (clay)	6.8×10^{-5}	8170	4300

[a]From Smedes et al. (1980).
[b]Computed for average current = 15 cm/sec: depth integrated, mean scalar speed for Buccaneer Gas and Oil Field current meter records, from Hamilton (1979).

found by adding the contributions by each size fraction. The resulting distributional pattern that might be expected for this example of continuous release of particulates at the rate of 1 m^3/day, and continuing for a year, is shown in Figure 8, for the area in which accumulated settling particle thickness is 1 micron or greater.

For the example depicted in Figure 8, accumulated thickness would be less than 1 micron at distances from the source of more than about 5 km downstream (toward the southwest), 4 km upstream (northeast), and 2 km to the southeast and northwest. Settling particles within the 1 micron thickness region of Figure 8 would be comprised almost entirely of coarse silts and larger fractions. The pattern of dispersion shown in Figure 8 and the results listed in Table 2 indicate that, except for coarse size material, sinking particulate contaminants would be distributed widely and at low concentrations.

Resuspended Sediments

Particulate contaminants that have settled to the bottom become available for redistribution and transport by resuspension. Divers in the Buccaneer Gas and Oil Field Study have reported fairly frequent instances of decreased transparency in the water because of resuspension of sediment material. As the

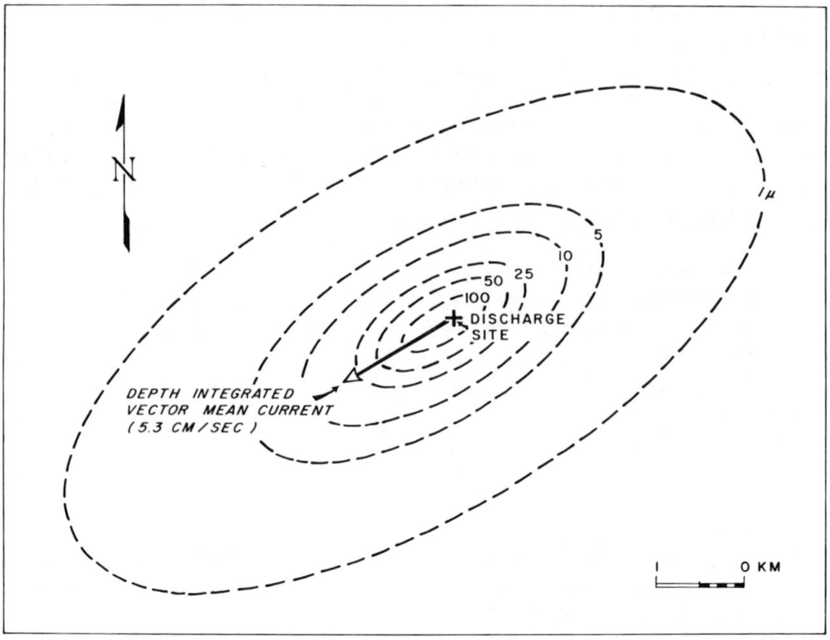

Figure 8. Accumulated thickness on the bottom of sinking particles after 1 year of continuous release at rate of 1 kg/day.

energy of water motion exceeds some threshold condition, sediment fractions would be brought back into the water column to be transported by the currents. As part of the investigation conducted by Hazleton Environmental Sciences Corp. in the Buccaneer Gas and Oil Field Study, resuspension dynamics by accelerating currents and by wave action were examined (Danek and Tomlinson, 1980). They concluded that currents, rather than waves, were the principal cause of resuspension in the Buccaneer Field although, under strong storm conditions, considerable reworking of sediments by wave action could occur. Based on sediment size distributions around the Field, they estimated that bottom currents in excess of about 26 cm/sec were necessary to accomplish resuspension of surficial sediment particles. Potential for bottom currents in excess of 26 cm/sec can be determined from current meter data recorded at 18 m in the Buccaneer Gas and Oil Field Study. Currents were measured near the bottom from March to October 1976 (Hamilton, 1979) and for a month each during July-August 1978 and February-March 1979. For all these records, currents were ≥26 cm/sec 9% of the time (hourly observations), with frequencies of occurrence of about 5% for February-March, 15% for April-June, near zero in July-September, and 33% in October.

Near-bottom currents at speeds ≥26 cm/sec typically persisted for periods of 6 hours or more with average speeds during the event of about 30-40 cm/sec. Except in summer, when bottom currents rarely exceeded 26 cm/sec, the currents were normally directed toward the southwest. Therefore, with these conditions during each resuspension event, sediment material would be transported about 6-8 km or more to the southwest before settling to the bottom again.

These results imply that, during summer, some accumulation of particulate contaminants might develop around the area but, during the rest of the year, any accumulations of other than coarse size material would be widely distributed and would be transported well away from the source.

SUMMARY

Results from analyses of data collected or acquired in the Buccaneer Gas and Oil Field Study were used to model and examine transport and dispersion characteristics of the waters for diluting potential releases of contaminants from the Buccaneer Field. Models were developed to consider distributional patterns and to compute dilutions that might be expected to result from continuous discharges. Because of differences in dynamic response, types of discharges were examined separately as dissolved and suspended materials, floating and surface film contaminants, sinking particulates, and as resuspended particulate matter.

Portrayals and nomographs of the distribution and dilution of dissolved and suspended contaminants were developed from areal growth rate models to represent conditions as water dynamics progress from a static situation (no current) to the case of dispersion with a steady current to the circumstance of varying currents. Models for the first two cases were developed from dye study results and the last (varying currents) from computations of equal frequency ellipses by bivariate analyses of current meter records. For the static case, which might occur briefly when, for example, tidal currents oppose and cancel ambient currents, results indicated slow radial dispersion with concentrations at fixed distances from the source increasing at a rate proportional to the natural log of the length of time of continuing discharge. With a steady current, discharge would stream along the current trajectory with diffusion along the current direction at a rate about three times greater than that tranvsverse to the current. With continuous discharge in a steady current, and since current speeds at the Buccaneer Field are typically five to ten times greater than the rate of diffusion along the current direction, dispersion would principally be along a narrow plume with the current trajectory and accumulation could not occur. Concentrations would decrease rapidly downstream. Current meter and drifting buoy records show that, beyond a few hours, considerable variability in speed and direction develops in the currents. Results of analyses of current records for discharges continuing for lengthy periods, considering current variability, show widespread dispersion particularly in the alongshre direction, and very rapid dilution. Dilution rates were somewhat lower near the bottom than through the rest of the water column. Charts of dilution magnitudes were constructed to portray distributions of contaminants resulting from long-term discharges and indicated weak net transport relative to dispersion by the varying currents, except in the surface layer where net transport would lead to a plume streaming toward the southwest.

Floating and surface film contaminants would be expected to be transported with the wind at about 3.5% of the wind speed. To represent overall distributional patterns and dilutions resulting from continuous long-term discharges, equal frequency ellipses were calculated from annual variance values of anemometer data recorded at the Buccaneer Gas and Oil Field, with magnitudes reduced by the 3.5% rule. Nomographs and dilution charts were derived for the case of long-term releases and showed very rapid dilution and widespread dispersion with somewhat of a northerly transport because of the prevalence of southerly winds.

Settling velocities for various size fractions of sinking particulates were determined from Stoke's law and, using the depth integrated mean current speed from current meter records at the Buccaneer Gas and Oil Field (15 cm/sec), horizontal displacements for each size fraction before settling to the bottom were calculated. Results indicated that, on the average, particles coarser than fine sands would quickly sink to the bottom, typically accumulating within the shadow of the platform from which they derived. Particles finer than medium silts would behave essentially as suspended materials. Bivariate analysis of current meter data was used to determine depth integrated average equal frequency ellipses to estimate distributions of accumulations of continuously discharged particulates after settling to the bottom. As an example, with total release of all size fractions of 1 kg/day and continuing for a year, accumulated sinking particulates would be about 1 micron thick at distances of 2-5 km from the source and, at 100 m from the source, thickness would be 200 microns or less.

Consideration of sediment distributions at the Buccaneer Field indicated that the occurrence of bottom currents in excess of 26 cm/sec would lead to resuspension of surficial sediments, including particulate contaminants that may have derived from Field operations and settled to the bottom. Wave action was concluded to be of only minor importance for resuspension, except during severe storms when wave energy might lead to considerable reworking of sediments. From hourly current meter observations at the Buccaneer Gas and Oil Field, near-bottom currents exceeded speeds of 26 cm/sec 9% of the time, with seasonal differences in frequency of occurrence indicating low liklihood of resuspension and transport of settled contaminants in summer and high liklihood during fall. Typical speeds and durations of events when near-bottom currents exceeded 26 cm/sec imply that resuspended particulates would be transported an average of 6 km or more during each resuspension event.

ACKNOWLEDGEMENTS

Material presented here is from a report prepared by the author on results in Work Unit 2.3.9 (Hydrodynamics) in the study entitled Environmental Assessment of an Active Gas and Oil Field in the Northwestern Gulf of Mexico, sponsored by the Environmental Proection Agency (Project No. EPA-IAG-D5-E693-EO) and conducted under the program management of the Galveston, Texas Laboratory of NMFS/NOAA.

Models developed in this study are derived from results of studies conducted by the Galveston Laboratory and the Atlantic Environmental Group, both of the National Marine Fisheries Service (NMFS)/NOAA, Hazelton Environmental Sciences Corp., Environmental Research and Technology, Inc., and Evans-Hamilton, Inc.

REFERENCES

Armstrong, R. S., 1979, "Describe seasonal circulation patterns in the oil field, part A," in: Jackson, W. B. (ed.), "Environmental Assessment of an Active Oil Field in the Northwestern Gulf of Mexico, 1977-1978, Vol. III: Chemical and Physical Investigations." NOAA, pp. 2.3.9-1 to 2.3.9-76. Available from: NTIS, Springfield, Virginia, Accession No. PB80107899.

Danek, L. J., and Tomlinson, M. S., 1980, "Currents and hydrography of the Buccaneer Field and adjacent waters," in: Jackson, W. B., and Wilkens, E. P. (eds.), "Environmental Assessment of the Buccaneer Gas and Oil Field in the Northwestern Gulf of Mexico, 1978-1979, Vol. VI." NOAA Technical Memorandum NMFS-SEFC-40, 33 pp. Available from: NTIS, Springfield, Virginia.

Hamilton, R. C., 1979, "Describe seasonal circulation patterns in the oil field, Part B," in: Jackson, W. B. (ed.), "Environmental Assessment of an Active Oil Field in the Northwestern Gulf of Mexico, 1977-1978, Vol. III: Chemical and Physical Investigations." NOAA, pp. 2.3.9-77 to 2.3.9-250. Available from: NTIS, Springfield, Virginia, Accession No. PB80107899.

Smedes, G., Calman, J., and Beebe, J., 1980, "Hydrodynamic modeling," in: Jackson, W. B., and Wilkens, E. P. (eds.), "Environmental Assessment of the Buccaneer Gas and Oil Field in the Northwestern Gulf of Mexico, 1978-1979, Vol. X." NOAA Technical Memorandum NMFS-SEFC-42, 57 pp. Available from: NTIS, Springfield, Virginia.

EPILOGUE

Brian S. Middleditch

Department of Biophysical Sciences
University of Houston
Houston, Texas 77004

INTRODUCTION

The preceding chapters contain details of the more significant findings resulting from the research that was conducted in the region of the Buccaneer Field between 1975 and 1980. It has not been possible in this small volume to allot space for the respective authors to present full details of all of their research on this project. A comprehensive bibliography has, therefore, been compiled for the reader who wishes to delve deeper into the subject.

This bibliography lists all of the technical reports which have been published by the National Marine Fisheries Service (together with NTIS accession numbers, where these have already been assigned). Each of the reports was submitted first to the NMFS in draft form, and then revised after peer review. The results of the pilot study (1975) and the first year of the main study (1976-1977) were each published in single volumes. The comprehensive technical reports on the second year of the study (1977-1978) appeared in two volumes, and these were accompanied by a synopsis volume which summarized the salient details. The report by each Principal Investigator on the third year of the study (1978-1979) was published separately, and the Principal Investigators of the final year of the study (1979-1980) prepared individual milestone reports covering all of the research related to their respective disciplines during the five-year period of the study.

Also listed in the bibliography are journal articles, theses, and a dissertation that resulted from this study.

BIBLIOGRAPHY

Anderson, J. B., Schwarzer, R. R., and Clark, H. C., 1977, "Sedimentology, geochemistry, and trace metal analysis," in: Jackson, W. B. (ed.), "Environmental Assessment of an Active Oil Field in the Northwestern Gulf of Mexico, 1976-1977." NOAA, pp. 89-174. Available from: NTIS, Springfield, Virginia, Accession No. PB283890.

Anderson, J. B., Schwarzer, R. R., Clark, H. C., Wheeler, R. B., Hokanson, C. L., Anderson, J. E., and Okafor, J., 1979, "Describe the fine sediments and nepheloid layer of the oil field, focusing upon their relationship to heavy metal adsorption/determine levels, pathways, and bioaccumulation of heavy metals in the marine ecosystem in the oil field," in: Jackson, W. B. (ed.), "Environmental Assessment of an Active Oil Field in the Northwestern Gulf of Mexico, 1977-1978. Volume III: Chemical and Physical Investigations." NOAA, pp. 2.3.2/2.4.2-1 to 2.3.2/2.4.2-118. Available from: NTIS, Springfield, Virginia, Accession No. PB80107899.

Armstrong, R. S., 1980, "Current patterns and hydrography," in: Jackson, W. B., and Wilkens E. P. (eds.), "Environmental Assessment of the Buccaneer Gas and Oil Field in the Northwestern Gulf of Mexico, 1975-1980, Vol. IV." NOAA Technical Memorandum NMFS-SEFC-50, 41 pp. Available from: NTIS, Springfield, Virginia.

Armstrong, R. S., and Hamilton, R. C., 1979, "Describe seasonal circulation patterns in the oil field," in: Jackson, W. B. (ed.), "Environmental Assessment of an Active Oil Field in the Northwestern Gulf of Mexico, 1977-1978. Volume III: Chemical and Physical investigations." NOAA, pp. 2.3.9-1 to 2.3.9-250. Available from: NTIS, Springfield, Virginia, Accession No. PB80107899.

Aumann, G. D., 1977, "The effect of structures on migratory and local marine birds," in: Jackson, W. B. (ed.), "Environmental Assessment of an Active Oil Field in the Northwestern Gulf of Mexico, 1976-1977." NOAA, pp. 551-574. Available from: NTIS, Springfield, Virginia, Accession No. PB283890.

Basile, B., 1978, "Environmental Effects of Offshore Oil Production." Ph.D. Dissertation, University of Houston, Houston, Texas.

Behrens, E. W., 1977, "Total organic carbon and carbon isotopes of sediments," in: Jackson, W. B. (ed.), "Environmental Assessment of an Active Oil Field in the Northwestern Gulf of Mexico, 1976-1977." NOAA, pp. 737-759. Available from: NTIS, Springfield, Virginia, Accession No. PB283890.

Boland, G. S., 1980, "Morphological Parameters of the Barnacle, *Balanus tintinnabulum antillensis*, as Indicators of Physiological and Environmental Conditions." M.S. Thesis, Texas A&M University, College Station, Texas, 69 pp.

Brooks, J. M., Estes, E. L., and Huang, W. H., 1980, "Investigations of surficial sediments and suspended particulates at Buccaneer field," in: Jackson, W. B., and Wilkens E. P. (eds.), "Environmental Assessment of the Buccaneer Gas and Oil Field in the Northwestern Gulf of Mexico, 1978-1979, Vol. II." NOAA Technical Memorandum NMFS-SEFC-36, 261 pp. Available from: NTIS, Springfield, Virginia.

Brooks, J. M., Estes, E. L., Wiesenburg, D. A., Schwab, C. R., and Abdel-Reheim, H. A., 1980, "Investigations of surficial sediments, suspended particulates, and volatile hydrocarbons at Buccaneer gas and oil field," in: Jackson, W. B., and Wilkens E. P. (eds.), "Environmental Assessment of the Buccaneer Gas and Oil Field in the Northwestern Gulf of Mexico, 1975-1980, Vol. I." NOAA Technical Memorandum NMFS-SEFC-47, 89 pp. Available from: NTIS, Springfield, Virginia.

Caillouet, C. W., Jackson, W. B., Gitschlag, G. R., Wilkens, E. P., and Faw, G. M., 1981, "Review of the environmental assessment of the Buccaneer gas and oil field in the northwestern Gulf of Mexico," Proc. Gulf Caribb. Fish. Inst., in press.

Chang, E. S., 1978, "Sterols in the Marine Food Web." M.S. Thesis, University of Houston, Houston, Texas.

Danek, L. J., and Tomlinson, M. S., 1980, "Currents and hydrography of the Buccaneer field and adjacent waters," in: Jackson, W. B., and Wilkens E. P. (eds.), "Environmental Assessment of the Buccaneer Gas and Oil Field in the Northwestern Gulf of Mexico, 1978-1979, Vol. VI." NOAA Technical Memorandum NMFS-SEFC-40, 33 pp. Available from: NTIS, Springfield, Virginia.

Devereux, R., and Sizemore, R. K., 1981, "Incidence of degradative plasmids in hydrocarbon-utilizing bacteria," Dev. Ind. Microbiol., in press.

Emiliani, D., Baxter, K. N., Jackson, W. B., Morgan, J., Zamora, G., and Lansford, L., 1977, "Demersal finfish and macro-crustaceans collected in the vicinity of Buccaneer oil field during 1976-1977," in: Jackson, W. B. (ed.), "Environmental Assessment of an Active Oil Field in the Northwestern Gulf of Mexico, 1976-1977." NOAA, pp. 275-310. Available from: NTIS, Springfield, Virginia, Accession No. PB283890.

Faw, G. M., 1981, "Crude Oil Biodegradation Studies in Buccaneer Gas and Oil Field." M.S. Thesis, University of Houston, Houston, Texas

Finucane, J. H., and Collins, L. A., 1977, "Ichthyoplankton," in: Jackson, W. B. (ed.), "Environmental Assessment of an Active Oil Field in the Northwestern Gulf of Mexico, 1976-1977." NOAA, pp. 339-486. Available from: NTIS, Springfield, Virginia, Accession No. PB283890.

Finucane, J. H., Collins, L. A., and Barger, L. E., 1979, "Determine the effects of discharges on seasonal abundance, distribution, and composition of ichthyoplankton in the oil field," in: Jackson, W. B. (ed.), "Environmental Assessment of an Active Oil Field in the Northwestern Gulf of Mexico, 1977-1978. Volume II: Data Management and Biological Investigations." NOAA, pp. 2.3.6-1 to 2.3.6-157. Available from: NTIS, Springfield, Virginia, Accession No. PB80165970.

Fotheringham, N., 1977, "Effects of offshore oil field structures on their biotic environment: benthos and plankton," in Jackson, W. B. (ed.), "Environmental Assessment of an Active Oil Field in the Northwestern Gulf of Mexico, 1976-1977." NOAA, pp. 487-549. Available from: NTIS, Springfield, Virginia, Accession No. PB283890.

Fucik, K., and Show, I. T., Jr., 1980, "Sources, fate and effects modeling," in: Jackson, W. B., and Wilkens E. P. (eds.), "Environmental Assessment of the Buccaneer Gas and Oil Field in the Northwestern Gulf of Mexico, 1978-1979, Vol IX." NOAA Technical Memorandum NMFS-SEFC-43, 105 pp. Available from: NTIS, Springfield, Virginia.

Gallaway, B. J., 1980, "Pelagic, reef and demersal fishes and macro-crustaceans/biofouling communities," in: Jackson, W. B., and Wilkens E. P. (eds.), "Environmental Assessment of the Buccaneer Gas and Oil Field in the Northwestern Gulf of Mexico, 1975-1980, Vol. II." NOAA Technical Memorandum NMFS-SEFC-48, 82 pp. Available from: NTIS, Springfield, Virginia.

Gallaway, B. J., and Margraf, F. J., 1979, "Simulation modeling of biological communities associated with a production platform in the Buccaneer oil and gas field," in: Jackson, W. B. (ed.), "Environmental Assessment of an Active Oil Field in the Northwestern Gulf of Mexico, 1977-1978. Volume II: Data Management and Biological Investigations." NOAA, pp. 2.5.1-1 to 2.5.1-67. Available from: NTIS, Springfield, Virginia, Accession No. PB80165970.

Gallaway, B. J., and Martin, L. R., 1980, "Effect of gas and oil field structures on pelagic and reef fishes, and demersal fishes and macrocrustaceans," in: Jackson, W. B., and Wilkens E. P. (eds.), "Environmental Assessment of the Buccaneer Gas and Oil Field in the Northwestern Gulf of Mexico, 1978-1979, Vol. III." NOAA Technical Memorandum NMFS-SEFC-37, 49 pp. Available from: NTIS, Springfield, Virginia.

Gallaway, B. J., Howard, R. L., Green, K., and Martin, L., 1976, "A Study Plan for the Buccaneer Oil Field Assessment Program." Report to National Marine Fisheries Service, Galveston, Texas from LGL Limited-U.S., Inc., Bryan, Texas, 69 pp.

Gallaway, B. J., Johnson, M. F., Howard, R. L., Martin, L. R., and Boland, G. S., 1979, "A study of the effects of Buccaneer Oil Field structures and associated effluents on biofouling communities and the Atlantic spadefish (*Chaetodipterus faber*)." in: Jackson, W. B. (ed.), "Environmental Assessment of an Active Oil Field in the Northwestern Gulf of Mexico, 1977-1978. Volume II: Data Management and Biological Investigations." NOAA, pp. 2.3.8-1 to 2.3.8-126. Available from: NTIS, Springfield, Virginia, Accession No. PB80165970.

Hada, H. S., and Sizemore, R. K., 1981, "Incidence of plasmids in marine *Vibrio* spp. isolated from an oil field in the northwestern Gulf of Mexico," Appl. Environ. Microbiol., in press.

Harper, D. E., Jr., 1977, "Distribution and abundance of macrobenthic and meiobenthic organisms," in: Jackson, W. B. (ed.), "Environmental Assessment of an Active Oil Field in the Northwestern Gulf of Mexico, 1976-1977." NOAA, pp. 175-273. Available from: NTIS, Springfield, Virginia, Accession No. PB283890.

Harper, D. E., Jr., Scrudato, R. J., and Giam, C. S., 1976, "Pilot Study of the Buccaneer Oil Field (Benthos and Sediments). A Preliminary Environmental Assessment of the Buccaneer Oil/Gas Field." Report to NMFS, Galveston, Texas from Texas A&M University, College Station, Texas, 63 pp. plus appendices.

Hollaway, S. F., Faw, G. M., and Sizemore, R. K., 1980, "The bacterial community composition of an active oil field in the northwestern Gulf of Mexico," Mar. Pollut. Bull., 11:153-156.

Holley, H., and Savastano, K., 1979, "Buccaneer oil field study data management and system development final report," in: Jackson, W. B. (ed.), "Environmental Assessment of an Active Oil Field in the Northwestern Gulf of Mexico, 1977-1978. Volume II: Data Management and Biological Investigations." NOAA, pp. 2.2.3-1 to 2.2.3-159. Available from: NTIS, Springfield, Virginia, Accession No. PB80165970.

Holley, H., and Savastano, K., 1977, "Data management system development and monitoring," in: Jackson, W. B. (ed.), "Environmental Assessment of an Active Oil Field in the Northwestern Gulf of Mexico, 1976-1977." NOAA, pp. 11-87. Available from: NTIS, Springfield, Virginia, Accession No. PB283890.

Howard, R. L., Boland, G. S., Gallaway, B. J., and Dennis, G. D., 1980, "Effects of gas and oil field structures and effluents on fouling community production and function," in: Jackson, W. B., and Wilkens E. P. (eds.), "Environmental Assessment of the Buccaneer Gas and Oil Field in the Northwestern Gulf of Mexico, 1978-1979, Vol V." NOAA Technical Memorandum NMFS-SEFC-39, 60 pp. Available from: NTIS, Springfield, Virginia.

Jackson, W. B. (ed.), 1977, "Environmental Assessment of an Active Oil Field in the Northwestern Gulf of Mexico, 1976-1977." NOAA, 759 pp. Available from: NTIS, Springfield, Virginia, Accession No. PB283890.

Jackson, W. B. (ed.), 1979, "Environmental Assessment of an Active Oil Field in the Northwestern Gulf of Mexico, 1977-1978. Volume I: Synopsis." NOAA, 62 pp. Available from: NTIS, Springfield, Virginia, Accession No. PB301339/AS.

Jackson, W. B. (ed.), 1979, "Environmental Assessment of an Active Oil Field in the Northwestern Gulf of Mexico, 1977-1978. Volume II: Data Management and Biological Investigations." NOAA, 799 pp. Available from: NTIS, Springfield, Virginia, Accession No. PB80165970.

Jackson, W. B. (ed.), 1979, "Environmental Assessment of an Active Oil Field in the Northwestern Gulf of Mexico, 1977-1978. Volume III: Chemical and Physical Investigations." NOAA, 722 pp. Available from: NTIS, Springfield, Virginia, Accession No. PB80107899.

Jackson, W. B., and Wilkens, E. P. (eds.), 1980, "Environmental Assessment of an Active Oil Field in the Northwestern Gulf of Mexico, 1978-1979." NOAA. Available from: NTIS, Springfield, Virginia.

Jackson, W. B., and Wilkens, E. P. (eds.), 1980, "Environmental Assessment of an Active Oil Field in the Northwestern Gulf of Mexico, 1975-1980." NOAA. Available from: NTIS, Springfield, Virginia.

Jackson, W. B., and Wilkens, E. P., 1980, "Synopsis." in: Jackson, W. B., and Wilkens, E. P. (eds.) "Environmental Assessment of an Active Oil Field in the Northwestern Gulf of Mexico, 1978-1979, Vol. I." NOAA Technical Memorandum NMFS-SEFC-35. Available from: NTIS, Springfield, Virginia.

Jackson, W. B., Baxter, K. N., and Caillouet, C. W., 1978, "Environmental assessment of the Buccaneer oil and gas field off Galveston, Texas: an overview," Proc. Offshore. Technol. Conf. (Houston, Texas), 3081:277-284.

Jackson, W. B., Baxter, K. N., and Caillouet, C. W., 1979, "Environmental assessment of an offshore oil field," Ecolibrium, 8:7-9.

McCall, J. O., 1979, "Isolation and Characterization of a bacteriocinogenic strain of the bioluminescent marine bacterium *Beneckea harveyi*." M.S. Thesis, University of Houston, Houston, Texas.

McCall, J. O., and Sizemore, R. K., 1979, "Description of a bacteriocinogenic plasmid in *Beneckea harveyi*," Appl. Environ. Microbiol., 38:974-979.

McKinney, L., 1981, "The genus *Photis* from the Texas coast with a description of a new species, *Photis melanicus*," Contrib. Mar. Sci., in press.

Martin, J. A., 1977, "Hydrography of the Buccaneer Oil Field and adjacent waters," in: Jackson, W. B. (ed.), "Environmental Assessment of an Active Oil Field in the Northwestern Gulf of Mexico, 1976-1977." NOAA, pp. 575-632. Available from: NTIS, Springfield, Virginia, Accession No. PB283890.

Middleditch, B. S., 1980, "Hydrocarbons, biocides, and sulfur," in: Jackson, W. B., and Wilkens E. P. (eds.), "Environmental Assessment of the Buccaneer Gas and Oil Field in the Northwestern Gulf of Mexico, 1975-1980, Vol. III." NOAA Technical Memorandum NMFS-SEFC-51, 140 pp. Available from: NTIS, Springfield, Virginia.

Middleditch, B. S., and Basile, B., 1976, "Deuteriated analogs as internal standards for the quantitation of environmental alkanes by gas chromatography," Analyt. Letters, 9:1031-1034.

Middleditch, B. S., and Basile, B., 1979, "Sulfur-34 as an internal standard for quantitation of environmental sulfur by combined gas chromatography - mass spectrometry," Analyt. Letters, 12A:777-781.

Middleditch, B. S., and Basile, B., 1980, "Alkanes in benthic organisms from the Buccaneer oilfield," Bull. Environ. Contam. Toxicol., 24:945-952.

Middleditch, B. S., and Basile, B., 1980, "Discharge of elemental sulfur and its distribution in the region of the Buccaneer oilfield," J. Chromatog., 199:161-169.

Middleditch, B. S., and West, D., 1979, "Bicyclohexyl interference in the quantitation of environmental alkanes by gas chromatography," Analyt. Letters, 12A:1279-1284.

Middleditch, B. S, and West, D., 1980, "Hydrocarbons, biocides, and sulfur," in Jackson, W. B., and Wilkens, E. P. (eds.), "Environmental Assessment of the Buccaneer Gas and Oil Field in the Northwestern Gulf of Mexico, 1978-1979, Vol. 8." NOAA Technical Memorandum NMFS-SEFC-41, 119 pp. Available from: NTIS, Springfield, Virginia.

Middleditch, B. S., Basile, B., and Chang, E. S., 1977, "Hydrocarbon levels and pathways in the major components of a marine ecosystem," in: Jackson, W. B. (ed.), "Environmental Assessment of an Active Oil Field in the Northwestern Gulf of Mexico, 1976-1977." NOAA, pp. 633-735. Available from: NTIS, Springfield, Virginia, Accession No. PB283890.

Middleditch, B. S., Basile, B., and Chang, E. S., 1979, "Environmental effects of offshore oil production: alkanes in the region of the Buccaneer oil field," J. Chromatog., 142:777-785.

Middleditch, B. S., Basile, B., and Chang, E. S., 1978, "Alkanes in surficial sediments from the region of the Buccaneer oilfield," J. Chromatog., 158:449-463.

Middleditch, B. S., Basile, B., and Chang, E. S., 1978, "Discharge of alkanes during offshore oil production in the Buccaneer oilfield," Bull. Environ. Contam. Toxicol., 20:59-65.

Middleditch, B. S., Basile, B., and Chang, E. S., 1979, "Alkanes in seawater in the vicinity of the Buccaneer oilfield," Bull. Environ. Contam. Toxicol., 21:413-420.

Middleditch, B. S., Basile, B., and Chang, E. S., 1981, "Alkanes in shrimp from the Buccaneer oil field," Bull. Environ. Contam. Toxicol., in press.

Middleditch, B. S., Basile, B., and Missler, S. R., 1979, "Determine levels, pathways, and bioaccumulation of selected discharge constituents (non-metals) in the marine ecosystem in the oil field/hydrocarbon modeling," in: Jackson, W. B. (ed.), "Environmental Assessment of an Active Oil Field in the Northwestern Gulf of Mexico, 1977-1978. Volume III: Chemical and Physical Investigations." NOAA, pp. 2.4.1-1 to 2.4.1-302. Available from: NTIS, Springfield, Virginia, Accession No. PB80107899.

Middleditch, B. S., Chang, E. S., and Basile, B., 1979, "Alkanes in plankton from the Buccaneer oilfield," Bull. Environ, Contam. Toxicol., 21:421-427.

Middleditch, B. S., Chang, E. S., and Basile, B., 1979, "Alkanes in barnacles (*Balanus tintinnabulum*) from the Buccaneer oilfield," Bull. Environ. Contam. Toxicol., 23:6-12.

Middleditch, B. S., Chang, E. S., Basile, B., and Missler, S. R., 1979, "Alkanes in fish from the Buccaneer oilfield," Bull. Environmental Contam. Toxicol., 22:249-257.

Mitchell, F., 1977, "Accessible data base management," in: Jackson, W. B. (ed.), "Environmental Assessment of an Active Oil Field in the Northwestern Gulf of Mexico, 1976-1977." NOAA, pp. 7-9. Available from: NTIS, Springfield, Virginia, Accession No. PB283890.

National Marine Fisheries Service, 1977, "Environmental Assessment of an Active Oil Field in the Northwestern Gulf of Mexico. A Report of NMFS/LGL Workshop I, 1977-1978." NOAA, 38 pp.

Olsen, K. D., and Sizemore, R. K., 1981, "Effects of an established offshore oil platform on the autochthonous bacteria community," Dev. Ind. Microbiol., in press.

Savastano, K., and Holley, H., 1980, "Implement, monitor, and modify data management system," in: Jackson, W. B., and Wilkens E. P. (eds.), "Environmental Assessment of the Buccaneer Gas and Oil Field in the Northwestern Gulf of Mexico, 1978-1979, Vol I." NOAA Technical Memorandum NMFS-SEFC-35. Available from: NTIS, Springfield, Virginia.

Show, I. T., Jr., 1979, "An application of compartmental models to meso-scale marine ecosystems," in: Matis, J. H., Patten, B. C., and White, G. C. (eds.), "Compartmental Analysis of Ecosystem Models." Vol. 10 of the Statistical Ecology Series, International Cooperative Publishing House, Fairland, Maryland, pp. 73-97.

Sizemore, R. K., and Olsen, K., 1980, "Bacterial communities," in: Jackson, W. B., and Wilkens E. P. (eds.), "Environmental Assessment of the Buccaneer Gas and Oil Field in the Northwestern Gulf of Mexico, 1978-1979, Vol. IV." NOAA Technical Memorandum NMFS-SEFC-38, 32 pp. Available from: NTIS, Springfield, Virginia.

Sizemore, R. K., and Olsen, 1980, "Bacteriology of a Gulf of Mexico gas and oil field," in: Jackson, W. B., and Wilkens E. P. (eds.), "Environmental Assessment of the Buccaneer Gas and Oil Field in the Northwestern Gulf of Mexico, 1975-1980, Vol. III." NOAA Technical Memorandum NMFS-SEFC-49, 21 pp. Available from: NTIS, Springfield, Virginia.

Sizemore, R. K., Hollaway, S. L., and Faw, G. M., 1979, "Determine and compare groups of bacteria from an active oil field and an unaltered area," in: Jackson, W. B. (ed.), "Environmental Assessment of an Active Oil Field in the Northwestern Gulf of Mexico, 1977-1978. Volume II: Data Management and Biological Investigations." NOAA, pp. 2.3.7-1 to 2.3.7-41. Available from: NTIS, Springfield, Virginia, Accession No. PB80165970.

Smedes, G., Calman, J., and Beebe, J., 1980, "Hydrodynamic modeling," in: Jackson, W. B., and Wilkens E. P. (eds.), "Environmental Assessment of the Buccaneer Gas and Oil Field in the Northwestern Gulf of Mexico, 1978-1979, Vol. X." NOAA Technical Memorandum NMFS-SEFC-44, 57 pp. Available from: NTIS, Springfield, Virginia.

Tillery, J. B., 1980, "Trace metals," in: Jackson, W. B., and Wilkens E. P. (eds.), "Environmental Assessment of the Buccaneer Gas and Oil Field in the Northwestern Gulf of Mexico, 1978-1979, Vol VIII." NOAA Technical Memorandum NMFS-SEFC-42, 93 pp. Available from: NTIS, Springfield, Virginia.

Tillery, J. B., 1980, "Trace metals," in: Jackson, W. B., and Wilkens E. P. (eds.), "Environmental Assessment of the Buccaneer Gas and Oil Field in the Northwestern Gulf of Mexico, 1975-1980, Vol. VI." NOAA Technical Memorandum NMFS-SEFC-52, 39 pp. Available from: NTIS, Springfield, Virginia.

Trent, L., Workman, I. K., Dime, S., and Jones, C., 1977, "Recreational fisheries and the distribution of predatory pelagic fishes," in: Jackson, W. B. (ed.), "Environmental Assessment of an Active Oil Field in the Northwestern Gulf of Mexico, 1976-1977." NOAA, pp. 311-337. Available from: NTIS, Springfield, Virginia, Accession No. PB283890.

West, D. L., 1979, "Hydrocarbons, Sulfur, and Acrolein in the Marine Ecosystem Associated with Offshore Oil Production," M.S. Thesis, University of Houston, Houston, Texas.

Wheeler, R. B., 1979, "Environmental Trace Metal Geochemistry of the Buccaneer Oil and Gas Field." M.S. Thesis, Rice University, Houston, Texas.

Wheeler, R. B., 1980, "Environmental trace metal geochemistry of the Buccaneer oil and gas field," Environ. Geol., in press.

Wheeler, R. B., Schwarzer, R. R., and Anderson, J. B., 1978, "Assessment of environmental impact of offshore production in Buccaneer oil field: sedimentologic and geochemical results," Proc. Offshore Technol. Conf. (Houston, Texas), 3082:285-290.

Workman, I. K., and Jones, C. E., 1979, "Determine effects of oil field discharges on species composition and abundance of pelagic fishes and demersal fishes and macro-crustaceans in the oil field," in: Jackson, W. B. (ed.), "Environmental Assessment of an Active Oil Field in the Northwestern Gulf of Mexico, 1977-1978. Volume II: Data Management and Biological Investigations." NOAA, pp. 2.3.5-1 to 2.3.5-149. Available from: NTIS, Springfield, Virginia, Accession No. PB80165970.

Zein-Eldin, Z. P., and Keney, P. M., 1979, "Bioassay of Buccaneer oil field effluents with penaeid shrimp," in: Jackson, W. B. (ed.), "Environmental Assessment of an Active Oil Field in the Northwestern Gulf of Mexico, 1977-1978. Volume II: Data Management and Biological Investigations." NOAA, pp. 2.3.4-1 to 2.3.4-25. Available from: NTIS, Springfield, Virginia, Accession No. PB80165970.

Houston Area Oxidants Study

Platform 296-B of the Buccaneer Field was also included as a study site in the Houston Area Oxidant Study (HAOS), a comprehensive study of air pollutants in the Houston area organized by the Houston Chamber of Commerce. Ozone levels were monitored (by Rockwell International) at this location continuously between June and October, 1977. Ozone concentrations greater than 80 ppb were recorded only during 1.1% of the hours monitored. Hydrocarbon levels were determined in air samples, also, at this location (by Research Triangle Institute). For more details see the following reports, all published by the Houston Chamber of Commerce, Houston, Texas:

"Houston Area Oxidants Study Program Summary," 1979.

"Ambient Ozone, Oxidants, and Nitrogen Oxides Measurements in the Houston Area," HAOS Publication OX-4, 1978.

"Detailed Hydrocarbon Analyses," HAOS Publication HCP-5, 1979.

INDEX